The Nephilim Chronicles
Fallen Angels in the Ohio Valley

Fritz Zimmerman

The Nephilim Chronicles
Fallen Angels in the Ohio Valley

Cover Illustration from, 12th Annual Report of the Bureau of Ethnology to the Secretary of the Smithsonian Institution, 1890-1891

Contents

Origins of the Nephilim
The Paleolithic 2,000,000 - 6,800 B.C.

The time period "Paleolithic" begins 2 million years ago with what is considered the first tool making and ends at the end of the Ice Age, about 68000 B.C. It is by far the longest time period that includes in the view of evolutionist, a period of transformation of primates from ape-like creatures to the more gracile and fully erect modern humans. Theoretically, each of these more "evolved" orders, spread out across the globe, replacing the more primitive species. However, in reality, species that have been around a million years or so, do not disappear as quickly as scientist would like us to believe.

This is their story. A species of man, defined as "archaic," characterized by the following physical traits, a receding lower jaw, projecting upper jaw, prominent brow ridge with a receding forehead, massive jaws and thick skulls walls. Many of the "archaic" men grew to an unusual stature, by even today's standards, with some as large as nine feet in height. Their remains have been found across the globe, but in no more greater density than in the Ohio Valley and associated with the Allegewi Hopewell mound builders.

The accepted progression of evolution, by academia, goes from Homo afrensis, H. habilis, H. erectus, H. sapiens (Archaic) that included Homo antecesser and H. Heidelbergensis, Neanderthal and then to the final sub species of modern man Cro-Magnon. It has been professed by anthropologist that a more modern, elegant variant of modern man interbred with this last "archaic" sub species of Cro-Magnon, resulting in todays populations. The last of the Cro-Magnon are said to have disappeared at the end of the Ice Age around 10,000 B.C. And yet, within the later Mesolithic, Neolithic and Atlantic periods (6,800-2250 B.C.) skeletal remains are still consistent with what would be considered Ice Age "archaic" or Cro-Magnon populations. The distinctness of an archaic type of skull, compared to that of a modern man being bigger teeth, projecting upper jaw, called prognathism, a massive lower jaw, a pronounced brow ridge, a receding forehead, and thick skull walls. The size of the skeletal remains in

of itself is not an attribute of an archaic human, but the immense size of the Cro-Magnon has to be part

of the gene pool that survives into the "Hunter and Fishers," the Amorites or Nephilim, the Beaker

People and eventually within the Allegewi and Hopewell populations in the Ohio Valley. So, unless we

surmise that de-evolution has taken place then the only explanation can be that what appears to be

"archaic" peoples survived much longer than scientist admit in the circumpolar regions extending

across the North American continent. The other population center of the Cro-Magnon was found in the

easter Mediterranean lands of present day Israel, Palestine and Syria. Legends of the last remnants of

these giants would be immortalized in the Bible, Eddas, Vedas and Greek mythology.

   Before exploring the geographic extent of the Cro-Magnons, we will examine their "evolutionary" or

"created" predecessors. This establishes the long duration of what has been classified as "archaic"

skull types. The discovery of skeletons of these early men also shows that heights exceeding six feet

was established early and continued into the Cro-Magnon era.

   It is the belief of evolutionist that Homo sapien's lineage started with Australopithicus afarenis, 4 to

2.7 million years ago and whose discovery, near Hadar, Ethiopia was nicknamed "Lucy."

Australopithicus head and brain were small, resembling the ancient

apes, but walked upright and had a skeletons similar to humans. The

next species in the evolutionary line was Australopithecus robustus who

lived from 2.2 to 1.6 million years ago and was similar in body size to

afrensis but was distinguished with a massive skull and teeth. They were

much like afrensis in that their face was flat very ape-like with no

Australopithecus

forehead. Contemporary to robustus around 2.5 to 1.6million years ago was Homo habilas,("Skillful

or Handy Man") who had a larger brain that allowed him to develop limited speech and make some

tools. However, Homo habilas, never improved the few tools he had and lived a life of a scavenger.

Habilis's face protrudes less than the earlier hominids, but was still very ape-like with a sagital crest or bony ridge that was evident on the top of the skull. This crest is where muscles were attached to their massive jaw that was used to crush their hard to chew and rough diet.

Sometime around 2 million B.C. the first true hunter-gatherer evolves or was "created" with Homo ergaster; what is considered by many scientists and creationists to be the first humans. The discovery was found in West Turkana, Kenya, when a complete skeleton of a young Homo ergaster male between 9 and 12 years old was uncovered. He was 5 feet four inches tall and it is estimated he would have been over 6 foot in height as an adult. His remains were dated at 1.7 million years old.

Primates had progressed from herds of herbivores and scavengers to hunters and warriors. It was originally believed that around 1.8 million years ago a number of H. ergasters migrated from Africa into Asia and gave rise to Homo erectus. Recent finds in China of Homo-erectus fossils have been dated to 1.9 million years. Another early Homo erectus skeleton was found by a German-Georgian archaeological team in Dmanisi, Georgia that dated to 1.7 million years, described as "tall and fully erect." These new finds led anthropologist to push Homo-ergasters emergence in Africa to 2.5 million years; being contemporaneous with Homo habilas and proof that habilas was an evolutionary "dead end."

Homo-erectus is believed to have lived between 1.9 million years to 300,000 years ago. Their brain, in the beginning was about 25 percent smaller than modern mans, but was soon equal in size. Homo erectus developed speech and language, their tool and weapon kits became more specialized and they learned to make fire. With these new technologies they were able to migrate north out of Africa and Southeast Asia into the colder European continent.

In 1994 Italian scientist uncovered a skull dating 900,000 year old in Caprina that resembled H. erectus. Their report was published in the October 2000 *Journal of Human Evolution.* The skull was described as having a massive brow ridge and a sharply angled occipital bone at the back of the skull.

In this time span of a million years Homo erectus changed very little. This fact was pointed out by A.W. (Bill) Mehlert, in 1994 *"Homo erectus 'to' Modern Man: Evolution or Human Variability"* "From an evolutionary viewpoint the small degree of change in erectus populations over an alleged period of one and a quarter million years must be disappointing, especially if the cranial capacities of the earliest and latest examples all lie within modern range of humans."

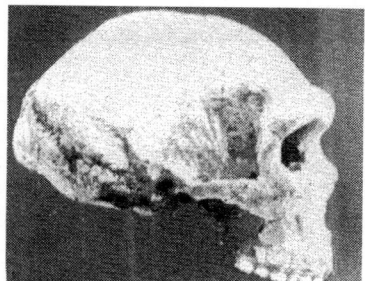

Homo erectus

Another apparent H. erectus skull was found in 1994 by a Spanish anthropologist in the Grand Dolina beds at Atapuerca, dating to 600,000 years old. Printed in the March 2001 edition of *Science Magazine*, The Gran Dolina frontal bone and parts of upper jaw and mid-face had a strange resemblance to modern humans. This resulted in the Dolina skull being called a new species called Homo antecessor or Archaic Homo sapiens. Another Archaic Homo sapien skull was discovered in Petralona, Greece that was dated between 250,000-500,000 years. Another new discovery found near Mauer Germany, was given to professor Otto Schoetensack from the University of Heidelberg, so naming this new contemporaneous species, called Homo heidelbergensis. Both H. antecessor and H. heidelbergensis are believed to have descended from the morphologically similar Homo ergaster. H. heidelbergensis had a larger brain and more developed tools, than its predecessors. More importantly is that H. ergaster, H. heidelbergensis were tall with average height being around 6 feet. These species was larger and more muscular than modern humans.

Homo heidelbergensis

This divergence of species is split within the Asian and European Heidelbergensis populations around 200,000 years ago giving rise to two separate species, the Neanderthals and a more evolved "archaic" homo sapiens. The Neanderthals lived in central Asia and Western Europe between 200,000 and 25,000 years ago. They were short, heavy-boned, with a low forehead, double arched brow ridge, large nasal opening and a weak chin. An occipital bun or bony area is the back of the head is present as in erectus.

Spy Neanderthal

Several skeletons discovered have had both "archaic" and "modern" features that are called Homo Sapiens (archaic). Homo-sapiens (archaic) are supposed to be the evolutionary link between Erectus and Modern Homo-sapiens sapiens. They had more modern features with a less prominent brow ridge and a more rounded skull. While these features were softened a bit from the more pronounced Homo erectus they still had all the features that is used to define an "archaic" skull. A receding chin, larger teeth for a rough diet, a pronounced brow-ridge with a receding forehead, a flattened top of the skull and a thick muscular neck.

What has been scientifically defined as modern Homo sapiens sapiens have been dated as early as 195,000 years. Some scientist believe that early modern man originated in East Africa and expanded into South Africa and Southwest Asia by 100,000 B.C. At the Omo site in Ethiopia, one skeleton was modern looking, the one next to it had more "archaic" type features. Modern humans are identified as having a more rounded skull with a less prominent brow-ridge and a projecting chin. These early finds still retained the prominent brow-ridge.

European and Asian Archaic Homo Sapiens are said to have disappeared by 28,00 B.C., with Neanderthal vanishing a little later at 25,000 B.C . The inexplicable time frame are those years between 40,000 and 25,000 BC when modern humans, Neanderthals and Archaic sapiens shared the territory we now know as northern Europe. While it is believed that the populations of Archaic Homo sapiens and Neanderthal were vanishing, their physical traits are visible within the Cro-magnon populations.

The interaction between Modern man and the Archaic forms of Homo sapiens is still being debated by scientist. The most accepted anthropological theory of the evolution of modern man is called the "Replacement model" which proposes that modern human evolved from archaic Homo sapiens 200,000 – 150,000 years ago only in Africa. Some of them migrated north into Asia and Europe replacing all Neanderthals and other late Archaic Homo sapiens around 100,000 BC. This theory proposes that all modern humans share a common ancestral link to Africa. This idea also suggests that all of the regional anatomical differences that we see amongst modern humans are the result of evolution in the last 40,000 years. This theory discounts any interbreeding between modern humans and Archaic Homo sapiens or Neanderthals

In the 1940s the "Multi-Regional Theory" was proposed, that emphasized, that the regional variations of modern humans (races) originated in geographically separated Homo erectus populations. Theoretically, Homo erectus had spread across the Old World, developing separately in to modern humans. This allowed for unique features to develop in humans in each geographic area; with considerable interbreeding and therefore genetic exchange between the various regional groups.

The newest theory and least accepted is the "Assimilation model." This model contends that human developed out of Africa and moved north and in some cases replaced archaic Homo-sapiens and Neanderthals but others interbred with the Archaic homo sapiens resulting in hybrid populations. This theory best explains the abrupt emergence of Cro-Magnon around 40,000 BC. Cro-Magnon shared Europe with Neanderthal for nearly 12,000 years. It is argued that these two people interbred creating a partial hybridized Cro-Magnon population.

There is mounting evidence that a hybrid Cro-Magnon-Neanderthal or Archaic hybrid did exist. Evidenced has been found within skeletons that display more "archaic" features than "modern." In *"The Archaeology of the U.S.S.R"* by Alexander Mongait writing for the Academy of Sciences of the U.S.S.R., 1959, "In 1953 science was presented with a new, important discovery. In Statoselye Cave near Bakhehisarai, A. A. Formozov found the burial of a one or two year old child of the Mousterain epoch. In direct proximity to the skeleton, he uncovered a hand-axe; scrapers, a point and the bones of a wild ass, an ox and a bear.

The skeleton was studied by anthropologist and it was established that it belongs to the modern type but has many features that approximate it to the Neanderthal type: large teeth, heavy cheekbones, ect. The Statoselye skeleton is one of the few finds of the remains of representatives of ancient man who occupies an intermediary position between the Neanderthal man and Cro-Magnon man."

A recent find in Lapedo Portugal was found that also shows traits of modern man, including the jaw and teeth and Neanderthal features, like the size of the femur and tibia. Carbon dating showed the skeleton to be 25,000 years old.

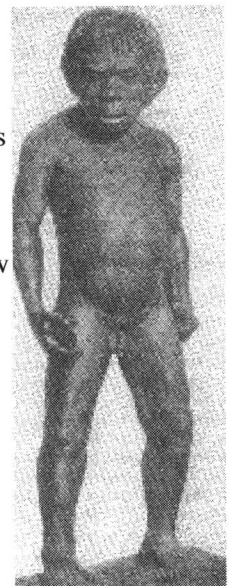

**Statoselye
skeleton**

In a Neanderthal blog by Hohn Hawks he cites a recent paper about a skeletal find in Pestera Romania by Andrei Soficaru, Adrian Dobos and Erick Trinkaus, 2006. the following is from the abstract by Soficaru.

"The early human remains from the Petera Muierii, Romania have been directly dated to 30,000 radiocarbon years before present. The Muierii fossils exhibit a suite of derived modern human features, including reduced maxillae with pronounced canine fossae, a narrow nasal aperture, small supercilliary arches, an arched parietal curve, zygomatic arch above the auditory porous, laterally bulbous mastoid, narrow mandibular corpus, reduced anterior dentition [...] However, these traits co-occur with contextually Archaic and or Neanderthal features, including a moderately low frontal arc, a large occipital bun, a high coronoid process and a asymmetrical notch. As with other early modern humans, the mosaic of modern human and Archaic/Neanderthal features, relative to their potential Middle Paleolithic ancestral populations indicates considerable modern human admixture."

At Mount Qafzeh in Israel, within a cave the remains of eleven individuals were discovered between 1933 and 1977, that date to 100,000 years old. The remains of what was designated as Skhul 9 was intriguing in that was a modern looking skull with Neanderthal features of prognathism and a prominent brow ridge. Some researches thought that these physical traits were the result of interbreeding between Neanderthals and a more modern form while others called the Skhul skull a forerunner of the Western European Cro-Magnons and proposed the term "proto-Cro Magnon" It is of interest to note that these skulls were described as "extremely large," with a height of five feet, nine inches, they were much larger than the average Neanderthal.

One of the Skhul skulls was reported to have had hyperdontia, or a double row of teeth. This dental anomaly has been reported in populations that have been of legendary great height. The Babylonian Talmud claimed that the Biblical giants, (Amorites) had in some cases, a double row of teeth. In The British Isles the legendary Fomorians (Muru or Amorites) also had double rows of teeth. This anomaly also occurs with some frequency in North America within the Shell Mound Culture, the Glacial Kame, and the later Allegewi Hopewell mound builders. All of these North American populations also display archaic facial features along with their great height.

Stephen Coons, reports in *"The Neanderthaloid Hybrids of Palestine"* that "The Neanderthal group was extremely variable, and showed within its ranks clear evidence of evolutionary change in a human direction." He sites the discoveries found in Palestine within the Mount Carmel caves where, "the skeleton of a small woman, fully Neanderthal and associated with it was a male mandible equal in size to that of Heidelberg, but possessed of that human feature, a chin."

These ancient hybridized people have been called Cro-Magnon by scientist that are the supposed forerunners to modern Europeans. Yet, Cro-Magnon also exhibit archaic skeletal features such as occipital buns, thick skull walls, heavy marking on bones from muscle attachements and heavy brow ridges that are not typical of other homo sapiens sapiens found in other parts of the globe. A.W. (Bill) Mehlert, also points out in, *"Homo erectus 'to' modern man: evolution or human variability*; "All H. Sapiens forms should be considered not as separate species but as a single human species encompassing a range of genetic and phenotypic diversity." The question he poses is "are erectus forms proof of an evolutionary progression from the apes, or are they simply temporal, regional, climatic, dietary or pathological variants of human beings?" A physically diverse archaic population of the world from which all modern humans derived seems more probable than current views that attach present populations to one species, Cro-Magnon."

H. heidelbergensis are believed to have descended from the morphologically similar Homo ergaster; both of which were tall. Itt is plausible the evolutionary progression went from ergaster to heidelbergensis to Cro-Magnon. This would include some of the "Mult-Regional theory" and part of the "Assimilation theory." Cro-Magnon was an evolutionary progression from heidelbergensis; that may have also included interbreeding with other populations.

It wasn't until about 40,000 B.C. that these early modern men (with archaic features) appeared in Europe, ushering in a new era known as the Upper Paleolithic from 40,000-12,000 B.C and a new defined species, Cro-Magnon. They were a tall people, the men being many times over six feet in height with a face broader across the cheekbones than it was long; he had a brain that was larger than his modern descendants. Their upper jaw projected slightly forward, the lower jaw was square, thick and strong. The brow-ridge was still prominent.

By twenty-five thousand years ago these ancestors of present men were roaming the cold wind swept plains of Southwestern Europe. It was the late-glacial epoch, when the severity of the climate reached its peak. In Europe and North Asia, cold steppes and tundra with patches of northern forest stretched over a larger expanse than previously.

Depiction of a Cro-Magnon campsite constructed from furs and mammoth tusks.

Cro-Magnon hunted reindeer, wooly mammoth, wooly rhinoceros, musk ox, arctic fox, arctic hare, wolverine, lemming, ibex, red deer, brown bear, wildcat, fox, wolf, otter, lynx, bison, cave lion and the ptarmigan and other Arctic birds, all of which were common in Southern Europe.

They chipped tools from flint, sharpening them skillfully by knocking fine chips from the edges on both sides called "Bi-facial knapping", They also made gravers, scrapers, drills, knives, awls and chisels with which he used to carve stone and bone and create sketches or base-reliefs of familiar animals, often well drawn and colored with red and yellow ochres. The most popular artistic motifs were geometric designs that were carved on bone tools and pebbles.

Their bone harpoons and fishing hooks strung with sinew were used to catch the fish that abounded in the lakes and rivers. Clams and mollusks in the shallows also supplied him with easily obtained food.

While scientist label the Cro-Magnon as the prototype of the European, they were in fact as racially diverse as modern men. Stephen Coons, in *"Chronological and Geographical Differentiation of the European Aurgnacian Group,"* (Aurignacian 25,000 B.C) "There is no type man more completely sapiens than a negro. If this is true, than the skin and facial differences in races have to be due to some

admixture with different stock such as the H. erectus that had left Africa some 1.8 million years ago. It

is also conceivable that during the Ice Age some of these H. erectus migrated back south into Africa,

again mixing with local populations." The Cro-Magnon skeletal remains found at Grimaldi were

described by Coons as representing "an early negro-white mixture, or a generalized proto-negroid in

the process of specialization." Coons also writes in *"Upper Palaeolithic man in China"* that in a Cro-

Magnon find near Chou Kou Tien that "One of these skulls, seems upon preliminary examination, to

resemble the European Upper Palaeolithic group very closely, it has also been compared to Ainu

crania." Ainu are believed to be a mix of European and Asian Cro-Magnon. It is to early to assign

racial designations to any of these populations. There should be no confusion that this publication is

about archaic species and not races.

The following historical accounts describe large skeletons that are classified as belonging to the

Cro-Magnon species. Important facts within these descriptions are the burial types and inclusions that

set an early precedent for burials of the later periods. Red ochre is found in graves along with the

presence of fire or ashes. These burial rites found within Cro-Magnon will reoccur in later burials that

also contain large skeletons that exhibit "archaic" type skull features.

Oakland Tribune, **May 30, 1909**
CORREZE SKULL IS ONE OF GREAT RACE ANCIENTS
Probably Belonged To A Cave Dweller of Western Europe-Owner Had No Dome.
Skull Fairly Complete
   Professor Boule introduced me to the wonderful skull, and placed it in my hands. It has been treated
by him with great skill so as to render the bone firm and hard, while detached portions have been fitted
into place, so that it is fairly complete. It will be remembered that this skull was found together with
most of the skeleton of the same indavidule, by two enthusiastic local archaeologist, buried at such
depth and in such position in the cave known as the Chapelle-aux-Saints is to leave no doubt as to its
belonging to one of a race of men contemporary with the mammoth and hairy rhinoceros-a race which
inhabited Europe in the great glacial period-which cannot be less than a hundred thousand years behind
us, and probably is more. The chief importance of the skull lies in the fact that it agrees in its very
peculiar form with the Neanderthals skull (from Bonn), the Spy skulls (Belgium), and the Gibraltar
skull. It in fact, confirms the conclusion that at this period the caves of Western Europe were inhabited
by a race of men with peculiar skulls

The fact was published some four months ago that the new Correzo skull agrees with the celebrated skull top of the Neanderthal in the extraordinary shallowness or absence of doing in the retreating forehead, the thick prominent eyebrow ridges, and in the excessive "lowness" or want of elavation of the back region. But further study of the new skull had enabled Prof. Boule to show that the outline of the new skull looked at from above exactly coincides with that of the Neanderthal skull- there is the same great length from the eyebrows to occipant and the great breadth at a series of corresponding regions.

Both of Enormous Size

The curious thing is that both these skulls are of enormous size-a good deal bigger in length and breadth than modern European skulls, and not as small and ape-like, though they are far shallower (that is, less high in the dome) than any skulls of living men. The ancient Neanderthal men's brain was not smaller, but actually a little bigger than that of modern Europeans, it was bigger in regions where tho modern Europeans is small, and smaller where that is large.

Oakland Tribune, **January 1, 1928**
ATTACKS ACCEPTED THEORIES ON PREHISTORIC EUROPEANS

Some widely accepted beliefs concerning prehistoric man in Europe tottered today when Dr. Alex Hrdlicka, Smithsonian anthropologist, presented results of his recent researches before the Anthropological Section of the American Association for the Advancement of Science.

One of the beliefs questioned concerned the existence of the so-called "Supermen" of Grimaldi as a distinct race. Some years ago in the caves of Grimaldi on the Mediterranean near Mentone, three gigantic skeletons, with some others were uncovered. These skeletons, all male, belonged to men considerably over 6 feet in height, two of them being 6 feet 4 inches, and of great strength. The skull capacities of the three "giants" were correspondingly large.

Coshocton, Ohio Tribune, **January 9, 1929,**
Algeria Finds Were the Bones of Our Ancestors

"The bones unearthed by the Beloit-Logan expedition were dug out of huge mounds of small shells. The hithero unknown pre-modern race at Gibralter and Sicily, Dr. Alfred Romer, University of Chicago paleontologist, reports that the animal bones found in the shell heaps are similar to those found in the northern Cro-Magnon sites and that modern people were evidently hunters on a small scale, the evidence being the presence of burnt bones of the giant ox, the wild bull and the lion in the dirt and shells of the heaps. The North African people lived in the open, the Cro-Magnon people in caves.

After intensive measurements of the skeletons in all possible features Dr. Cole concludes that they resemble modern Europeans far more than they do any other race. The anthropological measurements also indicate that these pre-modern inter-married with other races, which would in turn indicate that none of the present south European races are technically "pure." He believes that the skeletal evidence of that race shows inter-mixture with ancient brutish anthropoid-like Neanderthal men, who dominated the scene 50,000 years ago."

Origin and Character of the British People, **Charles Macnamara, 1900**

The skeleton of the tall race above referred to have been found all over Europe, but for our present purposes we would draw attention to skeletons discovered in a few well-known burial places; one of which is situated in the extreme south-east of France, and another in Dordogne, one of the western departments of that country.

In the natural cave of Baousse-Rousses, between Mentone and Ventimille, three skeletons were unearthed in the year 1892. Dr. R. Verneau has given a lucid account, and photographs of this discovery.

"The cave had been inhabited by animals of the early Paleolithic period, but the human skeletons found there were buried in graves dug out of the ancient formation which covered the floor of the cave to some depth. The bodies of these people had been laid upon, and were covered by layers of earth containing much iron pyrites, which had been carried into the cave, so that the bones composing these skeletons were stained by iron, and also with ochre. Surrounding the skulls and the necks of these skeletons, rows of perforated shells and bone were found, which had evidently been strung together and used as ornaments; the shells had regular and symmetrical patterns carved on them. Under the skulls and near the hands of the skeletons some finely flint instruments were found; these weapons are well-marked specimens of the type belonging to the early Neolithic period. There were numerous bones of various animals in the layers of earth which covered to a considerable depth the floor of the cave; but as much of this earth had been disturbed before Dr. Verneau examined the site of the skeletons, it is better not to place reliance on the relation of the remains of these animals to those of the human skeletons.

The skeletons were (1) that of an old, extremely powerful man, who measured 6 feet 6 inches in height; (2) that of a young woman, who was 6 feet, and (3) that of a youth, who was 6 feet 2 inches in height.

Phoenician Orgins of Britons and Scots, 1925

And it was presumably early pioneer stragglers of this same Nordic race at the end of the Old Stone Age who are represented by the "Red Man" of Paviland Cave, in the Gower peninsula of Wales, of the mammoth age, and the "Kneiss Chief " in the stone cist at Keiss (Kassi?) in Caithness. Both of these are interred with rude stone weapons, and are of the superior and artistic Cro-Magnon type of early men, which seems to have been the proto-Nordic or proto-Aryan. Indeed, the associate of the Keiss chief had a cranium described by Huxley: The Keiss chief is described by Laing as "tall man of very massive proportions", lying extended with face to the East. Huxley found his cranial index was 76, with projecting eyebrow ridges which gave the forehead a "receding" aspect and the forehead "low and narrow."

Davenport Daily Leader, November 25, 1894
GIANTS OF PREHISTORIC FRANCE

In a prehistoric cemetery recently uncovered at Montpelier, France, while workmen were excavating a waterworks reservoir, human skulls were found measuring 28, 31 and 33 inches in circumference. The bones which were found with the skulls were also of gigantic proportions. These relics were sent ot the Paris academy, and a learned "savant," who lectured on the find, says that they belonged to a race of men between 10 and 15 feet in height.

News, (Frederick Maryland)
Find Old Graves
Skeletons of Giant Warriors Unearthed in France
Men who lived 25,000 years ago Believed to have died fighting-Arrow found in head of one.

The discovery of 25,000 year- old graves containing well preserved skeletons of three prehistoric warriors has just been discovered at Solutre, a small village in the. [...] department of France, widely known for its prehistoric remains.

In the last 60 years remarkable specimens of remains of the prehistoric period have been found at Solutre, and a short time ago the Lyons faculty of science decide to undertake a methodical and scientific search of the district on its own account. The first search proved rather disappointing but the work was nerveless, energetically carried on, and after 20 days of patient toll the scientist in charge of the expedition discovered three men who lived in the latter Paleolithic or Aurignacien period, from 20,000 to 25,000 years ago.

The three skeletons were buried in the same position at a depth of three feet seven inches, five feet and six feet respectively the heads facing the rising sun. Lying on their backs the knees slightly raised, the hands clasped over the stomach, the skeletons were resting on beds of ashes. On either side of the head were two roughly hewn stones in the shape of a "cromlech," which it is believed indicate the exact position to be occupied by the body.

The skeletons evidently belonged to extremely powerful men, as the smallest of the three measured six feet two inches while the tallest measured six feet nine inches. The shape of the skull is remarkable, the forehead is rather low. The sockets of the eye are square and of large dimensions; the jawbone is prominent and the jaws are powerful, and still contain well preserved teeth.

As the great ice sheets began to melt approximately 10,000 BC, to 7,000 BC, the reindeer hunters appeared in the regions north of the Crimea, west of the Ural Mountains, extending into the Baltic. The Reindeer Hunters followed the migrated herds who stayed near the retreating ice, feeding on the new grasslands. Few physical remains have been found of the reindeer hunters in this region, but the top of a skull was discovered in western Lithuania and was described in, *The Prehistory of Eastern Europe, Mesolithic, Neolithic and Copper Age Cultures in Russia and the Baltic Area.* Marija Gimbutas, 1956 "The upper part of the skull was massive, dolichocephalic, with strong proclivity of the forehead, prominent and massive brow ridges and a narrow forehead. These traits suffice to show that the Kebelia man was sapiens, but had Neanderthaloid elements, in other words, was a Neanderthal-sapiens hybrid. Morphologically the Kebeliai skull is more primitive than that of the boreal period, discovered at Krisna, some 150 km. Southeast of Kebeliai. From this we can deduce that the Kebeliai skull probably belongs to the period between the last glaciation and the Baltic Boreal culture. A closely related fragment of a skull was discovered near Moscow in the Pleistocene alluvium of the Skhodia River. This was also a top of a skull, having the same traits as that of Kebeliai. The geological evidence made it possible to date the Skhodia man undoubtedly during the Late Glacial period."

A skeleton of a female between 25 and 30 years old was found in Denmark from Koelbjerg dated to about 7,000B.C. The Koelbjerg skeleton was said to represent a type closely related to the reindeer hunters.

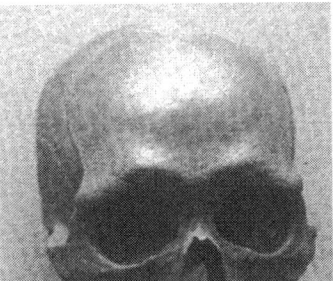

In Denmark near Koelbjerg was discovered the burial of a woman who was described as having a protruding brow ridge, massive jaw and prominent chin. The skull dating to 7,000 B.C., is dimilar to the Upper Paleolithic Cro-Magnon skulls.

The next time period of time, called Mesolithic, is generally accepted as beginning about 6,800 B.C. when the temperatures warmed and people became more settled in the newly formed woodlands. Cro-Magnon had not disappeared, their physical attributes defined by their height and their archaic facial features are still previlent within these later population populations.

Cro-Magnon in
the Boreal and Atlantic Periods
6800-2250 B.C.

In the post ice-age world there is the extinction of the giant sized herbivores, the mammoth, the wooly rhinoceros, horse, bison and saiga antelope. This extinction of many arctic animals was the result of global warming after the last glaciation. "The investigations of V.N. Suskachev, of the *Carnegie Institution of Washington* in 1928, found the turfs of the Karsky tundra (between the River Stchuchya and the Kara Sea) were the remains of firs, pines, larches, birches and alders there, where now the treeless tundra extends, and he suggest that formerly it was much warmer at that latitude." With the milder climate and the increased numbers of animals in the forest, the Reindeer Hunters no longer needed to follow the migrating herds. They turned to fishing as an alternative food source and lived in the forested plain that stretched across northern Europe. They lived in small groups along small streams and glacial lakes. In addition to fishing and the collection of shell-fish, they hunted small game animals, the reindeer remained the chief creature of the hunt.

According to Gumbutas in *The Prehistory Of Eastern Europe*, 1956; "This period is defined as the Boreal Period (6,800-5600BC) where a relatively homogeneous culture of hunters and fishers stretched from the western Baltic area to southwestern Finland, yielding harpoons, leaf shaped projectile points, fish hooks, ice picks, fish nets, net sinkers, gouges and stone axes. Patterns found on bone or horn tools had either incised marks or were made small holes or pits that arranged in regular shapes of chevrons, cross hatching or other geometric designs. These artistic motifs seemingly being a continuem of those found in the Upper Paleolithic. These tools and artistic designs were continued into Atlantic Period (5,600-2250 BC) when the culture of the Hunters and Fishers flourished during the warm, humid period."

Emerging from the Late Paleolithic there existed two cultural blocks in Europe. The culture of the Pontic area north and east of the Black Sea was linked with the Near East. This included the areas of the Northern Caucasus, Ukraine and southern Russia. The other area is the Baltic that included, Poland East Prussia, Lithuania, Estonia Latvia, Finland and northern and central Russia. The importance of the Pontic area are the cultural links with the Near East and the Biblical Lands. The early peoples in this region would later be known as the Amorites, Kurgan or Corded people, Scythians, and Magog.

In the Tardenosian Period, (6,800-5,600 B.C.) at a place called Murzak-Koba , within the Pontic area, harpoons and awls are found that are similar to those in the Baltic region. Many fish bones were found, showing the importance of fishing as a food source. Gimbutus described the burial, at Murzak-Koba as, "A double burial of a middle-aged man and a young woman both resting on the back in an extended position. The physical type of both individuals is stated to be like that of the Cro-Magnon men of the French Upper Paleolithic. The man was 180 cm. tall [5'9"] , long and large headed, and with heavy brow ridges, low wide orbits and a wide, long face."

The Baltic Forest Culture extended from the western Baltic Sea to southwestern Finland. A skeleton was discovered belonging to the Baltic Forest Culture in the southern Lithuania. The skull was found in a pit paved with stones. Bone harpoons and other stone tools accompanied the body. The skull was described as having a long and narrow face, similar to the Aurignacion skulls from the Upper Paleolithic. More specifically it was similar to the skull found at Combe Capelle and the Brunn skull of central Europe, both of which are Cro-Magnon.

The artistic, geometric designs found in Baltic Forest Culture was a continuem of those found in Upper Paleolithic. Artistic impressions are consistently found on bone or horn tools that had either incised marks, or were made displaying small holes or pits that were arranged in regular shapes of chevrons, cross hatching or other geometric designs.

Bone implement ornamented with a geometric pattern discovered near the Dneiper rapids from Marija Gimbutas in 1956 *"The Prehistory of Eastern Europe."*

The Hunters and Fishers continued to flourish over the forested belt, as the climate continued to warm into what is called the Atlantic period (5,600-2,250 B.C.). New technologies from the south are introduced that included the polishing and perforating stone tools and the manufacturing of pottery. Burials from this period have been found in glacial kames and shell mounds. Kame burials were placed in pits that were sometimes lined with bark. The use of red ochre on the bones or as an inclusion within the burials is commonly found.

In southern Scandinavia, the Erterbolle culture (5300-3950 B.C.) has been identified. They derived their living chiefly from the seas, where they hunted seals and whales from their dugouts. This has also been called the Kitchen Midden culture. A culture that was defined by the large heaps of oyster shells that contained burials with inclusions of bone, antler and flint artifacts.

Erterbolle skeletel remains that were found within the kitchen middens are still consistent with Cro-Magnon. The Cro-Magnon skull being defined as dolichocephalic (long), the jaw prognatheous (protruding), the nose flat and the suborbital ridges (brow ridges) pronounced. In 1975 a hill known as "Bogebakken" was being bulldozed in VedBaek Denmark for the construction of a school. During the initial excavations, some amateur archaeologist uncovered graves that subsequently led to an archaeological dig. The age of these burials dated to the "Atlantic" dating from 5,000B.C. to 3,000 B.C. A large number of polished axes of greenstone were found along with harpoons heads of roe antler and slotted bone points.

The excavations were described in, *Excavations of a Mesolithic Cemetery at VedBaek, Denmark* by Svend Erik Albrethsen and Erik Brinch Peterson, Copenhagen, 1975. "Carbon 14 dating of this site was 4,100 B.C. The burials were laid out in regular rows with most of the skeletons oriented east-west. Most of the pit graves contained only one individual, except one in which three skeletons were found. Almost all of the graves contained red ochre. In some of the graves remains of a bark coffin was observed."

Photograph of the glacial kame called Bogebakken, taken in 1924.

"One of the most elaborate of these graves was a young woman believed to be about 18 years old and by her right side was her new-born infant. Near her head was nearly 200 pendants from red deer teeth, the front teeth of wild pigs and perforated snail shells. Red ochre was also included in the burial. The baby was laid on a swans wing with red ochre sprinkled on and below the body. A large truncated blade was also with child, suggesting to the excavators that the young child was a boy."

"17 graves were unearthed that contained 5 children, 7 females and 10 males. The skulls featured prominent cheek bones and brow ridges and in several cases thick walls. Most of the skulls from Bogebakken, both males and female have prominent brow ridges that was regarded by archaeologist to have been a racial characteristic of this population. The skeletons they said, " show many primitive characteristics. The heavy facial features with prominent brow ridges and relatively low-set eye-sockets are typical of the Cro-Magnon race, so the population of Bogebakken may be regarded as remote cousins to the Cro-Magnon man."

Additional finds of Ertebolle skeletons were found from Dyrholm in Jutland and from Korsor in Zealand as reported by Steven Coons, in *The Stone Age.* "A fragment of a skull from Dyrhom was described as being extremely thick with pronounced torus occipitalis. The Korsor skeleton was a well preserved skeleton of an adult male. The cranium was described as "large and massive," with protruding brow ridges and a sloping forehead and heavy mandible, that was characteristic of the Cro-Magnon race."

In an earthen burial mounds in Sweden dating from 4500-3800 B.C. were cremations, internments in a sitting position, double graves with men and women and a rich child graves that included dogs. Dogs within graves occurs in many cases with children; the belief being that they would protect them in the afterlife as they did in this. Timber structures over two graves indicated the remains of a charnel house. At the Popovo burial site that dates to 5,000 BC "The cemetery was situated on the top of a sand-and-gravel hill. The bodies were placed in two rows along the river bank. In five instances the head was facing to the east, and in one instance to the west. The graves were .035-0.7 m deep. The dead were placed in were extended position and were covered, together with small lumps of coal.

East of the Baltic at Lake Onega within the former Soviet Union, similar skeletons were discovered and reported in *Archaeology in the U.S.S.R. Alexander Mongait 1959,* "There is a very interesting Neolithic burial ground on the South Oleny Island in the northwestern part of Lake Onega, near the Tranonega Peninsula. Over 170 burials (excavations by I.V. Ravodonikas, 1936-38) were found in it. This was the cemetery of the folk who lived on the shores of the lake. The dead were brought to it in boats or on rafts and buried in shallow pits where they were painted with raddle [red ochre]. Articles like stone and bone implements and ornaments were placed in the graves. On some of the male skeletons there were stone javelin points and arrow-heads, bone daggers and harpoons, schist axes and knives, necklaces and pendants made from the teeth of wild animals mostly, incisors of the elk and bears fangs. There are interesting burials of women with babies: the skeletons of the babies lay next to the women's skeletons or between their legs. Some of the graves are in deep vertical pits in which skeletons stood upright. In one of the burials the skeleton had on its chest a large bone dagger with sharp flint flakes fitted into a groove along the edge.

Photograph is from "*Archaeology of the USSR,*" 1959. The skull from Oleny Island with the remains of a head-dress made from beaver incisors. Note the protruding brow-ridge and a forehead that is completely lacking along with the massive jaw of this female skull. These physical traits are more attributable to "archaic" skulls than "modern." Another "archaic" physical trait that is evident in this skull is the position of the mandible foramen. This is the hole in the jaw where the nerve goes through. The mandibular foramen is an opening on the inside of the vertical part of the mandible for the branch of the mandibular nerve that reaches the teeth. In modern skulls the mandible foramen is located on the chin and in archaic forms of humans it is located underneath the molars.

"Ontogenetic Migration of the Mental Foremen in Neanderthals and Modern Humans" by Frank L'Engle Williams and Gail Krovitz, originally published in the *2004 Journal of Human Evolution"* "Mental foramen position in modern human adults appears to be more anterior with respect to its position in Neanderthals. Nineteenth century researchers were the first to note the unusual position of the mental foramen in Neanderthal manibles. These researchers compared the position encountered in Neanderthal mandibles frequently located below the first mandibular molar (M 1), to its position in

modern humans most often positioned interior to the first or second premolar (p 3 or p4), or under the interdental septum."

Gimbutus commented on the Olen Island skeletal remains, stating "Skeletal material from northwestern Russia also evinces the presence of Europoid and Mongoloid traits. The skeletons from southern Olen Island in Lake Onega belong to two different types, Cro-Magnon like Europoid and Mongoloid. The Europoid type from Olen Island cemetery was close to an East Baltic or Ladogan type."

Looking further to the east In Siberia the tool kits and weapons are similar to those found to the west, in the Baltic region. In a report done by the *Carnegie Institution of Washington* in 1928, Neolithic artifacts from Siberia were described, "There are points of lancet-leaf, willow leaf types and more slender lancet-like form. Lance heads were as variable in form as arrow points, but of larger size. This same may be said of scraper, awls and drills were also of different types as to size, form and hafting. Two kinds of stone fishing sinkers were found: grooved balls and natural flat pebbles with notches on two sides."

"The Siberian Neolithic man also manufactured some large stone implements, in the shape of slender chisels, wide gouges, bent gouges, axe-wedges and axes with ears. These implements were fastened to wooden handles by means of thongs or ropes made of vegetable fibres. The large stone implements were intended for work on wood and to dig canoes out of tree trunks."

A few burials were discovered in the region, but no good description of the skull types were given except that they were long headed and craniometric measurements showed that they were identical to skulls found in the Kurgan (burial mounds) of south Russia who were identified as the blue-eyed and blonde Usuns or Wusuns and the Dinlins of the Chinese annals of the second century B.C. They were legened to have lived in north Mongolia and were of the same stock as with Sacea, the western Scythians, and with the Nordics of northern Europe.

Shell mounds are found as far east as the Japanese archipelago and as far north as the Karelian Islands. The dates of the shell mounds range from 7,500 to 2,000 B.C. Their tool kits are similar to those in the west including polished plummets, bone combs and slate spear heads. This group has been called the Joman Culture. "Joman" is derived from pottery associated with the early Joman that had "cord markings." This type of pottery is also found to the west of the Ural Mountains and in the Baltic region, associated with shell mounds and the later Bell Beaker People. Tool kits, such as the toggle harpoon are evidence that the Joman were deep sea fishing far off the the Japanese coast.

Skeletal remains of the Joman, also known as the Ainu, indicate that in eastern Asia as well as in Europe, was already racially complex, and that peoples of the Cro-Magnon type stretched across the entire width of the northern half of the Eurasiatic continent. Stevens Coons, in *Upper Paleolithic Man in China, writes* "By means of this knowledge we may explain, at least in part, the enigma of the Ainu, a large-headed, broad- faced Europoid-Mongoloid hybrid, living on the outer periphery of eastern Asia. At the same time fresh light is thrown upon the human materials which may have taken part in the early peopling of America,"

The following article describes large skeletons in northern Japan. Besides the large size of the skeleton, another physical attribute described was flat shin bones. This is also found with the Allegewi Hopewell mound builders of the Ohio Valley. The copper arrowheads would date the find to sometime after 2700 B.C.

Mansfield News, **July 13, 1918**
STONE AGE SKELETONS
Bones of Early Japanese show Their Great Height
Fifteen human skeletons were unearthed in the provinve of Kawachi. This is considered the birthplace of Japanese Civilization. Of the relics of the Japanese stone age discovered by Professor Okushi three of the skeletons were in perfect preservation all bones being intact. *East and West News* says it rarely happens according to scientific records that so many perfect skeletons are discovered in one place.

Among indications that people of that period lived on uncooked food is the fact that upper and lower teeth are evenly worn down. Decayed teeth are not found. The bony structure of the skeletons are massive, shin bones in most cases are somewhat flat. Some of these skeletons stand seven feet high and even shorter ones are over six feet. Skeletons were found in a lying position with knees drawn up. Without doubt these people belonged to the stone age in Japan.

While making the excavation stone implements, earthenware and two copper arrowheads were found. Two white jade earrings were discovered which may be Chinese in origin.

Marija Gimbutas in 1956 "The Prehistory of Eastern Europe. "Up to the beginning of the second millennium B.C. The culture of Mesolithic appearance continued in the East Baltic area and northwestern Russia. At the end of the third millennium B.C. its pottery was predominantly decorated by comb-like impressions. Hence, it can be designated as a "Comb-marked Pottery" complex. Its carriers were Europeans with massive skulls and wide faces bearing a certain Cro-Magnon-like implication."

"In the broad sense, the culture of north-eastern Europe in the hunting-fishing stage is a counterpart of the Eurasian culture which stretches across the whole forested zone of northern Europe and Asia to North America. Northern Scandinavia, northeastern Europe and western, central and even Siberia indicate related traits. The Scandinavian Arctic culture has characteristics common with the East Baltic and northern Russia. On the other hand, the area beyond the Urals in Asia shows an equally clear affinity to northeastern Europe and indeed the whole circumpolar world possesses related cultural characteristics."

"The only domesticated animal was the dog, and its remains have been found in nearly all sites. Bone was used for making awls, needles, knives, arrowheads, hooks and harpoons, ice picks, etc. Wood was used for tools, as finds in peat bogs attest."

"Articles of slate and quartz including large celts and gouges, net sinkers, scrapers, strait arrowheads, knife blades, spears daggers are characteristic of northeastern Baltic and and northwestern Russia and particularly of Karelia." A particular concentration of sites is known from the lakes region of Finland, Karelia and northwestern Russia, especially around Lake Onega. The situation of the settlements, as well as the mass of fish bones, harpoons, conical projectile points, hooks and net sinkers [plummets] attest fishing as one of the main sources of livelihood."

Net sinker or plummet from the Baltic region. Some of the plummets were highly polished, implying that they were not only utilitarian, but also had a spiritual significance.

In *"The Races of Europe"* by Carleton Stevens Coon, in describing the Upper Palaeolithic man in Europe, writes " In totality of facial features, with few exceptions, the Upper Paleolithic people may be said to have resembled modern white men. Some however, probably looked like a certain type of American Indian, notably that of the North American Plains."

Migrations of the Cro-Magnon, Hunters and Fishers to the North American, Pacific Coast

The "Berengia Theory" states that the Americas were populated after several migrations from Asia, across the Berengia land bridge that was briefly open when the ice sheets began their retreat from about 10,000 B.C. This theory is under continued scrutiny, based on new evidence suggesting that the earliest immigrants were more genetically diverse than previously thought and didn't walk here, but came in boats. The earliest skeletal remains found in western, North America resemble the prehistoric Joman people of Japan and their closest modern descendants, the Ainu. The Joman were an extension of the circumpolar complex of Hunter and Fishers who stretched across Siberia to the northern Japanese islands.

First Americans-Origins of Man, "LookSmart.com" February 1999,. "When the seafaring theory was proposed in the mid- 1970s, it sank for lack of evidence. Any shoreline outposts of an ancient maritime culture would have been submerged when sea levels rose some 300 feet about 12,000 years ago at the end of the Ice Age. But as the timeline for the new-world occupation has changed, the theory seems downright sensible, if not quite provable. The Pacific Rim has vast resources of salmon and sea mammals, and people need only the simplest of tools to exploit them: nets, weirs, clubs, knives. Whereas ancient landlubbers would have had to reinvent their means of hunting, foraging, and housing as they passed through different trains, ancient mariners could have had smooth sailing through relatively unchanging coastal environments. And recent geological studies show that even when glaciers stretched down into North America, there were thawed pockets of coastline in the northwest where people could take refuge and gather provisions. "Most archaeologist have a continental mindset," says Robson Bonnichsen, an anthropologist at the Center for the Study of the First Americans at the University of Oregon in Eugene. "But the peopling of America is likely to be tied very much to the development and spread of maritime adaption."

The Joman and Ainu have skull and facial characteristics more similar to Europeans than to mainland Asians. The most celebrated of the Joman people is not from Japan, but Washington State and is known as the Kennewick man who dates to 7,000 B.C. Additional skeletal remains have been found that are also an apparent extension of the Cro-Magnon Hunters and Fishers, both Spirit Cave Man and two Minnesota skulls, one dating 7,900 and the other 8,700 years old have been analyzed and showed European facial characteristics.

Kennewick Man, with archaic facial features of a prominent brow ridge and sloping forehead.

As a homogeneous Maritime Culture we should find on both the Atlantic and Pacific coasts of North America, similar artifacts and skeletal remains from the years 7,000-2000 B.C. as those found in Europe and Asia. Tool kits should include bone harpoons and awls, implements of slate and quartz, leaf shaped projectile points, fish hooks, ice picks, fish nets, net sinkers or plummets, gouges and stone axes. Burials should be found in glacial kames, near lakes and within shell mounds. Red ochre being included many of the graves in addition to evidence of charcoal or ashes. A bed within the burial pit may be lined with bark or stones. Artifacts within the grave may be ritualistically broken. Skeletal remains showing archaic features, with facial prognathism, protruding brow ridges with a low sloping forehead, thick skulls walls and some instances of occipital buns. The most most compelling evidence are the large skeletons, that is a physical trait of only one species of man, Cro-Magnon.

Several interesting historical accounts were found that describe ancient burials in the Aleutian Islands that reveals evidence of the migrations of the Joman to the North American Pacific coasts. In the first report the skeleton was found in a kitchen midden or shell heap, and had artifacts that were ritualistically broken. The second report describes a large seven foot skeleton, found in a ridge or kame overlooking the ocean. Polished slate implements were also found in the Kuril Islands associated with the Joman, and with the Maritime peoples in the Baltic region and the Northeast American Atlantic coasts. The bodies were placed like spokes of a wheel. This type of arrangement of the dead is common within the later Allegewi and Hopewell Sioux, Iroquois and Cherokee peoples.

Indiana Evening Gazette **April 4, 1950**
Aleutian Skeleton Found

State College, Pa., April 4-The skeleton of an ancient Aleutian woman, excavated by an Army Air Forces chaplain on the island of Architka, Aleutian Islands, has been presented to the Pennsylvania State College.

Capt. Cecil C. Cowder, of Bigler, Clearfield County, said he excavated the skeleton on November 9 while he was stationed at Shenya Air Force Base.

The body had been buried in an ancient "kitchen mitten," a refuse heap, at a depth of five feet. With it were a number of bone harpoons, several bone fish shanks, stone blades, a bird, a fish and numerous pieces of hard bone and rough stone which probably were intended for use in the future life. The harpoons had been deliberately broken and then placed carefully on top of the body."

The island of Architka is halfway between the coast of Kamchatka and the northern Japanese Kuril Islands; (the home of the Joman), and mainland Alaska's Bristol Bay. It can be found on a globe or map where the international dateline makes a large "V" to incorporate this island.

Washington Post, **Sept. 16, 1944**
Major Finds Grave of Giant Aleutian

An advanced Aleutian Base (U.P.) Site of a strange burial of a prehistoric giant was discovered on an Aleutian Island recently by Major. E. E. Chittenden, Kearney Neb. The ancient Aleut, who had been at least 7 feet tall, has been buried on a low ridge overlooking the ocean, and in the same shallow grave with him were the skeletons of five women, placed to form a geometrical pattern.

Major Chittenden found the burial site while excavations for a military installation were being made, and he states the six skeletons had been placed with their heads together, so that trunk and leg bones extended outward like the spokes of a wheel. In the unusual grave were carved ivory ornaments and weapons made of polished slate.

The idea of a Maritime migrations occurring before people began walking over the frozen Berengia Land Bridge was crystallized when the oldest skull found in North America was discovered on an island off the California coast. The find was reported in Infoplease.com *Oldest Human Remains in North America Found* "In 1959, the partial skeletal remains of an ancient woman estimated to be 10,000 years old were unearthed in Arlington Springs on Santa Rosa Island, one of the eight channel islands off the southern California coast. Her remains dated to 11,000 BC. The earlier date and the location of the womans remains adds weight to the alternative theory that some early settlers may have constructed boats and migrated from Asia by sailing down."

The oldest skeleton found within the American continent is that of the Penon woman who was uncovered in Mexico City. The skeletal remains identical to the Hunters and Fishers of Europe and Asia. The find was reported in, *CNN.com/Science and Space*, December 4, 2002. "Penon woman was found while digging a well near the Mexico City International Airport. Scientist believe that Penon woman died anywhere between 10,700 and 11,000 BC at the age of 27. Silvia Gonzalez from Oxford University believes, "the bones of the Penon woman belong not to Native Americans, but to the Ainu people of Japan"

Gonzalez also stated in *The Independent* by Steve Connor, Science Editor December 2002. Dr. Gonzalez told BBC News Online: "Studies of Native Americans early indicated a link with modern-day Asains, supporting the idea of a mass migration across the Bering land bridge. But one DNA study also pointed to at least some shared features with Europeans that could have derived from a relatively common ancestor who lived perhaps 15,000 years ago."

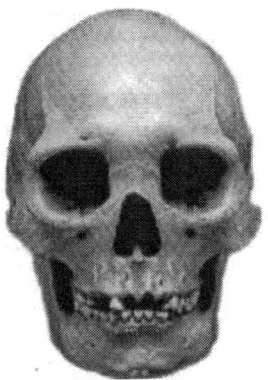

Skull of the Penon woman with protruding brow ridge and an accentuated temporal line. The mandible foreman is also under the molars, similar to the skull found at Oleny Island, in northern Russia.

A few of the tools kits and burial practices on the west coast are identical to these found in the eastern Woodlands, that are associated with the Meadowood and Point Peninsula Iroquois Cultures. DNA studies have found a genetic link with the northwest coast's, Yakima Indians and the Ohio Hopewell. Were the Yakima a detached Iroquois tribe? Daniel S. Meatte wrote in 1990 in *Prehistory of the Western Snake River Basin,* " Between 4,500 and 4,000 B.P., with possible extensions until 3,500 B.P. Identified cultural attributes include massive turkey-tail and cache blades, caches or obsidian blank/preforms, large side notched projectile points, flexed or semi-flexed inhumations, possible cremations, and candid skull interments. Additional characteristic include the use of red ochre. Human burials are placed in unmarked cemeteries with a preference for high sandy knolls along river terraces."

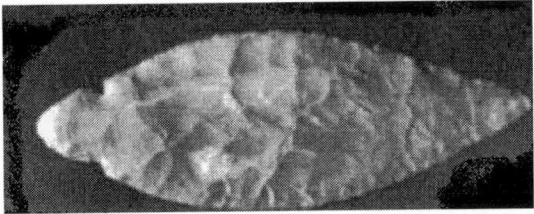

Turkey Tail points were mined from Harrison County, Indiana grey flint and associated with the Red Ochre Iroquoians.

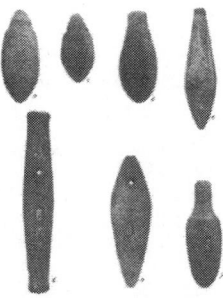

Artifacts found in Northern California are identical to the burial inclusions found in the Great Lakes region, associated with the Red Ochre, Glacial Kame and Meadowood Iroquois. Artifacts consist of plummets, a bar amulet and trapezoidal one-holed pendant.

The burials and artifacts on the Snake River are consistent with those associated with the Glacial Kame and Red Ochre burials found in the Great Lakes region. Additional evidence of the the Iroquoians, trading or inhabiting the extent of the continent, are the Pacific coast items found as burial inclusions in the Midwest.

History of Darke County, Ohio, 1905

The locality in and around Nashville, German Township has some interesting information. One or two mounds have been opened yielding a lot of relics, skeletons, etc. Two large shells native of the Pacific Coast, were taken from one of the mounds. The inside had been cut out of them leaving a large cavity capable of holding about one gallon, and make a very beautiful addition to the furniture of the ancient people of the stone age.

One of the most interesting burial spots was discovered on the farm of Jesse Woods in German Township. In digging the cellar under the house where he lives, Mr. Woods discovered a skeleton in a sitting posture. It was covered with plates of Mica and was the central figure in a group of other skeletons arranged in a circle around it. The skeletons in the circle were lying at full length.

The following articles describe skeletal remains found along the coastal regions of the Pacific and a few of the western states. Slate spears, bone awls, stone smoking tubes and plummets are the most common artifacts found; indicating a people that was reliant on fishing as their principle subsistence.

The skeletal remains described in the following article are large, in addition to the skulls being described as having "primitive" characteristics. It is reasonable to assume that the remains described are of the same stock as the Penon Woman and Kennewick Man, who both had "primitive" characteristics.

Another physical characteristic that is evident within this population is the physical abnormality of possessing a double row of teeth. While a large skeleton would appear to be rare, in combination with a double row of teeth would imply that a single people is being represented.

Bancroft's *Native Races,* **1882,**

Further Bacroft writes, "Mr. Taylor heard from a resident of San Buenaventura that " in recent stay on Santa Rosa Island, in 1861, he often met with entire skeletons of Indians in the caves. The signs of their rancherias were very frequent, and the remains of metates, mortars, earthen pots, and other utensils, and other utensils very common. Extensive caves were met with which seemed to serve as burial places of the Indians, as entire skeletons and numerous skulls were plentifully scattered about in their recesses." Some very wonderful skulls are also reported as having been found on the islands, furnished with *double teeth all the way round the jaw."*

Near Comox, one hundred and thirty miles northwest of Victoria, a group of mounds were examined in 1872-3, and found to be built of sea sand and black mold, mixed with some shells. They were from five to fifty yards in circumference. In one by the side of a very large skull was deposited a piece of coal; and in another with a very peculiar flattened skull was a child's tooth.

Evening News, **(Ada, Oklahoma) November 8, 1912**
PRIMITIVE MEN OF GIGANTIC STATURE

Eleven skeletons of primitive men, with foreheads sloping directly back from the eyes, and *two rows of teeth* in the front of the upper jaw, have been uncovered at Craigshill, at Ellensburg, Wash. They were found about twenty feet below the surface, twenty feet back from the face of the slope, in a cement rock formation, over which was a layer of shale. The rock was perfectly dry. The jaw bones, which easily break, are so large that they will go around the face of the man today. The other bones are also much larger than those of the ordinary man. The femur is twenty inches long, indicating a man of eighty inches tall [6' 8"] The teeth in front are worn almost down to the jaw bones, due, it is believed, to eating uncooked foods and crushing substances with the teeth. The sloping skull shows an extreme low order of intelligence.

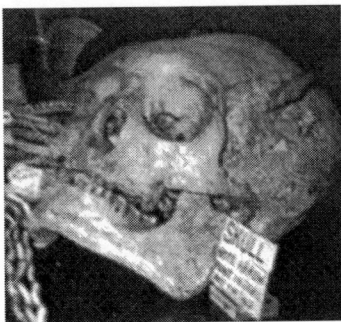

This photo appeared on *www.viewzone* **in February, 2002, described only as ancient skeleton, that was photographed in window of a Seattle, Washington shoppe. Is this the photograph of one of the skulls from Craigshill or Ellensburg, Washington?**

Decatur Weekly Republican, **April, 9, 1923**
New Link in Man History Is Found on West Coast
Santa Barbara Mound Yield Remains of People Older Than Neanderthal

SANTA BARBARA, Cal., Oct 27.-All doubt as to the greater age of the skulls of the "Santa Barbara man" uncovered here this week, as compared with the Neanderthal man of Central Europe, has been dispelled in the minds of scientist doing excavation work on the Burton Mound fronting the Santa Barbara ocean beach according to J.P. Harrington of the Smithsonian Institution in a formal statement tonight.

Dr. Harrington, who has been in charge of southern California archaeological work for the Smithsonian Institution for several months, is certain that a new link in the Anthropological chain has been established definitely by the excavations of the last few days. Further examination of the gorilla-like skulls unearthed on Burton Mound, he asserts, has definitely proven that the Santa Barbara man existed in a period far earlier than the era of Neanderthal man. Not only that, but he possessed a culture which far exceeded that of the Neanderthal.

Burton Mound in Santa Barbara from a sketch done in 1809. The mound has since been levelled.

Tools are Found

Artifacts found in the hardpan which gave up the skulls showed Santa Barbara man used tools and implements, which although crude were greatly in advance of those supposed to have been employed by the Neanderthal man in the dawn of the world's civilization. Instruments resembling pestels, crude barbless fish-hooks and other relics encrusted in the protecting calcerous soil point almost unmistakably to that conclusion, Dr Harrington said tonight.

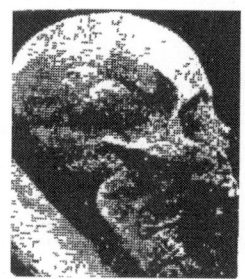

The skulls of the Santa Barbara men were carefully cleaned today in order that more minute investigation might be made. This led to the discovery that the primitive owner of the skeleton remains, possessed a mouth larger than any man of modern or ancient times. The mouth of one of the skulls was widely opened, as if the early man had died in great pain or fear. The jaws measured nearly seven inches. The same pronounced suprorbital ridge existed in both skulls with the same lack of forehead and other scientific evidences of primitive existence.

Skulls Thick

The thickness of the skulls is twice as great as those of Indians found in the burial grounds, known to be 1,000 years old or more. The average thickness of each skull is approximately three-quarters of an inch.

Dr. Jesse Walter Fewkes, chief curator of the Smithsonian Institution telegraphed Dr. Harrington today for a complete report of the discoveries made here. An authoritative and official statement has been dispatched to him.

Newark Daily Advocate, (Newark, Ohio) July 12, 1897
Ghastly Indian Relics
Bones of a Giant Race on San Nicolas Islands

After nearly three weeks' sojourn on the barren island of San Nicolas a party of relic hunters reached Long Beach, Cal., loaded with skeletons, skulls and ancient implements and ornaments of stone and shells, the remains of Indian tribes which inhabited the now almost desert waste in bygone ages.

There were 11 in the party which left Long Beach in the gasoline schooner San Clemente for San Nicolas Island, which lies 65 miles off the coast from Santa Barbara. Four days were occupied in the journey to the island owing to the dense fogs, and after landing the party the schooner returned to Long Beach and the explorers were left to their work.

The party found 87 skulls buried in the sand of the island, but were only able to secure three entire. They made one excavation 20 feet square in which they found nine skeletons in a crouching attitude, as though men, women and children had been buried alive. In another place they found the remains of hundreds of bodies that had been burned, and some of the party believe that cremation was practiced by the ancient people of the island.

Positive evidence was found that the island was inhabited by two or more different races in the dim past, one of which was of great size, a peculiar characteristic being gigantic jawbones. Some of the specimens of the latter brought by the party are almost large enough to slip over the head of an ordinary man. Mr. Longfellow, one of the party, speaking of the trip, said:

"One of the most interesting relics brought back by us was part of a skeleton of a large man in whose bones a long bone spear point was sticking. In the shattered skull was a big round stone used as a war implement. The spear passed near the heart and entirely through the shoulder blade. I am sure that two different races fought and died on the island, as most of the bodies were of moderate size while some were almost giants. The latter were always in isolated graves. We found many implements and weapons of stone, but all are very crude and show almost no ornamentation-San Francisco Chronicle.

The Dubuque Herald, December 5, 1900
THE CHANNEL ISLANDS
THEY HAVE LONG BEEN A MYSTERY
Rich Field for Anthropological and Archaeological Investigations
Lies Off Coast of Southern California

San Francisco, Dec. 4-An expedition of scientific students has been made up in Los Angelas and Pasadena to explore thoroughly the Channel islands off that part of the coast of California known as Santa Barbara and San Buenaventura counties during the next six months. It ss to be sustained by Stanford university largely, and to a less extent by several denominational colleges in southern California. Anthropological and archaeological students who have spent several weeks each on these islands, say they are one of the very richest fields for work in that department of knowledge on the pacific coast. The channel islands constitute California's only archipelago, with the possible exception of the rocky and scanty Farallone islands. They have been objects of romance, legends, curiosity and mystery for a generation or more.

St. Nicholas island lies eighty miles immediately opposite the little city of San Buenaventura and is the most interesting of all the Channel islands from the many points of view. As far back as the memory of any person in southern California runs hundreds of white skeletons have dotted the valleys and hillsides. Strange utensils of serpentine sandstone and steatite are found there among the human bones and the island and its erstwhile inhabitants have a history so curious that it is difficult of comprehension.

As far as the eye can trace there are barren levels with innumerable circular depressions, showing where primitive dwellings once stood. Not a vestige remains of the materials used in the construction of these rancherias. Hundreds of shell mounds are scattered about and are found to consist of astonishing numbers of mollusks, the bones of every species of fish found in the channels, skeletons of seals, sea elephants, whales, sea otter, the island fox and various aquatic birds. Without question these animals were used for food by the tribe that once thronged those boundrys.

An examination of some of the mounds discloses all sorts of curious utensils-stone cooking pots, mortars, pestles, drills, bone needles and fish hooks, shell beads, charm stones [plummets], pipes, cups and a few arrowheads, spear points and swords made of bone.

The most grewsome of all the sights on this strange island is to be seen on the broad plateau south of the Chinese camp at Coral harbor. Here acres of the naked sand are littered with hundreds of disjointed skeletons, and present the most reckless illustration of the "ground plan" of humanity that imagination can picture. Measurements have been made by several scientists of the thigh, leg and arm bones, and literally bushels of skulls and other parts of the human frame have been brought to Los Angeles from St. Nicholas islands for investigation. The general opinion is that the Indian race that swarmed over the islands was much larger than any civilized race of today and that some of the men must have been seven feet three inches high. The skull of this extinct tribe often measure several inches more than some of the other large skulls of today. Many skulls found lying about on the island show that their possessors must have suffered death from a club or blunt battle axe.

Woodland Daily Democrat, (Woodland Ca.) **February 13, 1922**
Bones of Giant Found in South
Best Preserved Skeleton of Extinct Tribe Hauled from Channel

San Francisco, June 10- Up to three hundred years ago, a giant race of Indians inhabited the coastal regions of California. Remains of these have been discovered in the islands of the Santa Barbara Channel. To William Altman, assistant curator of the Golden Gate Park Museum belongs the honor of discovering one of the tallest and best-preserved skeletons of this extinct tribe.

The giant skeleton found was ten feet from the surface, and around it were a large number of mortars and pestles, charm stones and obsidian arrow heads. One of the skeleton measured seven feet four inches."

Artisans and Artifacts of Vanished Races, **T. Dickerson, 1915**
Bones Show Giant Man Many Years Ago
Remains Found In California Believed To Date Before The Mound Builders

Los Angeles-Men working under the personal direction of Frank S. Daggert, director of the Museum of History and Science and Art, are excavating on La Brea ranch and have discovered what they have been searching for-the bones of a man.

Several bones resembling those of a huge man, have been dug from the clay beds and are carefully guarded until all parts of the skeleton have been found. If, as if hoped, the find is genuine, it will startle scientist throughout the world and will prove that America, long regarded as a continent barren of humanity prior to the age of the Mound Builders, was peopled with a race of thousands of years before these builders appeared.

"There is little doubt," said one man who is familiar with the work, "that the bones are those of a man. The question to be determined is how long ago he walked the earth. This must be disclosed by geologist. The formation of the bones indicates that they are not those of primitive man, and it is possible that he was one of a race who inhabited this land in the Pliocene Age."

Anthropologist of the University of California will, it is believed, regard the discovery as further proof of their contention that North America and Europe were once joined together and that the back bone of this peninsular is the Aleutian Islands. Scientist believe that the two-toed horses, remains of which have been found in Nevada, made their way across the neck of land from Western Europe.

If the bones now being taken from La Breas' pit bear a close resemblance to the fossil remains of the early races if Europe, proof will be established that man also used the pathway between the two continents thousands of centuries before Columbus was born.

Fresno Bee, November, 26, 1937
Skeletons Of Vanished Race Predating Indians are Found Near Lodi

San Francisco, Nov. 26- The horizon of earliest mankind in California has been moved back many thousands of years through the discovery of fossilized skeletons in the Lodi region

The University of California anthropological department made the Maidu and Miwok Indian tribes of startling announcement today that the skeletons probably of people antedating the American Indian.

Professors at the university thus far have been unable to identify the age of the skeletons but stated there is little doubt they represent the earliest California man of which there is any knowledge.

Before Bow and Arrow

Many of the artifacts revealed by the excavations are apparently peculiar to this early culture alone and there is evidence that it antedated even the bow and arrow, indicating a great lapse of time between that period and that of the earliest Indians whose burial mounds have been found in the same region heretofore.

The newly discovered burial mound had been built upon by another later race now virtually extinct, from which sprang the Maidu and Miwok Indian tribes of the Sacramento Valley.

Found By Ditch Diggers

The fossilized skeletons were unearthed by deep ditch digging operations in thick alluvial deposits near Lodi several months ago. They were called to the attention of the Sacramento Junior College, which made an extensive study of the remains under the direction of Dr. J. B. Lillard, the college president. Later, they were forwarded to the University of California for the identification and further study.

Professor A. L. Kroeber, chairman of the university's anthropological department has pronounced it the most important discovery in the universities history and one that will throw new light on all American antiquity.

Little Relation

Kroeber says the skeletons present features entirely new from a scientific standpoint, and bear little specific relation to the latter cultures, particularly the historic Indians. Many still had the remnants of the fureral finery about their necks and bodies in the form of beads.

Mansfield News, Mansfield Ohio, July 25, 1911
Important Historical Find In California

From an excavation made by workman in the employ of the Por Costa Water company have been found a large number of Indian relics of great age including the specimens of crude pottery already mentioned and the skeleton of an Indian giant more than seven feet tall. The skeleton is in the possession of Dr. Neff of Concord, who is mounting it for an exhibition. The pottery specimens consist of charm stones of baked clay of spindle shape and pierced so that they may be suspended from the neck by cords. In addition there are a large number of knives and arrowheads of obsidian."

Native Races, **Vol. IV. 1882 Hubert Howe Bancroft**

Dr. Yates sent to the Smithsonian Institute, in 1869, a collection of relics taken from mounds in Alameda County. It is not expressly stated that these were shell mounds, although I have heard of the existence of several in that county. This collection included, "stone pestles, perforators or awls, sinkers, a phallus, spindles, a soapstone ladle, stone mortar and pestle, pipe bowls, shell and perforated stone ornaments, an ancient awl and serrated implements of bone."

Quite a number of mounds are known to exist on the peninsula of San Francisco, several being in the vicinity of the silk factory on the SanBruno Road. One of them covered an area of two acres, was at least twenty-five feet deep, and from it was taken arrowheads, hammers, and many other relics. One of these shell mounds, near the old Bay View racetrack is being opened by Chinamen engaged in preparation for some building, as I write this chapter. Mr. James Deans of whose explorations I shall have more to say when treating of the antiquities of British Columbia, has brought me a large number of stone and bone relics taken from this deposit, the different classes of which are illustrated in the accompanying cut.

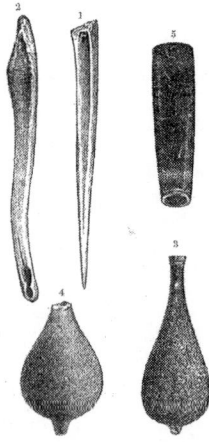

Figure 1 is an awl of deer-bone, and fig 2 is another implement of the same material, curiously grooved at one end. These bone implements occur by the thousands, being from three to eight inches in length. Figures 3, 4, are perhaps stone sinkers, or as is thought by some, weights from diorite, the latter from sandstone, and not polished. Figure 3 is four inches long, and an inch and a half in its greatest diameter. Hundreds of these pear shaped weights are found in the mounds, but the end is usually broken off, as is the case with fig. 4 Fig. 5 is an implement carved from black clayey slate, and is brightly polished surface. It is four inches long, and one inch at the smaller. It is hollow, but the bore diminishes in size regularly from each end, until at a point about an inch and a half smaller end it is only a quarter of an inch in diameter.

Oakland Tribune, **August 24, 1896**
SEVEN SKELETONS
Remarkable Discovery Made at Shell Mound Park
SKULLS RESEMBLING APES

An interesting discovery has been made at Shell Mound Park, where the skeleton of a prehistoric race of Indians was excavated. These skeletons are of a race unknown at present and are undoubtedly of great antiquity.

They were discovered in a shell mound on the west side of the racetrack. The mound is the usual kind, formed of shells, and is about ten feet high and 100 yards in length. Men were digging in the ground in order to investigate the soil, when their spades struck against bones.

A skull was laid bare. The skull was of the most unusual formation and appearance. Professor John Merriam of the State University was immediately sent for and work was suspended until he arrived.

Professor Merriam went to work himself and very carefully dug for the remains. In a short space of time three skeletons were discovered. The skeletons are of ordinary size and the most extra-ordinary characteristic about them is the shape of the skulls. They are more like the skulls of apes than human beings, and present a type of an unusually degraded and depraved race of Indians. There is little or no forehead and the lower part of the skull is shaped like an ape's.

Professor Merriam declares he has never seen any skulls like these, although they bear a strong resemblance to the heads of the Flat Head Indians, former residents of the more northern coast.

The skeletons are undoubtedly of great antiquity, as a careful study shows that they have laid beneath the mound for many hundreds of years. The mound itself is formed like other shell mounds. The peculiar race of Indians, who lived along the coast, camping on the very shore of the ocean, existed principally on shellfish. As soon as they devoured the fishes they probably threw the shells beside their campfire. As time went by, these piles grew into small mounds of various heights. The mound recently excavated is at least ten feet in height and at one time probably stood close to the waves of the ocean.

That the skeletons were there before the mound was built is proved upon examination. The remains were found below the skulls and beneath strata of ashes. The ashes were undisturbed and formed a two-inch covering over the bones.

The question now arises, how the ashes came there? The bodies were probably placed two or three inches below the surface and a covering of ashes above them as an additional protection from the air, or it may have been that the ashes were thrown there in an entirely accidental manner, as the result of a camp fire. The fact that the ashes were there however proves that the relics are ancient. Seven or eight skeletons have been taken out, but they are badly broken condition. They are in the possession of Prof. Merriam.

This photo of the Emeryville shell mound, near San Francisco was originally taken in 1902 and printed in 1907 in *"American Archaeology and Ethnology."* A large dance pavilion was built on top of the mound. The raceway is visible to the left of the pavilion.

San Antonio Light, **September 19, 1932**
Huge Skeletons of Indians Found

STOCKTON, Calif., Sept 24- Three skeletons, one seven feet long have been uncovered by Harry T. Sanford, college of the Pacific archaeologist, and crew of men engaged in excavating an Indian burial mound near Garwood Ferry bridge. They are believed to be the largest Indian skeletons ever unearthed in California.

Wisconsin Rapids Daily Tribune, **November 26, 1937**
Find Trace of Race Which Preceded Ancient Indians

Berkely, Calif. Nov. 26-(AP)-University of California anthropologist announced today they had uncovered the fossilized bones of some people who may have been the original native sons of this golden state. The scientist believe the beetle-browed, bulldog-jawed skulls and sturdy skeletons are those of a primitive human race that peopled the coastal plains long before the Indians. In one instance they found such bones beneath the burial mounds of a later race from which sprang the present-day Miwok and Maidu Indians. A profound geologic change, which completely buried the earlier men, their villages and burial mounds, separated them from the later race.

The bones, and numerous weapons and utensils, however are not as old as others found elsewhere on this continent, the scientist said. The remains were uncovered accidently during ditching operations at Lodi, 60 miles east of here. Sacramento Junior College anthropologist declined to estimate the bage of the bones but said they were "thousands of years" old. Artifacts in the mounds included charm-stones made of abelone shells, numerous quartz crystals, ashaltum objects, some creations resembling slate pencils, shell beads, barbless bone points, points chipped from stone and stone grinding mortars.

The Daily Citizen **(Iowa City, Iowa) November 25, 1892**

A few years ago some remarkable skulls were found in Oregon by a scientist who is now connected with the Metropolitan museum in Central park, New York. They resemble the skulls of sheep, and yet there is no doubt that they were human. An examination of their form led to the idea that they came nearer to the connecting link between man and the brute creature than any other skulls which had been dug out of the earth.

Marion Daily Star **(Marion, Ohio) April 14, 1923**
RELICS OF LOST RACE ARE FOUND
Cemetery, 1,500 Years Old, Is Bared in Oregon
MANY SKULLS AND IMPLEMENTS FOUND
Ancient Race of Mound Builders Taught by Asiatics, is Theory Held.

Portland, Ore., April 14.-Pages of history were turned back to the year 500A. D., when Buddhist missionaries taught the dwellers along the Pacific slope, in the discovery near Albany, Oregon, of what is pronounced a burial ground of an ancient race of mound builders.

A farmer, Clyde Peacock, unearthed the relics while plowing his fields, the plowshare was caught by a rock and after digging the rock out it was found to cover a fine specimen of mortar. Further digging unearthed skulls, knives, skeletons, charcoal, pestles and additional mortars.

Thirteen well-preserved skulls have been removed from the excavation, which is fifty feet long, twenty feet wide and about two feet deep. Disintegrated skeletons, pestles and mortars were found with most of the skulls.

The people of the race were of great size and strength, according to J. G. Crawford, local authority on prehistoric specimens. He examined the skulls and bones carefully. They vary in weight and some are thinner than others. The teeth in a few are badly worn, while others are in nearly perfect condition.

Crawford believes that the mound builders covered the bodies of their dead with earth and built fires over them, thus baking the forms. He is of the opinion that the custom was derived from Buddhistic missionaries 500 years after the birth of Christ, and preceded the time when the North American Indians inhabited the section of the country.

Amateur archaeologist of the vicinity advanced the theory that the plot was once the burial ground of the Indians. Crawford, however, declared the skulls to be those of mound builders after closely comparing them with other specimens of mound builders and Indians he had in his laboratory.

As many as thirty mounds have been opened in the district surrounding the present discovery within the last few years. Their contents have been distributed among various museums. From evidence obtained so far it is held possible that there may have been communication between the mound builders and people who inhabited the mesas and Mexico. Color is added to this theory by the type of decorations found upon relics in the mounds and from the fact that the mound builders evidently incorporated sun worship with their phallic ceremonies. In some instances remnants of clay alters have been found in the center of the mound surrounded by cluttered heaps of human bones.

That the mounds precede the advent of the white man is believed to be conclusively proved by the fact that no beads or metal such as the Indians traded have ever been found.

Bancroft's *Native Races*, 1882

Shell mounds are described as very abundant throughout Vancouver Island, and also on the mainland, and all are composed of species of shells still common in the coast waters. One at Comex covers three acres, and is from two to fourteen feet deep. The relics discovered in mounds of this class include stone hammers; arrow points of flint, slate, and of a hard green stone; spear-heads, knives, needles, and awls, of stone and bone, one of the knives being sixteen inches long and of whale-bone; bone wedges, sometimes grooved; and finally stone mortars, comparatively few in number, since acorns and seeds were not apparently a favorite article of food. Human skeletons also occur in the shell mounds. At Comex a skeleton is said to have been found with a bone knife broken off in one of the bones. Mr. Deans believes he can distinguish two distinct types of skulls in Vancouver Island-the 'long-headed' in the older cairns, and the 'broad-headed' in the shell mounds and modern graves: and this distinction is independent of artificial flattening, which it seems was practiced in a majority of cases on skulls of both types.

In addition to the mounds, Mr. Dean states that earth-works very similar to those found in the eastern states are found at many localities in British Columbia. Indeed, he has sent me several plans, cut from Squire's work on the antiquities of New York, which by a simple change in the names of creeks and in the scale would represent equally well the northwestern works

Fitchburg Sentinel, (Fitchburg Massachusetts) January 16, 1904
A Gigantic Skeleton
Bones of a Man Eleven Feet in Height

Winnemucca, Nevada, Jan. 16-Workmen engaged in digging gravel here have uncovered a number of bones that once were parts of the skeleton of a gigantic human being. Dr. Camels pronounced them bones of a man who must have been 11 feet in height.

The Standard (Albert Lea, Minnesota) November 11, 1882

There were giants, or, at least, there was one giant in those days, when the imprint of a foot nineteen inches long and six inches broad, recently discovered in Nevada, was made. As there are marks of bristles along the edges, it is not believed by all that it was a human monster that trod there, but what sort of creature it was is still a matter of doubt.

Helena Independent, **December 6, 1925**
Nevada Once Home Of Tribe Of Giants
Huge Skeletons Found In Burial Mounds Of "Lost City"

Reno-Announcement of an appropriation of $25,000 by the Haye Foundation for thorough investigation of Nevada's "lost city" has stimulated wide interest in this buried metropolis of a vanished people. Located in the triangle formed by Arizona, California and the mighty Colorado river, the ruins have been explored sufficiently to prove existence of a far flung city, centuries before Columbus sailed for the New World.

Scores of ancient houses and buildings have been uncovered, the broad streets of a well-ordered city exposed and hundreds of skeletons found in burial mounds.

Race of Giants Indicated

The explorers have recovered treasures in pottery, Indian jewelry, ancient weapons and utensils of domestic life. Skeletons over seven feet long indicate that a giant race once held sway over the Southwest. Weapons, bead work, pottery and other articles indicate a high state of culture.

The preliminary excavation show houses strongly built of stone and clay, rectangular in design and of Pueblo character. Rooms were well furnished and in larger buildings were places where families evidently gathered to feast and hold counsel. Well woven blankets, skins finely tanned and many products of agriculture indicate a civilization infinitely surpassing anything found in this country by early colonist.

Still Older City Beneath

The city apparently was built on the ruins of a more ancient metropolis, and this deeply buried debris of the past may shed light on the peoples who populated the Western Continent when Rome dominated the nations of the Old World.

Exploration may also reveal the story of the catastrophe that overwhelmed the city and blasted its people to dust. Did an invading tribe level the buildings and stamp out the inhabitants? Or was destruction wrought by the heavy hand of nature? Why was no evidence of the culture of the lost city handed down in tradition.

Change in Climate Seen

The general impression of the explorers is that the city flourished in a vast fertile area, and that pronounced climatic changes converted a region of abundant moisture into arid desert. Possibly volcanic eruptions buried rivers and springs.

A few miles from the ruins pictures show a beautiful girl being sacrificed by a high priest of the tribe, and hieroglyphs explain that the lovely maiden was offered in vain to the rain god after seasons of heat had blasted crops and dried up springs and streams.

Did an invading tribe level the buildings and stamp out the inhabitants? Or was destruction wrought by the heavy hand of nature? Why was no evidence of the culture of the lost city handed down in tradition.

Archaeologist to Send Expedition to Explore Graveyards in New Mexico
Where Bodies Were Unearthed
Special to the New York Times

Los Angeles, Cal., Feb. 10. -Owing to the discovery of the remains of a race of giants in Guadalupe, N. M., antiquarians and archaeologist are preparing an expedition further to explore that region. This determination is based on the excitement that exist among the people of a scope of country near Mesa Rico, about 200 miles southeast of Las Vegas, where an old burial ground has been discovered that has yielded skeletons of enormous size.

Luciana Quintana, on whose ranch the ancient burial plot is located, discovered two stones that bore curious inscriptions, and beneath these were found in shallow excavations the bones of a frame that could not have been less than 12 feet in length. The men who opened the grave sat the forearm was 4 feet long and that in a well preserved jaw the lower teeth ranged from the size of a hickory nut to that of the largest walnut in size. The chest of the being is reported as having the circumference of seven feet.

Quintanaq, who has uncovered many other burial places, expresses the opinion that perhaps thousands of skeletons of a race of giants long extinct will be found. This supposition is based on traditions handed down from the early Spanish invasion that have detailed knowledge of the existence of a race of giants that inhabited the plains of what now is Eastern New Mexico. Indian legends and carvings also in the same section indicate the existence of such a race.

Indiana Progress, **(Indiana Pennsylvania) October 14, 1891**

The gigantic skeleton of a man who measured eight feet six inches in height was recently dug up by some laborers near the Jordan River, just outside Salt Lake City, Utah.

New York Times, **March 17, 1924**
Find Skeleton of Giant
Idaho Road Men Dig Up Bones of Prehistoric Herbivorous Woman

Lewiston, Idaho, March 16 (Associated Press)- A huge skeleton, believed to be that of a prehistoric human being, has been discovered in the Salmon River country, south of here, by two members of the State Highway Department who have brought their find to this city.

The lower jaw and vertebrae will be sent to the Smithsonian Institution at Washington, D. C., for analysis as to the probable date of existence.

The bones were found in the side of a cliff at a depth estimated to be fifty feet. Nearly the entire skeleton was recovered.

Measuring more than eight feet in height and possessing numerous strange features, the skeleton has aroused widespread interest. Three physicians pronounced it to be that of a woman.

Belief that the person was of a herbivorous race had been expressed, owing to the peculiar formation of the jaws and teeth. Both the upper and lower have only ten teeth each and all are intact.

Malcom Rogers who was the former director of the Museum of Man in San Diego, named the

remains in the shell mounds as the "La Jollans." At several campsites near burial cairns he observed many broken artifacts, and reported in, *Ancient Hunters of the Far West,* 1964, Malcom Rogers "He concluded that these implements had been "sacrificed" in some ceremony, perhaps in the vaguest sense of religion, or at death, or merely an act of departure." "Dr. Spencer Rogers, Chairman of the Department of Anthropology at San Diego, "There is an impressive similarity in many physical characteristics between La Jollan and the prehistoric populations of the Island of Kyushu, Japan."

Migrations of the Cro-Magnon, Hunters and Fishers to the North American Atlantic Coast

Cultural and physical similarities are substantial in comparing the remains of the Maritime peoples of the northeastern Atlantic coast of North America and those from the European Atlantic period (5,600-2,250 B.C.). Burials from this period in Northern Europe have been found in glacial kames and shell mounds. Burials were found within pits where fires were evident. Other northern European burial pits have been found that were lined with bark. The use of red ochre on the bones or as an inclusion within the burials were commonly found within graves. Some of the burial mounds have showed evidence of a charnel house. Also present within graves are implements of flint and slate that were sometimes intentionally broken before placed with the burials. Some of the bodies were oriented to the east, that may be evidence of solar worship. Bodies were also interred in a sitting position. Some of the skulls were described as "large" or "massive" with well developed eye ridges, large jaws and a sloping forehead. The Maritime populations have been determined by scientists as a mix of Cro-Magnon, European and Mongolian.

"The Prehistory of Eastern Europe, Mesolithic, Neolithic and Copper Age Cultures in Russia and the Baltic Area." By Marija Gimbutas, 1956. "The abilities in art of the hunters-fishers are shown by geometric and biomorphic designs on objects of daily use and on amulets and also carvings in amber and by pick engravings. The pattern on bone and horn tools were either incised or made up of small holes orbits of regular shape, probably made by a bow drill."

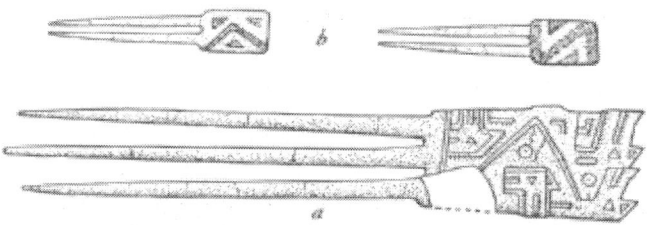

Antler Combs or Hairpins from Cape Cod shell heap. Both sides of "b," are shown, and the incised geometrical lines were originally colored. Image is from "Archaeology of Maine" 1922 by Wm. Moorehead.

Bone spear with tally marks from "Archaeology of Maine" 1922 by Wm. Moorehead.

A migration of the European Maritime Hunters and Fishers to the east coast has been argued by several historians including, F. Ridley who stated in *"Transatlantic Contacts of Primitive Man, Eastern Canada and Northwestern Russia,* "For several years, similarities between the Archaic implement complex of the St. Lawrence Valley of northeastern North Americas and those of the Atlantic and Arctic cultures of northwestern Europe have been recognized (Spaulding, 1946; Gjessing, 1948). The Atlantic culture dated for a period of 5000 B.C. to 2500 B.C. and the northeastern North America Archaic dated approximately 3000 to 1000 B.C., made common use of the polished gouge, polished celt, ground slate knives and ground slate projectile points (Ritchie, 1951). The Arctic culture of Scandinavia, succeeding the Atlantic culture, continued use of these implements and added the mortuary custom of covering the body with red paint, a custom common in the northeastern North America Archaic cultures. Gjessing (1948) states that large series of barbed slate points found in Northern Norway are identical with those of the Red Paint Archaic culture of Nova Scotia and New England. The comparison in northern Europe extends eastward across Russia to the northern Ural Mountains, for here in use at 2500 B.C. were stemmed and corner-removed points, beveled adzes, leaf shaped knives plus the implements listed above. In summary, the primitive people of the St. Lawrence

Valley in North America and those on Northern Europe from Scandanavia to the Ural Mountains, shared a tool and mortuary complex which appeared first in Europe."

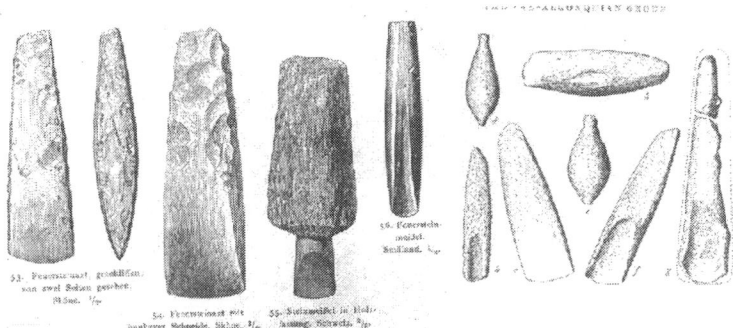

A series of tools from the Maritime people from Sweden on the left and Maritime Archaic from Maine on the right. The most prolific tools being the celt and the gouge that was used in the construction of boats or dugouts. Plummets are also found in abundance in the Baltic and the eastern and western shores of North America associated with Maritime peoples. They are also found in burial mounds in the Ohio Valley.

The Maritime Archaic people were initially located within coastal regions of Maine, New Brunswick, Quebec, Labrador and Newfoundland. They are also called the Red Paint People because of their penchant to include red ochre within their burials. They were originally seafarers who utilized deep-water fishing for large game fish including swordfish as part of their diet, along with harbor seals, auk, and other seabirds. They also hunted Caribou in the tundra of the northeast. Their implements were, plummets, ground slate and bone spears and lances, ulus (semi-circular knives), bone toggling harpoons and awls, and some large corner notched and leaf shaped stone and slate points and some made from exotic quartzite from Ramah Bay in northern Labrador, where they traveled over a thousand miles by boat. The most prolific artifact that ties these circumpolar people together are the plummets.

Several plummets from Maine that were found in "Red Paint" graves *"Archaeology of Maine"* **1922 by Wm. Moorehead.**

The most wide spread burial type in the Late Archaic that is found across the American continent are the shell mounds. The use of shell mounds to inter the dead began as early as 6,000 B.C. and was continued into historic times by Native Americans. Within the lower levels of the shell mounds are found, plummets, adzes, whale-tail batons, strait and notched stemmed spearheads of slate, flint and copper, stone ornaments, antler and bone combs. Many times these grave offerings are found "broken" within the grave.

Mussels provided an generous food source to these early setters whom discarded the shells in heaps that over time grew into large mounds. Shell mounds were both house sites and the repositories for the dead. The dead were placed within a pit either in a flexed, extended or sitting position. Children have been found accompanied by a dog in these shell heaps, to protect them in the afterlife as they did in this.

Shell mounds with identical artifacts and skeletal remains are found throughout the extent of coastal North America, from the Atlantic to the Gulf of Mexico to the Pacific and represent the first pan-American civilization. Shell mounds on the Eastern seaboard, are similar to those on the west coast, with slate spears, plummets, red ochre, and large skeletons resting, in some cases, on a bed of ashes.

Stephen Coons and other archaeologist have remarked on the similarities of the skulls from

Denmark, that looked like a certain type of American Indian, notably that of the North American Plains." There is a genetic marker called Haplo X that occurs in some Native American populations that was derived from European Caucasians around 7,000 B.C. Haplo X has been found in tribes that at one time inhabited the northern tier of North America. These include the Sioux and the Yakima on the northwest Pacific coast. The Great Lakes, Ojibwa also carry this genetic marker.

The following articles describe large skeletons found on the eastern seaboard. The large skeleton discovered on Amelia Island, Florida and at other locations along the Atlantic seaboard with a double row of teeth implies a genetic connection to those on the Pacific Coast with the same dental abnormality.

Annual Report of the Board of Regents of the Smithsonian Institution, 1874
Antiquities of Florida, by Augustus Mitchell, M.D., of St. Mary's, Georgia

While in the South during the winter of 1848, pursuing the study and collecting specimens of ornithology, I was impelled by curiosity to examine a mound of moderate size situated on the southern portion of Amelia Island, Florida, being kindly furnished with colored laborers, and aided by Dr. R. Harrison.

This mound was about 15 feet in height, and 30 feet in diameter at the base, flattened and worn by attrition for ages; there having been two growths of love oak upon it, as stated by an old Spanish inhabitant of the place. The soil composing the mound was of a light sandy, yellowish loam.

We commenced the examination by cutting a trench 4 feet wide directly through the center, from the apex to the base, and then another trench at right angles to the former. The excavation revealed a number of relics, and the mode of burial of the mound builders. They must have commenced by digging into the surface of the ground about 2 feet, then, partially filling the excavation with oyster shells, they placed their dead on these in a sitting posture, their legs bent under them, with their faces to the east, and their arms crossed upon the breast, and next spread over them a stratum of earth. The confirmation of the crania found in this mound appears to differ somewhat from that of the present Indians.

The teeth of many crania of this mound were, without exceptions, in a perfect state of preservation, the vitrified enamel of these organs being capable of resisting exposure for centuries. Not one carious tooth was found among the hundreds in the mound. Many were entire in the lower jaw, the whole compactly and firmly set. *In some the second set was observed, while one jaw had evident signs of a third set, a nucleus of a tooth being seen beneath the neck of a tooth of a very old jaw.*

Pursuing my investigations, and excavating farther toward the southeast face of the mound, I came upon the largest sized stone ax I have ever seen or that had ever been found in that section of the country. Close to it was the largest and most perfect cranium of the mound, not crushed by pressure of the earth, complete in its form, quite dry, and no sand in its cavity, contiguous to this was nearly a quart of red ochre, and quite the same quantity of what seemed to be pulverized charcoal, as materials of war paint. Anticipating a perfect specimen in this skull, I was doomed to disappointment, for, after taking it out of the earth and setting it up so that I could view the fleshless face of this gigantic savage, in the space of two hours it crumbled to pieces, except small portions. According to the measurements of the bones of this skeleton its height must have been quite seven feet.

Coal was freely diffused throughout the mound, which contained but little pottery. Two stone hatchets were found, and a small stone ax, in addition to the large one described. This instrument bore evident marks of fire.

There is one large mound around on the eastern end of Amelia Island, Florida, and two mounds on the central portion of Cumberland Island, Georgia, likewise, most of the Islands on that coast, from which could be obtained large collections of materials for the advancement of ethnological science.

Historical Collections of Virginia, 1845

On the Wappatomaka have been found numerous Indian relics, among which was highly a finished pipe, representing a snake coiled around the bowl. There was also discovered the under jaw bone of a human being (says Keucheval) of great size, which contained *eight jaw teeth in each side,* of enormous size; and what is more remarkable the teeth stood transversely in the jawbone. It would pass over any man's face with entire ease.

The Washington Post, February 10, 1890
Indians Seven Feet Tall
Skeletons of a Race of Giants Exhumed at Pleasantville, N. J.

May's Landing, N. J., Feb. 9-Forever a week past crowds have been flocking to the site of an unearthed Indian graveyard near Edgewater avenue in Pleasantville. The first lot of skeletons unearthed was about 1,000 yards from the city post office, and embraced eight bodies, closely laid together in a deep chamber, snugly packed in with tortoise, oyster, and clam shells. One of this number had bead and shell decorations, which together with its extreme height points to the fact that it must have been the powerful old chief Kineawaugha, whose descendants still own farms along the shore.

Prof. C. H. Farrel of Baltimore: Charles K. Simpson, of New York: John H. Cooley, Jr., of New Haven, Conn., and several gentlemen from the University of Pennsylvania immediately went to the scene. Messrs, Risley and Farr, the owners of the land gave to the Archaeological Association of the University of Pennsylvania the right to search for relics on their land. These researches have been watched by thousands of people with great interest. Besides weapons of war savage ornamental war decorations and numerous valuable shells, stones, &c., over fifty skeletons have been exhumed.

Dr. Charles R. Abbott, curator of the association, is continuing the search, and the skeletons are to be shipped to the university at once. They run in size from small child to several of seven feet in height, and one, supposed to be an old medicine man, Wauneck, must have been at least eight feet in height. About fifty students were upon the ground this morning and continued their search until stopped by rain.

The citizen's gaze in silent wonder on these relics of a race that at one time ruled the land. For seven miles along the shore can be seen large mounds of clam and oyster shells left here by Indians who used to congregate by hundreds to open oysters for winter food, and it is near these shell mounds that the great number of skeletons have been taken up. In some instances weapons of war made of stone and flint have been found lying beside some of the exceedingly large skeletons. The relics will be put on exhibition at the museum of the university in Philadelphia.

Alton Evening Telegraph, **May 3, 1934**
Jersey Farmer Plows up Prehistoric Giant's Axe

JERSEYVILLE, May 3, (Special) One of the largest axes of prehistoric origin in the memory of residents here was uncovered the past week by Louis Houseman on the farm where he reisides, seven miles northwest of Jerseyville.

The axe was weighed at the post office and lacked but several ounces of 10 pounds. The field where the axe was uncovered had been in cultivation for a number of years, but Housman has a reputation for plowing several inches deeper than the average farmer, and it was to his practice in this respect that the ax was brought to the surface. Houseman recently began farming the place where the find was made.

The ax had been scratched on a former occasion by a plow share, a mark on one of its sides showed. The relic was brought to Jerseyville by Houseman and left at the Munsterman filling station on South State street. He has received several offers for his find, but has refused them.

Much speculation has arisen relative to the physique of the man who carried such a heavy weapon or implement. Such a tool, corresponds to some of the unusually large skeletons of prehistoric men that have been unearthed in western and southwestern Jersey county.

History of the Colony of New Haven, Before and After the Union with Connecticut, **1838**

At the settlement of the English, the Indians in the center of the place retired to Indian Point, lying between East River and the Sound. Here they had a burying ground, the traits of which are now to be seen. (The house of Daniel Buckingham, Esq., stands on one side of the burying ground. In digging the cellar of the house, a number of skeletons were exhumed, one of which was near eight feet in length. They were buried in a horizontal position, and appeared to have been laid on a bed of charcoal, and covered with the same.)

The News, **(Frederick Maryland) September 28, 1897**
Skeletons of Indian Warriors

Mr. John Widgeon, curator of the Maryland Academy of Natural Sciences, Baltimore, Washington county on Saturday for the purpose of securing an Indian skeleton from the land on which the Delaware and Catawba tribes are supposed to have fought a battle 160 years ago. Mr. Widgeon found a skeleton seven and one-half feet long near Sharpsburg, where the Antietam creek empties into the Potomac River. The skeleton was shipped to Baltimore. About two months ago Mr. Otho Gray, a resident of Antietam, unearthed in the same locality a gigantic skeleton, which was sent to the academy of Natural Sciences.

The Archaeologi Mystified at Finding Skeletons of Men Who Were 7 Feet Tall

Portsmouth (Ohio) Times, October 2, 1936

Georgia Sand Dunes Yield Startling Proof of a Prehistoric Race of Giants

The Archaeologist Were Mystified at Finding Skeletons of Men Who Were 7 Feet Tall

Professor Holder, archaeologist, is directing the excavation work, which is been sponsored by the Smithsonian Institution. Slowly, painstakingly is endeavoring to place together the slender threads that will lead him into the past. He has explained the options the Smithsonian enterprise will throw important information on a thus far unrecorded tribe and perhaps establish a new link in the history of mankind in North America.

The Golden Isles extend in a chain from Savannah as far south as Fernandina. Today, they are today inhabited by the wealthy Americans.

Today, only one of all the islands still remains open to the public. It is called Saint Simons and Sea Island. And had it not been for the never ceasing studies of modern civilization, it might well be that the new proof of America's prehistoric giants might never have been found. For it was the ground breaking for Georgia's new Glynn County airport-which will be constructed on Sea Island. -That received the first evidences of the find, which has since brought archaeologist fairly tumbling over one another.

Workers on the proposed new airport hadn't set off more than two or three charges of dynamite when they were amazed to find a number of shattered skulls and skeletons scattered about. One of the nation's leading archaeologists, Dr. F. M. Setzler, of the United States National Museum, was dispatched to the scene. One look, and Dr. Setzler was convinced that the earth beneath the sand dunes would bear importantly upon the history and habits of southern coastal aborigines.

So systematic work began. Some of the first skulls to be disinterred by Preston Holder have been examined at the Smithsonian Institute by Dr. Alex Hedlicka, foremost authority on North American types. According to Dr. Hedlicka, the Sea Island skulls follow closely the Timucuan characteristics, while the pottery, implements and adornments uncovered in and about the village and burial grounds indicate Hichiti or Creek, affinities. That they were an early type of North American Indian is little doubt.

And as science continues to spread eager fingers over the moss-laced quietness of Sea Island, gradually the secrets of those early inhabitants of gigantic stature are being pieced together. In one mound delved into by Holder he found evidence, which led him to believe he had stumbled upon the site of a temporary camp, rather than a permanent village such as that which was located at the airport site.

The mound was composed of at least three layers of shell, each six inches to a foot thick, separated by layers of clean sand one to three feet thick. Very little midden, or garbage was found in the shell, which established Holden's belief that the site was not permanently occupied. The mound was fifty feet in diameter, with a six-foot raise. Burials were found to have been made immediately beneath the layers of shell.

It was in this mound that the archaeologist made the important discovery of a couple of complete skeletons of a young man, believed to have still been in his teens at the time of his death. From tip to tip it measured exactly six and one-half feet. Every detail of the burial indicated that he had been an important member of the tribe-probably a chieftain, or at least the son of a chieftain.

His bones were arranged with exceeding care. And between his right arm and his side were found three small bone awls, three large deer bone awls and three split and worked bones in the process of being made into implements or weapons. Over his left shoulder were four mussel shell pendants and a chipped stone spear-point, while fastened about his left knee was a string of sea-snail beads, numbering about 80 beads in all. The method of burial disclosed other bones- those of an older man, probably buried previously, had been recklessly disturbed by the giants during their burial of the young chieftain. These had been scattered back into the grave, over those of the younger man, with an abandon which archaeologist say is not all characteristic of the Florida Indians.

Of the first four internments made in this mound, all were of the full-flexed type, or curled up with knees close to the chin. Two of these were children, buried close together in "spoon fashion." They were heavily covered with hematite paint, a red pigment used by these Indians. One of the skeletons still wore an apron woven of 225 olivelia shell beads. Other burials yielded by the mound were all of the prone or full-extended type. Skulls were lacking from these.

Because of the generally disturbed condition of the contents of the mound, and the lack of order in which the bodies were placed, the excavators surmised that the burials had been made at various times-probably on fishing expeditions which were undertaken from time to time.

At the village under the airport site, Holden and his workers uncovered approximately 4,000 sherds or pieces of tribal pottery and cooking utensils. While a great deal of the pottery was plain ware, and in general quite crude, there were a few pieces, which were somewhat decorative. Colors ranged from black, through gray and red, to buff. The decorated ware showed at least five types of stamped design including the "check" stamped, the "delta" stamp, and a "herring bone" stamp. In addition there were discovered three distinct types of cord-marked ware, three types of thong-marked ware, and examples of rare incised and punctuate sherds.

New York Times, **February 15, 1925**
Find Florida Giant's Bones
Road Workers Unearth the Skeleton Parts of 7-foot Man

Boca Grande, Fla., Feb. 14-Discovery of a skull one-fourth larger than that of the normal modern, together with bones indicating a probable height of not less than seven feet, led to speculation today over theories of a giant race believed to have once inhabited Florida.

The portions of the skeleton were found yesterday by the workmen grading road near the Charlotte and Lee County lines. The bones, which are believed to be those of a male, are to be shipped to the Smithsonian Institution.

Portsmouth New Hampshire Gazette, **July 7, 1927**
FIND GIANT SKELETONS IN FLORIDA

Tamp, Fla., July 21-Giant Indians who roamed Florida swamps 500 years or more ago, living on shell foods which they cracked with their teeth is a picture unfolded by archaeologist who have delved into a burial ground on the gulf island near here.

The skeletons were discovered on a small section of land, where a lone fisherman has lived for years. Scientists estimated the bones are at least 500 years old and are remains of a tribe known as the Caribs, natives of the West Indies. They are believed to have inhabited the stae and adjacent islands before the arrival of the Spaniards in Florida.

The skulls, larger than those of current history, battered and crushed, indicated tribal battles. The jaw and teeth are unusually large. Likewise are the body bones, indicating the Indians of past ages were vertible giants in comparison with those of today.

Mounds similar to the one in which the bones were unearthed are common in the state. The bones have been sent to the Smithsonian Institution for further examination.

Smithsonian Institution's, *Bureau of Ethnology 44th Annual Report*
Shell Heap, Colbert County, Alabama

At the junction of the creek with the river, in the extreme northeast corner of Colbert County, is a large shell heap or kitchen midden composed entirely of mussel and periwinkle shells of several varieties, but all of them such as are now to be found in the river.

Scattered promiscuously among the debris were the usual objects found on Indian village sites; A large number of flint implements, more than a bushel, mostly knives or spearheads, the majority of them broken; cooking stones in abundance, usually cracked or shattered, but some showing only slight traces of heat; cupstones, none with more then five or six impressions; a few mortars; quantities of stones showing marks of use of hammers, others apparently pestles or rubbing stones, nearly all used in their natural shape or showing but slight marks of a dressing tool; hundreds of pointed bone implements, such as are usually called "needles," "awls," or "perforators," among them many spines from dorsal fins of large catfish and drumfish; numerous flaking tools and other implements made of antler, some of which holes drilled in the ends for inserting flint or bone points; only a few fragments of pottery; mammal and bird bones, with a large preponderance of those from deer in small pieces, and of various species of fish...

At 114 feet, in the east bank, a foot above the bottom, were the fragmentary bones of two infants, nearly the same size and neither apparently over two years of age, possibly less; the bodies were closely folded, heads in contact, and bones intermingles.

At the same distance, in the center of the trench, was a hole 3 feet in diameter, dug a foot into the soil. On the bottom lay some rough flat rocks on which was the closely folded skeleton of a man much above the average size. It lay on the right side, head south. The teeth were worn down to the gums; on some, the entire enamel was gone. The bones fell to pieces at a touch. Among the bones were several broken flints and two unfinished ones. Lining the margin of the depression were water worn boulders of quartzite from 5 to 50 pounds in weight. Altogether, at the bottom and around the side of the grave were 13 of these large stones.

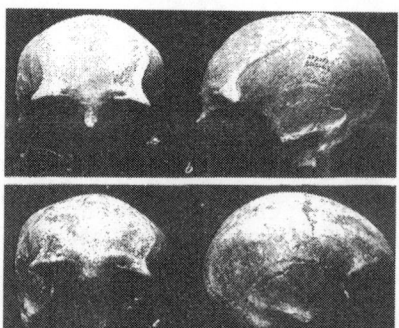

Skulls from the Colbert County, Alabama shell mound with protruding brow ridge and a sloping forehead. From the Smithsonian Institution's, *Bureau of Ethnology 44th Annual Report.*

By 2000 B.C the shell mounds become more numerous within the interior of the eastern continent. The tribes known as the Sioux, Cherokee and Iroquois are defined from of the early Maritime peoples. Their historic homelands are established at this early date. The Iroquois in the northern Great Lakes region, the Sioux, ranging from the Atlantic to the Ohio Valley and west to the Mississippi and the Cherokee in the southeast.

Maritime Origins of the Hopewell Sioux and Cherokee Mound Builders

"According to Native American traditions, the only people that have claimed heredity to the Hopewell mounds and earthworks are the Dakota Sioux Nations." was reported in the *The Prehistoric Aborigenes Of Minnesota And Their Migrations,* N. H. Winchel, *Popular Science Monthly,* September, 1908. "The Dakota, or Siouan, family comprised the following Indian nations, arranged approximately in order of apparent derivation: Biloxi, Tutelo, Waccon, Catawba, Huron Iroquois?, Cherokee?, Winnebago, Omaha, Osage, Issati, Mandan, Missouri, Dakota, Iowa, Ottoe, Hidatsa (and Crows), Blackfeet, and numerous subtribes, viz.,Ogala, Quapaw, Ponka, Assinboin, Akansea, Kansa, and others."

There is considerable evidence that the Cherokee and the Iroquois have a common ancestral origin with the Sioux. Linguistically, the Iroquois, Sioux, and Cherokee are similar, along with many of their customs. In "*A Brief History of the Cherokee,*" Mary Evelyn Rogers writes, "Linguistic studies show the Cherokees had been separate from the Iroquois, their closest linguistic relative, for at least 3500 years, based on a 1961 report per Duane King in introduction to "*The Cherokee Nation.*"

The historic locations of the Iroquois in the northeast and the Cherokee in the southeast are legened to be their ancestral homeland. This would account for the unhindered trade of the Hopewell and the diffusion of cultural traits across such a wide expanse. The Sioux, Iroquois and Cherokee all share similar legends that their earliest homelands were in the northeast. According to Chief Attakullakulla's speech to the Cherokee Nation in 1750, "we traveled here from rising sun, before the time of the stone age man." The Cherokee have in their language, names for whales and sea serpents; to which it can be conjectured that they once lived on the shores of an ocean.

Siouan Tribes in the Ohio Valley, American Anthropologist, 1943

"All of the traditions [of these tribes] speak of a movement from east to west covering a long period of time. The primordial habitat of this stock lies hidden in the mystery that still enshrouds the beginnings of the ancient American race; it seems to have situated, however, among the Appalachian mountains, and all their legends indicate that the people had knowledge of a large body of water in the vicinity of their early home. This water may have been the Atlantic Ocean, for, as shown on the map, remnants of Siouan tribes survived near the mountains in the regions of Virginia, North Carolina, and South Carolina until after the coming of the white race."

These traditions can be supported by the cultural similarities with the Maritime Archaic, dating as early as 6,000 B.C., with more considerable evidence found in the Late Archaic. The "Late Archaic" dates from, 3,500-1,000 B.C. This is when populations increased, trade routes were established and secured across the continent, seasonal camps became more permanent and the mortuary practices and artifacts of the Maritime Archaic spread across North America. The period around 1,500 B.C is when the Iroquois and Cherokee split and become linguistically distinct.

The people of the Maritime Archaic are also the first people to build mounds over their dead, dating as early as 5,500 B.C. at L'Anse Amour, located on the north shore of the Gulf of St. Lawrence, close to the Quebec-Labrador border. The L'Anse Amour mound was a circular, covered with large flagstones, which contained the skeleton of a boy. Within the burial mound were stone and bone spearheads, adzes, gouges, axes, pyrite fire kits, slate tools, chipped stone, antler and bone, swordfish bills, ulus, gouges, harpoons, ceremonial paint objects, a bird bone whistle and plummets. The Morrill mound near the mouth of the Merrimack River dated to 5200 B.C. while another burial mound located in southern Labrador dated to 4900 B.C.

Watson Break

The earliest mound and earthwork combination are located at Monroe Louisiana at the Watson Break site. This 135-foot diameter earthen circle connects eleven mounds, several exceeding twenty feet in height. Watson Break has been dated between 3,400 and 3,000 B.C. Many of the strait stemmed biface spearheads are similar to those of the northeast. The most persuasive evidence that ties the Watson Break to the Maritime Archaic are the plummets, found within the mounds.

Plummet uncovered at Watson Break mound and earthwork site.

Poverty Point

Another earthwork that occurs prior to the "Woodland Period," (1,000 B.C-500 A.D) is at Poverty Point Louisiana, dating 1,500 B.C. The enormous earthwork consisted of six concentric rings in a C-shaped design, cut by five avenues that are believed to be celestially oriented.

Artifacts recovered from Poverty Point are similar to the Laurentian or the Maritime Archaic who had moved from the coastal areas in to the interior. Several of the stemmed bi-faces discovered were identical to those found at the Morrill Mound in Maine dating to 4,000 B.C. Great quantities of plummets were found and picked up around the earthwork. Steatite bowls were also found similar to what were found in shell mounds on both the Pacific and Atlantic coasts.

Plummet from Poverty Point

Similarities exists between the Early Woodland earthen henges in the Ohio Valley and the Late Archaic shell rings found along the southern Atlantic and Gulf coast. At Sapelo Island Georgia, the diameter of the shell ring is 210 feet. This is significant because 210 feet diameter or a circumference of 660 feet is the most common size of the henges attributed to the Allegewi in the Ohio Valley. The coastal shell rings may represent the earliest astronomical temples in North America.

Also found within historic Cherokee homelands is this mound and earthwork complex that featured a henge 210 feet in diameter or 660 feet in circumference at Camden, South Carolina. John Bullick, *Bureau of American Ethnology, Bulletin 180,* 1961, proposed that the Cherokee culture in the Southern Appalachians had been in that region for more than 2000 years.

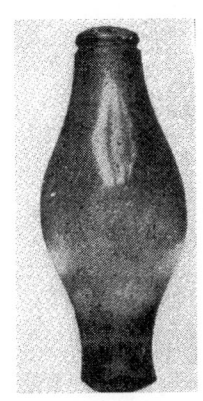

Plummet from the Crystal River mound

In northeast Florida along the St. Johns River were numerous sand and shell mounds. The two mounds at Tick Island and Mt. Royal were 555 feet in circumference. Mound internments in this area were similar to the Ohio Hopewell, with burials found in a spoked or sun burst position along with Hopewellian artifacts of triple tubes of copper, copper crescents and plummets. Some of these burials date as early as 500 B.C., 300 years earlier than the first Hopewell burial mounds in the Ohio Valley.

Shell mounds in Florida continued to be constructed into the Woodland Period, with artifacts also resembling the Ohio Hopewell. Shetrone reported in *The Mound Builders,* 1941 that the gulf coast shell mound at Crystal River, Florida had plummets that were found with burials throughout the mound. "It is significant that a deposit of plummets or pendants comprising similar materials and forms was taken from the Seip Mound of the Ohio Hopewell culture."

Similarities between the Maritime Archaic and the Hopewell were noted in, *"Bone Implements from Shell Heaps around Frenchmans Bay, Maine,"* Hadlock, 1943, "In all the Shell Heaps worked by the Robert Abbe Museum, artifacts similar to those of the so-called Red Paint Culture have been found throughout numerous levels. In most instances they were in a poor state of preservation and showed evidence of extensive use. The presence of these artifacts in the shell heaps shows that the inhabitants were familiar with their use and manufacture, and these implements are not a culture other than the Eastern Woodland Indians."

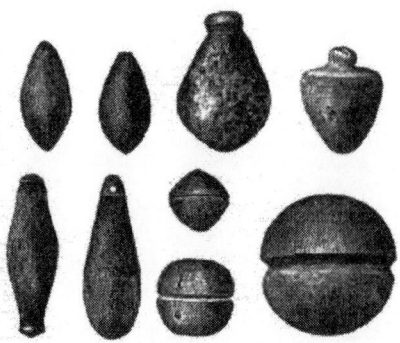

Series of plummets found within Fairfield County, Ohio and diagrammed in the *History Of Fairfield County, Ohio* 1905.

The first evidence of the Ohio Valley Sioux is within the Late Archaic as the Maritime people began moving into the interior and the earliest people known in western New York, called the Lamoka Focus. *The Archeology of the Northeastern United States*, 1952 Richard S. MacNeish, writes, "Various authors have pointed out the relationship of this focus [Lamoka] with those of the shell mound people in the central part of the United States, and fairly large number number of specific resemblance occur with the McCain site in Dubois County, Indiana, as well as the lower levels of the Annis Mound in Butler County, Kentucky"

In *The History of the Osage*, Louis F. Burns added, "recent archaeological findings seem to indicate that both the Dhegiha Sioux and Chewere Sioux were the Shell Mound Culture of Kentucky, [Indiana] and Tennessee." Skeletal remains found in these shell mounds are identical to the later Hopewell. The shell mounds in the interior reveal Maritime-type artifacts, and substantiate a connection with the Atlantic coastal area.

The earliest evidence of the Hopewell Sioux in the Midwest are the shell mounds that are found most extensively in southern Ohio, Indiana, Kentucky and Tennessee. Shell mounds within these areas date from 3,500-1,000 B.C. One of the larger shell mounds was described in the 1876, *Indiana Geological Survey*, at Clarksville, Indiana, "Just below the falls of the Ohio, in Clarke County, there is a shell heap extending for a mile up and down the river."

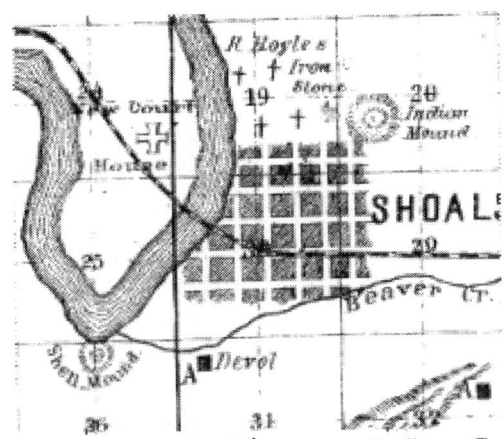

Shell and earthen mound are depicted in this 19ᵗʰ century *Indiana Geological Survey,* map of Shoals, Indiana. The proximity of these two types of mounds is evidence of the transition from shell to earthen mounds by the early Sioux people.

While the early Sioux become more populous in the Ohio Valley, the Iroquois spread their emerging culture across the northern teir of the Great Lakes. From the combined confederation of the Sioux, Iroquois and Cherokee, the great mound building nation would be formed in the Ohio Valley. All three of these tribes were involved and though becoming distinct, they shared a common lineage and culture that would become known as the Hopewell Culture.

Maritime Origins of the Iroquois

As another contingent of the Maritime culture spread into the Great Lakes region, they are given another archaeological designation, the Lake Forest Tradition, which included the Meadowood, Glacial Kame and Red Ochre Phases. Tool kits with the Lake Forest Tradition are similar to those in the Maritime regions with bone awls, plummets, drills, pestles, stone tubular pipes, winged bannerstones and projectile points made from ground slate and flint. The subsistence of the Lake Forest tradition was hunting and gathering with an emphasis on fishing. This is an important aspect with the thousands of glaciated lakes that were formed within the region of northeast Indiana, Northwest Ohio and southern Michigan.

The Meadowood, Glacial Kame and Red Ochre phases, represent different tribes of Iroquoians. Robert Converse writes in *The Glacial Kame Indians,* "Glacial Kame "phase" are associated with a burial cult which includes two other cultures, the Red Ochre "phase" (Wisconsin, Northern Illinois, Northern Indiana and the central lower Peninsula of Michigan) and the Meadowood "phase" (Western New York and Southern Ontario) There are areas of overlap."

"Common among all three (Glacial Kame, Red Ochre, and Meadowood) copper beads, shell beads, tubular pipes, birdstones, trapezoidal gorgets, pop-eyed birdstones and objects of bone and antler."

Bar amulets are often associated with the transitional "Turkey Tail Phase" of the Late Archaic Red Ochre Culture."

Throughout northeast Indiana and extending into southern Michigan in the lakes region, Meadowood points and pop-eyed birdstones are found. This birdstone called "Old Ringneck" was uncovered in Steuben County, Indiana, near Alvarado. Also near this site was once a circular earthwork.

Meadowood habitation sites have been dated as early as 1,250 B.C. Seven habitation sites of the Meadowood people have been found in the Maritime Provinces. In Nova Scotia, Meadowood-style arrows have been identified on Lake Kejimkujik and the Mersey River. In this region, eel weirs have also been discovered. These eel weirs are a series of triangular-shaped, stone fish weirs along the Mersey River which were used to capture eels in the fall. A similar weir is located in Laketon, Indiana, on the Eel River.

Fish weir on the Eel River in Laketon, Indiana. From *The Nephilim Chronicles, A Travel Guide to the Ancient Ruins in the Ohio Valley*. 2010.

Meadowood Iroquois artifacts are found from the Maritime regions through northwestern New York and west into lower Canada and the Great Lakes regions of Northwest Ohio, Southern Michigan and Northern Indiana. Along the shores of lakes within this region, were mound pods, consisting of 3 or 5 mounds in a group which are usually small and domed shaped. The mounds along the lake shores may represent some of the earliest burial mounds in the Great Lakes region. All of these have been destroyed except for one mound pod in Williams County, Ohio, at Nettles Lake and another at Croton Dam in Michigan. These two sites were similar in that five, domed shaped mounds were on the south side of the lakeshores, with conical mounds nearby. Croton Dam's 5 domed shaped mounds have been nearly erased, with the two larger conical mounds still visible despite recent university archaeological destruction.

Early Iroquoian, conical mound at Croton Dam in Newaygo County, Michigan after being excavated. In these mounds were found subsurface cremations, copper spear points, stemmed points, copper beads, beaver incisors, stone drills, copper needles, red ochre, fire kits, and a child's burial accompanied by a dog. Burial traits and artifacts resemble the Late Archaic more than the subsequent Woodland period. From *The Nephilim Chronicles, A Travel Guide to the Ancient Ruins in the Ohio Valley.* 2010.

Evidence of the Maritime people moving into the interior lakes region can be found in the similarities of artifacts. Mark Schurr from Indiana University conducted an archaeological survey of Lagrange County, in northeast Indiana and concluded that prior to 1,500 B.C., the cultural influence of the county was from the northeast. Cameron Parks, a local collector of artifacts in Northeast Indiana submitted *"Slate Artifacts from Dekalb County, Indiana"* to the Indiana Historical Society. Parks photographed several slate points, and realized that they were identical to points found in the northeast, associated with the Maritime or Red Paint people.

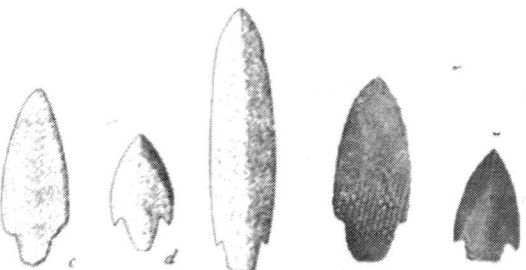

The slate points on the left are from Maine and were illustrated in *The Lost Red Paint People of Maine,* by Walter Brown Smith, 1930. To the right are identical slate points found in Dekalb County, Indiana that were photographed by Cameron Parks.

Typical Red Paint or Maritime graves were described n *Maritime Antiquities of the New England Indians with Notes on the Ancient Cultures of the Adjacent Territory* by Charles C. Willoughby

"The soil of the burial places at Bucksport and Orland was a coarse gravel while that at Ellsworth was a finer gravel with area of sand which showed quite distinctly the outlines of the graves, an enabled accurate cross-sections to be made. These show three methods of the disposition of the body. The usual method was to dig a basin-shaped hole about five feet in diameter at the top, and approximately three to four feet deep. In this the body was placed in a flexed position, together with fire-making outfit, a bag or package of red ochre, adzes and other implements. The grave was then filled with earth, and a fire built over it. This fire apparently did not come in contact with the body, otherwise carbonized bone doubtless would be found. Another type of interment the body was not flexed, but laid horizontally in a grave twenty one inches deep in the sand. In a few graves a small amount of buff colored powder was found which showed a decided reaction when acid was applied."

The act of burning an area prior to burial, or building fires over the grave was also evident in many of the burial mounds in the western Great Lakes region. Ashes and burnt earth was found at the base of mounds that have been excavated in northeast Indiana. Reports also described mounds where a mantle of earth had been burnt brick hard with a layer of ashes over this. It was believed that the cremation fires had been covered with earth, followed by the surrounding structure being lit and collapsing over the mound area.

The Meadowood, Glacial Kame and Red Ochre Phases are followed by the Early and Middle Woodland Period, (1000 B.C-500 A.D.) Point Peninsula Focus, who are defined as possessing, Vinette I and II pottery with exterior cord-marks, arrow points that were thin semi-lozenge, turkey-tailed, broad stemmed, broad side notched with expanded base or round ends, and double side notched, clay tubular pipes, trapezoidal convex-topped rectangular broad gorgets, birdstones of two types (bar and broad base with transversal ridges) narrow rectangular round-ended gorgets, thin elongated objects of slate, flat copper celts, cylindrical beads of rolled sheet coper, sinew stones of the ovoid flat pebble type, expanded base drills with scraper edge, cremation (both single and multiple) and bundle burials.

The Archeology of the Northeastern United States, 1952, Richard S. MacNeish, writes, "Point Peninsula culture in its earlier stages shows more resemblance to Adena [Allegewi] than to Hopewell, but in its later stages Point Peninsula certainly becomes more Hopewell-like. Point Peninsula II and ultimately Point Peninsula III, as I shall point out, there are indications that Point Peninsula III is ancestral to Owasco and that some Owasco is ancestral to some Iroquois groups. In other words there is evidence for cultural continuity from Early Woodland through Point Peninsula I, to historic Iroquois. Thus the proto-Iroquois may have entered New York bearing Point Peninsula culture in Early Woodland time."

Point Peninsula burials have been found along the extent of the Great Lakes region from New York to Wisconsin. A Point Peninsula burial mound was excavated near Walkerton, in Porter County, Indiana. Bodies were described as being in the flexed position and surrounded by red ochre. Elbow and platform pipes were within the mound that are distinctive of Point Peninsula. Pendants with one perforation and bone harpoons were also found within the mound that are common in Early and Middle Woodland Point Peninsula and also in the Late Woodland, "Intrusive Mound Culture."

The Point Peninsula people have been linked with the Late Woodland, "Intrusive Mound Culture." These are burials that have been found in the upper region of burial mounds that were placed there at a later date. Why would the Iroquois do this unless they thought that these graves contained their ancestors? Intrusive graves have been found at Portsmouth, Mound City and other mounds sites. We know that the ancient Iroquois were trading with the Ohio Hopewell Sioux and shared similar religious and burial rituals. It is likely that they were involved in the construction of the great earthworks in the Ohio Valley.

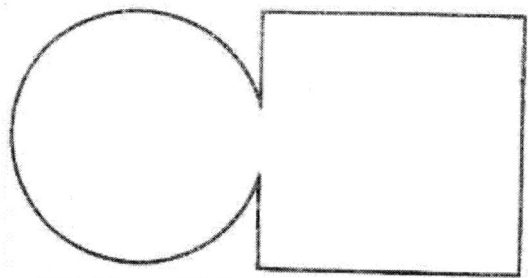

From *Archaeological Atlas of Michigan,* 1927, "The outstanding feature of the county was an earthwork quite unusual in its outline for Michigan, but common in Ohio. In the north part of the city of Tecumseh, near the band of the River Raisin, at what was once known as Brownsville, there was circle joined by a passageway to a square."

Evidence of an early Iroquoian presence in the Ohio Valley was found near Bainbridge, on Paint Creek, where a cache of Turkeytails were discovered. *Another Red Ocher Discovery in Ross County,* by Barry Grandstaff and Gary Davis reported; "This is not the first evidence of a Red Ocher presence in Ross County. There is the well known Spetnagel cache of over 200 ceremonially "killed" Turkey Tails found in the excavation of a house basement in nearby Chillicothe in the 1920s (Converse 1980). There is also the find made by Clark Johnson near Bourneville within a short distance of our site consisting of Turkeytails, gneiss bar amulets and rectangular gorgets in close proximity to Adena material. Our discovery is further evidence of a strong Red Ochre presence in the area at a time when Adena should have been in its early stages."

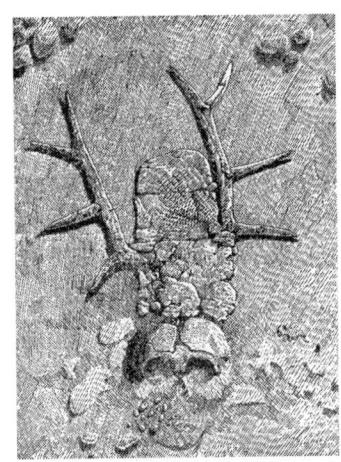

One similarity the Point Peninsula Iroquois shared with the earlier Maritime Archaic and the Hopewell Sioux were animal head-dresses. On a sandy ridge in the south shore of Ridinger Lake in Kosciusko County, Indiana, Point Peninsula Hopewell burials were unearthed, and reported in, *Hopewell Artifacts from a Burial in Kosciusko Co., Indiana: The Ridinger Lake Site,* **Dr. R. M. Gramly, Ph. D., Buffalo New York. Found within this burial was a copper head-plate that was described by Dr. Gramly, "The head-plate is dished and would have conformed to the wearer's head. Head plates or "helmets," as they are sometimes called, supported dear horns or wooden imitations and were ceremonial regalia. Iroquois Indians of New York State to this day still refer to the "horns of office" and a chief who has fallen from grace may be deprived of his powers or"dishorned," so to speak."**

E. A. Allen wrote in *The Prehistoric World, 1885,* "Early explorers have left abundant testimony to show that in many cases the Iroquois Indians resorted to mound burial. Thus, it seems that it was the custom of the Iroquois every eighth or tenth year, or when ever they were about to abandon a locality, to gather together the bones of their dead and rear over them a mound."

The Prehistoric Aborigines of Minnesota and Their Migrations, 1908, "The mounds that are common in southern Michigan and along the Lake Huron shore northward from Detroit, as well as those in northern Ohio [and Indiana], western New York are attributable to the Iroquois or to some of their kindred tribes, viz; the Hurons, Eries and Neutrals. The Iroquois dominion extended to the north shore of lake Huron even in historic times. Indeed there is good reason for believing that the Iroquoian and Siouan stocks at this time possessed the whole country east of the Mississippi River and south of the Great Lakes to Northern Georgia, constituting together the great Ohio dynasty of the mound builders."

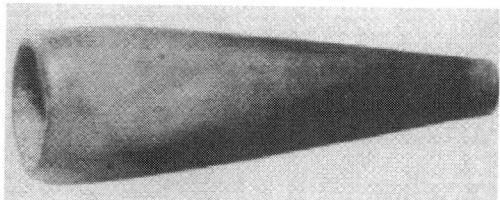

Tan mottled pipestone tubular pipe from a gravel kame near Oxford, Ohio from *The Ancient Ohioans,* **Raymond Vietzen, 1946.**

Skeletal remains found within burial mounds and glacial kame burials in the southern teir of the western Great Lakes are described as having prominant brow ridges, thick skull walls and other attributes of the Upper Paleolithic Cro-Magnon. Position of the skeletons within mounds and kames vary; some are in a sitting position, others were placed in a spoked position similar to those found in the Ohio Valley. Some mounds contained the remains of many skeletons. In most cases there is evidence of fires, either baked earth on the floor of the mound, or a layer of ashes above or below the burial.

Burial procedures across the entire Great Lakes are similar to Allegewi-Hopewell with a few noteble differences. Spoked burials occur with the Allegewi-Hopewell, but what is different with the Iroquois is that they were placed in a sitting position facing to or away from a central fire pit or sea shell.

The following is a list of spoked burials that extend from Ontario through New York, Pennsylvania, Ohio, Indiana, Illinois to Wisconsin. Included are skulls that are reported to have had protruding brow ridges, with "primitive" characteristics.

Annual Archaeological Report for Ontario, 1896-1897"

"One group exhibits circular to oval mounds from 25 to 40 feet in diameter and heights from 31/2 to 5 feet. Burials are on the original ground level, which has been prepared by burning in certain areas. In one case the skeletons radiate outwards from the center, apparently in extended positions. In others they were possibly the same, but one of them contained a number of partial burials as well as an entire one. In this structure were found a few pieces of mica, a rough stone sinker, a few mussel shells, a large slate knife or chisel, a small rough stone axe, a trapezoidal slate gorget with one perforation and roughly made chert points, both triangular un-notched and triangular notched."

Memorials of a Half Century (Michigan) 1885

By far the finest group of mounds that has come to my to my knowledge occurs on the bands of the Grand River, three miles south of Grand Rapids. They were still perfect when the writer had the satisfaction of seeing them in 1874.

Of the smaller mounds, six were opened. In all skeletons were found, generally one only in each, and all were so decayed that it was impossible to preserve them. They were of ordinary size, except one, which is pronounced gigantic, the proportions "indicating a stature of seven feet." All were in a sitting posture, and faced different points.

County of Williams Ohio, 1905

On the south half of the northwest quarter of Section 10 in Brady Township, on the land now owned by James F. Smith, was a solitary mound of considerable magnitude. On opening it, six full-developed skeletons were found and one of a child about eight to ten years of age. They were lying in a circle with their heads in the center, in close proximity to each other. Dr. Frank O. Hart, of West Unity, now deceased, secured the skulls from this mound and described them in a written article as follows:

"They were very thick. The brow ridge is very prominent. The orbital processes are profoundly marked. Average distance between temporal ridges of frontal bone, three and half inches; from temporal ridge of frontal bone to occipital joint, nine inches; length from beginning of frontal bone to occipital joint, twelve inches; from occipital joint to foramen magnum, three inches."

History of Northeast Indiana, 1905

A number of years ago M.F. Owen excavated a mound situated in a piece of woodland, on the east shore of the first "West Lake: on the north side of the old highway. On this mound had grown a large white oak tree, which, having just been felled, showed a growth of between 300 and 400 years. Among the roots of the tree were unearthed a skeleton in a sitting posture, facing west, the bones of which crumbled rapidly when exposed to the air. There was found and preserved a root which had grown apparently into the ear orifice of the skull, afterwards emerging through the eye, and firmly attached thereto is a well preserved piece of the frontal bone, showing great development above the eyes.

Counties of Whitley and Noble Indiana, 1882

A member of the historical force opened a mound in the Salem Church Cemetery, Washington Township, but discovered nothing save a considerable quantity of charcoal. Mr. Denney opened two mounds on the farm of Samuel Myers, Orange Township, both containing nothing but Charcoal; he also opened three more near there, on the farm of Otis Grannis, one of them being eight feet in height and about eighty feet in diameter at the base. Three quite well preserved skeletons were taken from the mound, one of the skulls being almost in entirety, and having a much *better frontal development than the average.* On this mound was an oak tree four feet in diameter and probably more than three hundred years old. This mound is probably the largest in the county. Two other mounds near it, of average size, contained a bed of charcoal each. Mr. Denney, assisted by his brother Orville, opened three more on the bank of Skinner's Lake, Jefferson Township, and took from one a quantity of human bones; but this mound had been opened a number of years ago by novices in the neighborhood, who used no particular care either to observe or preserve, and the number of individuals buried there is unknown, though these were several. The other two mounds contained charcoal.

Mound in Noble County, Indiana was once situated on an ancient lake shore. The large hole in the center is the result from excavations in the past and recently by universities. Two additional small mounds, described in the previous history, are also still visible. This mound site and 82 others in Indiana are not listed, nor recognized as historical sites. From *The Nephilim Chronicles, A Travel Guide to the Ancient Ruins in the Ohio Valley*, 2010.

American Antiquarian, **April 1878**
Lake County Illinois

   Mr. W.B. Gray, of Highland Park, also mentions the discovery of a skull in a mound near Fox Lake, in Lake County, Illinois. This skull is certainly very remarkable; the frontal lobe or arch seems to be entirely wanting; the large projecting eye-brows, deep set eye sockets, the low, receding forehead, and the long, narrow and flat shape of the crown rendered it a very animal-looking skull. If it was not a posthumous deformation it certainly is a remarkable skull and might well pass for the "missing link." It was found in a mound six feet below the surface, in company with thirteen other skeletons. The skeletons were found lying with their heads to the center and their feet arranged in a circle around this point.

Smithsonian Institutes Bureau of Ethnology, **1890- 1891**
Sheboygan County, Wisconsin

 About 2 miles west of this, (a mound containing a large skeleton), on a bluff overlooking the marsh, was another mound of simular form and slightly larger, which had been previously opened by Mr.

Hoissen of Sheboygan. It was found literally filled, to the depth of two and a half feet with human skeletons, many of which were well preserved and evidently those of modern Indians, as with them were the usual modern weapons and ornaments. Beneath these was a mass of rounded bowlders aggregating several wagon loads, below which were 40 or 50 skeletons in a sitting posture, in a circle around and facing a very large sea shell.

The most compelling evidence that ties the Great Lakes burials with those found in the coastal regions are the gigantic size of the skeletons. Descriptions of the skulls are consistent with Upper Paleolithic Cro-Magnon, with furrowed brow, sloping forehead, massive jaws and thick skull walls.

Supporting evidence that the people buried in the glacial kames around the Great Lakes share the same origins as those on the coasts, are the reports of skulls with a double row of teeth.

Perforated Skulls from Michigan, *American Antiquarian Vol., XII, No., 1* **Jan., 1890**
Mr. Bates and two friends made an excavation in a vacant lot located within the area of this old cemetery. they came upon a curious and interesting burial spot, at a depth of two feet, five skulls were found, lying in a circle, facing the center. Within the circle were ashes and charcoal,--evidence of fire; but the bones were not all burned. On the perforated skulls two had *"double teeth"* in front. Mr. Bates says the third may have had also.

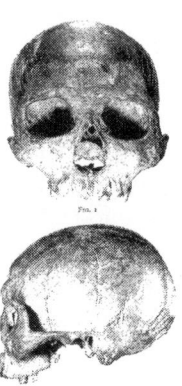

A Study of the Glacial Kame Culture,in Michigan, Ohio, and Indiana, **1948 by Wilbur M. Cunningham**
The Burch Site [Michigan]
During the year 1905 or 1906, Clark Burch, a farmer, while digging gravel to fill low spots in his barnyard, unearthed five human skeletons on his farm near Coldwater in Township 6 South, Range 7 West, Branch County, Michigan, less than twenty miles from the state line of Indiana. These skeletons were in a gravel deposit defined as glacial outwash and probably the remnant of a strong sandbar.
The skeletons were taken to the farmyard where they were examined, photographed, and measured by a physician who has long since died and whose records cannot now be located. They were then

placed in a corn crib and, with the exception of two skulls, one of which was loaned to a student of dentistry, were destroyed when that building later burned.

The physician who measured the bones is said to have made the statement that two of the skeletons were unusually large persons. Clark Burch, the finder of the skeletons, is still living on the same farm, and his testimony indicates that the burials were undoubtedly in a sitting position, with most of the artifacts "in their laps." With or near the skeletons he found much red paint, charcoal, some pieces two feet long, and many shell disk bead; the beads were so numerous that he and his helpers did not bother to pick them up all of them. The bones were not covered with red paint, as they are in some burials. The teeth is some skeletons were in a good condition and in others badly worn. Mr. Burch insisted that at least *one of the adult skeletons was equipped with two rows of teeth!* Reports of large skeletons and unusual tooth arrangement seem to be common in conjunction with finds of this character.

History of Medina County, Ohio 1881

In digging the cellar of the house, nine human skeletons were found, and like such specimens from other ancient mounds of the country. They showed that the mound builders were men of large stature.

The skeletons were not found lying in such a manner as would indicate any arrangement of the bodies on the part of entombers. In describing the tomb, Mr. Albert Harris said: "It looked as if the bodies had been dumped into a ditch. Some were buried deeper than others, the lower one being about seven feet below the surface. When the skeletons were found, Mr. Harris was twenty years of age, yet he states that he could put one of the skulls over his head, and let it rest upon his shoulders, while wearing a fur cap at the same time. The large size of all the bones was remarked, and the teeth were described as *"double all the way around."*

History of Hardin County, Ohio
Ancient Burial Mound and its Contents, Hardin County, Ohio, John B. Matson, M.D. to Judge John Barr, Cleveland, December 10, 1869

Dear Sir, --In the fall of 1856, in Hardin County, Ohio, near the Bellefontaine and Indiana Railway, between Mt. Victory and Ridgeway, I commenced removing a gravel bank for the purpose of ballasting a part of the above named railroad. I learned shortly after my arrival there, that the bank was an ancient burial ground. This information caused me to examine the ground, and note discoveries.

The mound covered an area of one and a half acres; being covered with an orchard of apple trees, then in bearing...The mound was what I would call double; the larger and higher part to the west. About two-thirds of the mound was embraced in this part. The eastern part, presenting the appearance of a smaller hill having been pressed against the other, leaving a depression between them of three or four feet, below the highest point of the smaller and five or six feet below a corresponding point of the larger.

On the north side of the eastern portion, under an oak tree stump (150 years old by growths) was the remains of the largest human bones I have ever seen. The joints of the vertebra seemed as large as those of a horse... I found in this part of the mound the remains of at least fifty children, under the age of eight years; *some with two, others with four incisors; some with eight, and others with no teeth.*

History of Jefferson County, N.Y., 1878

Near the north-west corner of Rodman, on lot number two, on the farm of Jared Freeman, was formerly an interesting work, of which no trace remains, except a boulder of gneiss, worn smooth by grinding. Before the place had been cultivated, it is said to have shown an oval double bank, with an intervening crescent-shaped space, and a short bank running down a gentle slope to a small stream, on of the sources of Stony Creek, that flows near. Several hundred bushels of burnt corn were turned out, over an area one rod by eight, showing that this must have been an immense magazine of food. On the farm of Jacob Heath, on lot No. 25, near the west line of Rodman, and on the north bank of North Sandy Creek, a short distance above the confluence of the two main branches of that stream, there formerly existed an enclosure of the same class. It included about three acres, was overgrown with heavy timber, and furnished within and without, when plowed, a great quantity and variety of terra cotta, in fragments, but not metallic relics. Under the roots of a large maple was dug up the bones of a man of great stature, and furnished with *entire rows of double teeth.*

History of Jefferson County, N.Y., 1878

One of the most conclusive evidences of ancient military occupation and conflict, occurs in Rutland, near the residence of Abner Tamblin, one mile from the western line of the town, and two miles from the river. It is on the summit of the Trenton limestone terrace, which forms a bold escarpment, extending down the river, and passing across the southern part of Watertown. There here occurs a slight embankment, and ditch irregularly oval, with several gateways; and along the ditch, in several places, have been found great numbers of skeletons, almost entirely of males and lying in great confusion, as if they had been slain in defending it. Among these bones were those of a man of colossal size, and like nine-tenths of the others, furnished with *a row of double teeth in each jaw.* This singular peculiarity, with that of broad flat jaws, retreating forehead, and great prominence of the occiput, which was common to most of these skulls, may hereafter afford some clue to their history.

Earthworks in the Iroquois lands around the Great Lakes consisted of oblong and circular earthworks and others that were perfect squares. The oblong and circular works are believed to be fortifications. The square works in one locale was described as being similar to those in Indiana and Ohio that were constructed by the Allegewi-Hopewell. However, the Iroquois squares along the Great Lakes are generally smaller in size. Descriptions of the square works with avenues going down to the adjoining creek and burial mounds within their interior are more descriptive of a ceremonial center. The ceremonies within the works were likely similar to those to the south, that were aligned to solar or lunar events.

A series of square earthworks were found from New York State to Michigan. The occurrences of the square earthworks are in the same general locations as the burials that contained large skeletons. This

is only a partial list of the square earthworks to give an idea of their size and their geographic extent.

Antiquities of Long Island, 1874

On the subject of this fortification: When this part of Long Island was first settled by the Europeans they found two fortifications in this neighborhood, upon a neck of land ever since called, from that circumstance, Fort Neck. One of them, the remains of which are yet very conspicuous, is on the southernmost point of land on the neck, adjoining the salt meadows. It is nearly, if not exactly a square, each side of which is about thirty yards in length. The breastwork or parapet is of earth, and there is a ditch on the outside which appears to have been about six feet wide. The other was on the southernmost point of the Salt Meadow, adjoining the bay, and consisted of palisades, set in the meadow.

This last described work was a true Indian fort, as is shown by all plates and sketches of such works as accompanying, Smiths "History of Virginia", Debrys "Voyages" and all the early works in this country, but no instance has ever been shown of the North American Indians having either in ancient or modern times, erected for the purpose of defense or for any other purpose, a four square fort of earth, with regular wall and ditch.

All this view of the case brings us to the conclusion that the two forts upon Fort Neck were constructed at different periods of time, and it may be far remote from each other; that the one first described, regular in its form, and built of earth, was the work of a people entirely different in the modes of living and in other respects from the aboriginal race found here by our forefathers.

Otzinachson: A History of the West Branch Valley of the Susquehanna, 1889

That the valley of the West Branch was inhabited by a superior race, of whom we have no account, appears to be evident. Traces of peculiar fortifications, resembling, those found in some of the Western States, are yet to be pointed out. One of these existed on the farm of Mr. Shoemaker, on the north side of Muncy Creek, it was square, and consisted of embankments thrown up in regular order, covering about one fourth of an acre. A simular one existed on the farm of Gov. Shultz, below Williamsport. On the other side of the river, nearly opposite the mouth of Lycoming Creek, was found another, resembling the one on Muncy Creek, traces of which can probably be seen to this day. Mr. Shoemaker of Muncy, now an old man, but with memory bright and mind unimpaired, informs me that years ago he made a personal examination of this latter fortification, and found all the embankments well defined. Large trees were growing upon them, however, the concentric circles of which would indicate many hundred years of growth, and entirely preclude the idea of their having been thrown up by the Anglo Saxon race. Mr. Shoemaker also states that many years ago he made inquiry of an old Seneca Chief concerning them, but all the information the old Indian could give, was that he had it from his ancestors that they were erected by White Indians. Of them we have no definite knowledge whatever.

History of Sandusky County, 1882

There are evidences of another fort of the same kind above the Williams Reserve a short distance, on the high bank of the river, In section thirteen, township five, range fifteen, (Sandusky Township) This work is different in form from those heretofore mentioned, being nearly square, and is supposed to include about three acres of land. It is situated at a place where there was once an Indian village called Muncietown, about three miles below the city of Fremont.

History of St. Joseph County Michigan, **1880**

Within three hours ride of Colon Village, there are no less than six fortifications of these ancient people. One of them is distinctly visible yet, and is in a square form, fronting the St. Joseph River, with an avenue leading to the rear to Bear Creek.

A Place Called the Thumb, Sanilac County Michigan, **1880**

In 1931, Wilbert B. Hinsdale in his "Archaeological Atlas of Michigan" located for a record, Indian villages sites, mounds, garden beds, circular and square enclosures, and other remains. In Watertown Township in Sanilac County an unusual construction consisted of a square enclosure with an interior mound. In the southwest corner of Worth Township, Hinsdale located a rectangular enclosure with an opening to the north, and a mound in front of the opening.

Civil History of Michigan, **1895**

Forts of the square or rectangular kind are sometimes found. There is said to be one or two miles below the village of Marshall, one in the township of Prarie Ronde, several on the Kalamazoo, and in other places. In Bruce Township in the county of Macomb, on the north fork of the Clinton, are several...eight miles from Lake St. Clair. In sinking the cellar of a building for a missionary, sixteen baskets full of human bones were found of a remarkable size. Near the mouth of this river, on the east bank, are ancient works representing a fortress, with wall of earth thrown up simular to those in Ohio and Indiana

The following list consists of giant skeletons that were found in both subsurface burials and within burial mounds. The geographic extent is the Great Lakes region. Some of the burials mounds are similar to the Allegewi in southern Ohio, with an earthwork that surrounds the mound. This is evidence that the early Iroquois shared many religious and burial aspects with the Allegewi. To what extent the Allegewi were intermarrying with the Iroquois is unknown, but it is possible that these similarities represent the Allegewi being absorbed by the Iroquois, as they were with the Sioux in southern Ohio. Distinct to these burials from those in southern Ohio is the placement of the interred in a sitting position.

New York

Mound at Tonawanda Island, from *Harpers New Monthly Magazine*, **May, 1860**

Notes on the Iroquois **by Henry Schoolcraft, 1847**
Ancient Entrenchments on Fort Hill

The work occurs on an elevated point of land formed by the junction of a small stream, called Fordhams brook, with Allens Creek, a tributary of the Genessee River. Its position is about three miles north of the village of LeRoy, and some ten or twelve northeast of Batavia. The best view of the hill, as one of the natural features of the country, is obtained a short distance north of it, on the road from Bergen to Leroy.

But the most remarkable and distinctive tract connected with its archaeology is the discovery of human bones denoting an uncommon stature and development, which are mentioned in the same communication. A Humerus or shoulder bone, which is preserved, denotes a stature one-third larger than the present race, and there is also a lower jaw bone, preserved by a physician at Batavia, from the vicinity, which indicates the same gigantic measure of increase.

Notes on the Iroquois, **Henry R. Schoolcraft, 1847**

Skeletons found about Fort Hill (Auburn, N.Y.) and its vicinity, sustain the impression that the former occupants of their military station, were of a larger and more powerful race of men than ourselves. I learned that the skeletons generally indicated a stouter and larger frame. A humerus or shoulder bone, of which one has been preserved, may safely be said to be one-third larger or stouter than any now swung by the living. A resident of Batavia, Thomas T. Everertt, M.D., has in his cabinet, a portion of a lower jaw bone, full one-third larger than any possessed by the present race of men, which was found in a hill near Le Roy, some two years since.

History of the Holland Purchase, **1849**

A mile north of Aurora Village, in Erie County, there are several small lakes or ponds, around and between which, there are knobs or elevations, thickly covered with a tall growth of pine; upon them, are several mounds, where many human bones have been excavated... There are in the village and vicinity few gardens and fields where ancient and Indian relics are not found at each successive ploughing. Few cellars are excavated without discovering them. In digging a cellar a few years since upon the farm of P. Piersen, a skeleton was exhumed, the thigh bones of which would indicate great height; exceeding by several inches, that of the tallest of our own race.

History of the Holland Purchase, 1849

The ancient works at Fort Hill, LeRoy, are especially worthy of observation in connection with this interesting branch of history… Forty years ago an entrenchment ten feet deep, and some twelve or fifteen feet wide, extended from the west to the east end, along the north or front part, and continued up each side about twenty rods, where it crossed over, and joining made the circuit of entrenchment complete. It would seem that this fortification was arranged more for protection against invasion from the north, this direction being evidently its most commanding position. Near the northwest corner, piles of rounded stones, have, at different times, been collected of hard consistence, which are supposed to have been used as weapons of defense by the besieged against the besiegers. Such skeletons as have been found in and about this locality, indicate a race of men averaging one third larger than the present race; so adjudged an anatomist.

History of St. Lawrence and Franklin Counties, New York, 1853

In the town of MaComb, St. Lawrence County, are found traces of three trench enclosures, and several places where beds of ashes mark the site of ancient hearths. One of these was on the farm of William Houghton, on the bank of Birch Creek, and enclosed the premises now used as a mill yard. It is somewhat in the form of a semi-circle; the two ends resting on the creek, and might have enclosed half and acre… On an adjoining hill, now partly occupied by an orchard, traces of an ancient work formerly existed, but this has also been obliterated.
In the pond adjoining there was found, many years since, a skeleton, said to have been of great size.

History of Seneca County, New York, 1876

There were several mounds on the Calver place, and we often plowed up bones and ancient crockery. In 1850 we opened one of these mounds, and found a very large skeleton, with a well-shaped skull, and a stone pitcher near the head. The pitcher seemed to have been made of sand and clay. Small vessels of the same material, filled with clam-shells, were placed inside of the elbows. Some of these pitchers would hold half a gallon. We gave them to Gen. Brish. These things were as wonderful to the Indians as to us.

History of Niagara County, New York, 1878

"About one and a half miles west of Shelby Centre, Orleans County, is an ancient work. A broad ditch encloses in a form nearly circular about three acres of land: the ditch is at this day well defined, several feet deep. Some skeletons, almost entire, have been exhumed, many of giant size, not less than seven to eight feet in length. The skulls are large, and well developed in the anterior lobe, abroad between the ears and flattened in the coronal region."

History of Niagara County, New York, 1878
Town of Cambria

A search enabled them to come to a pit, but a slight distance from the surface. The top of the pit was covered with slabs of the Medina Sandstone, and was twenty-four feet square by four and a half in depth- the planes agreeing with the four cardinal points. It was filled with human bones of both sexes and all ages. They dug down at one extremity, and found the same layers to extend to the bottom, which was the same dry loam, and from their calculations they deduced that at least four thousand souls had perished one great massacre. In one skull two flint arrowheads were found, and many had the appearance of having been fractured and cleft open by a sudden blow. They were piled in regular layers, but with no regard to size or sex... One hundred and fifty persons a day visited this spot the first season, and carried off the bones. They are now nearly all gone and the pit ploughed over. The remains of a wall were traced near the vault. Some of the bones found in the latter were of unusual size. One of these was a thighbone that had been healed of an oblique fracture. One was the upper half of a skull so large that that of a common man would not fill it.

History of Cattaraugus County, Ellis, 1879
Limestone Creek Area

Twenty years later, Mr. Older revisited the place, and found the work much changed by the hand of the white man. The smaller circle had been leveled, and a lumberman's road crossed its enclosure. Some workmen were attempting to remove a stump of about two feet diameter, which stood at the two circles, and interfered with the excavation of a cellar which had been marked out. Oxen were attached and the stump was easily turned out from its ancient bed, disclosing a mass of human bones, from which the earth had been entirely removed by the small fibrous roots. 'On examination,' writes Mr. Older, 'They proved to be skeletons entire, having been deposited there three or four in depth, with faces downward and heads to the east. A remarkable characteristic of these skeletons was their enormous proportions. Compared with my own stature and physical formation they must have been giants indeed! I am five feet eleven and a half inches in height, and I wear a hat seven and a half in size by hatters measure; but a skull of one of these skeletons would sit loosely on my head; a rib bone would pass round me from spine to colon, outside my garments, including an overcoat; a shin-bone would reach from my ankle two or three inches above the top of my knee joint; a thigh-bone reached from my knee to the upper part of the hip bone; and the sub-maxillary would encase my jaw like an easy-fitting mask. The teeth were enormous, particularly the molars. An attempt was made to preserve portions of these remains, but by exposure to the atmosphere they crumbled to a fine powder. These people must have been at least eight feet high, with other proportions corresponding.' The earthwork here mentioned is still visible. Its location is on the land now owned by Job Moses.

History of Cattaraugus County, 1879

About two miles south of the village of Rutledge, in the town of Connewango, on lot No. 45, at a point about sixty rods east of Connewango Creek and near the residence of Norman E. Ct. Cowen, there was discovered by the first pioneers of this section a sepulcher mound, nearly circular in form, and having an entire circumference of one hundred and seventy feet. The height of the mound was about twelve feet. Mr. Cheney spoke of this work as "having some appearance of being constructed with the ditch or vallum outside of the mound as in Druid Burrows..."

Within the mound there were discovered nine human skeletons, which had been buried in a sitting posture and at regular intervals of space, in the form of a circle, and facing towards a common center. There was some sight appearance that a frame-work had enclosed the dead at the time of their internment. The skeletons were so far decayed as to crumble upon exposure to the atmosphere, but were all of very large size. An osfemur (the largest found here) was twenty-eight inches in length.

Daily Gazette from Ft. Wayne Indiana, May 1, 1885
The Mound Builders, Many Evidences of a First Race In and About Randolph.

"Between Lake Erie and Dayton, Chatauqua County, are the remains of a circular sepulcher mound. The mound has an elevation of 80 feet, and is 120 feet in circumference. According to antiquaries, this mound must have been the mausoleum of some great warrior. In the towns of Lear and Conewago excavations were made some years ago in several of these mounds. In one of them eight skeletons were found. They were in a sitting posture and arranged in a circle. Large blocks of mica were found in some of the mounds among the skeletons. This material is so frequently found in these burial mounds that it is believed to have been regarded as a sacred relic by the extinct race whose work still remains in the most gigantic earthworks all over the continent.

Near the station of the New York, Pennsylvania and Ohio railroad is a peculiar earth formation, which was designed by those who fashioned it thousands of years ago to represent a serpent, according to the conclusion of those who have read the customs of the mound-builders by the monuments they left. This particular formation is 425 feet long, and enthusiastic antiquaries who visit it are unanimous in the resemblance it presents to a snake basking in the sunshine."

Smithsonian Institute Bureau of Ethnology 1898-1899 (New York)

An exceptional example of the burial mound was described by Mr. T. A. Cheney. It was in Conewango township, Cattaraugus County, and on the brow of a hill. The account is not perfectly clear, but is here given in Mr. Cheney's own words:

The form of the tumulus is of intermediate character between the ellipse and the parallelogram; the interior mound, at its base, has a major axis of 65 feet, while the minor axis is 61 feet, with an altitude above the first platform or embankment of 10 feet, or an entire elevation of some 13 feet. This embankment, with an entrance or gateway upon the east side 30 feet in width, has an entire circumference of 170 feet... In making an excavation, eight skeletons, buried in a sitting posture and at regular intervals of space, so as to form a circle within the mound, were disinterred. Some slight appearance yet existed to show that framework had enclosed the dead at time of interment. These osteological remains were of very large size, but were so much decomposed that they mostly crumbled to dust. The relics of art here disclosed were also of a peculiar and interesting character--amulets, chisels, etc., of elaborate workmanship, resembling the Mexican and Peruvian antiquities.

New York Times, February 11, 1902
Find Giant Indian's Bones
Workmen on Harlem Road Unearth Relics of Teekus Tribe
Special to the New York Times

Katonah, N. Y., Sept 6-While a gang of men in the employ of the New York and Harlem Railroad were taking sand from an immense mound near Purdy's Station to day fill an excavation, they unearthed several skeletons of unusual size.

The bones are believed to be those of Indians who once lived in this vicinity and belonged to a tribe that was led by the great Chief Teekus, from whom the Titicus Valley, now a part of the New York watershed, takes its name. Besides finding the bones, the workmen also exhumed a score or more of arrowheads, hatchets, and copper implements. It is believed that the large mound in which the relics were found was once the burying ground of the Teekus Indians. The last Indians were seen in the valley a short time after the Revolutionary War.

The bones found to-day were brought to Katonah and will be reinterred in the local cemetery.

Canada

Pioneer Society of Michigan, 1904

We frequently hear of the discovery of the skeletons of a gigantic race, and we are therefore the more puzzled to know to what race the mound builders belonged, for although we are called a new country, comparatively speaking, we may be the oldest.

A few years ago an article appeared in the Toronto Telegraph stating that in the township of Cayuga in the Grand River, on the farm of Daniel Fredenburg, five or six feet below the surface, were found two hundred skeletons nearly perfect, in a string of beads around the neck of each, stone pipes in the jaws of several of them, and many stone axes and skinners scattered around in the dirt. The skeletons were gigantic, some of them measuring nine feet, and few of them less than seven. Some of the thigh bones were six inches longer that any now known. The farm had been cultivated a century and was originally covered with a growth of pine. There was evidence from the crushed bones that a battled had been fought and these were some of the slain. Were these the remains of Indians or some other race? Who filled this ghastly pit?

Warren Evening Mirror, (Warren, Pennsylvania) May 17, 1910
Grim Relics of Early Fight
Manitoba Workman Unearth Skeletons In Common Grave

Snowflake, Man., May. 17-Workman digging on the brow of a hill on the Charles Sims homestead unearthed the skeletons of 20 human beings, which had been buried in all sorts of positions. The bones were those of men, women and children who in ages past, had been slaughtered, evidently in some battle between warring tribes of redskins.

The spot could have been no peaceful Indian burying ground, as the bodies were thrown in every position, some on top of the others. The skeletons show the men to be of gigantic stature.

Beads and other adornments for Indian women are quite plentiful. To Dr. Corbett of Snowflake belongs the credit of making the discovery.

Pennsylvania

History of Bradford County, Pennsylvania, 1878

In 1822, while digging a cellar on the farm of Gen. McKean, the excavation came to what was a supposed to be "an impenetrable rock, but striking it with a crow, it gave forth a hollow sound." They redoubled their efforts, and at last the stone broke and fell into a vault. And now, with visions of long-buried treasure flitting through their minds, they carefully removed the earth from the arch, speculating the while as to the probable extent of the "treasure trove" and the amount of salvage the general would be likely to claim. On removing the cap they found "not what they sought" but a sepulcher. A careful examination of the sarcophagus revealed it flagged at the bottom, the sides artistically built up, and a flat stone laid on the top. The sarcophagus measured nine feet in length, two feet six inches in width, and two feet deep. In it was found a skeleton, measuring, as it lay, eight feet two inches in length (this measurement was made by Dr. Williams, late of Troy, now deceased). The teeth were sound, but the bones were soft and easily broken. There were two of these sepulchers within the space of the cellars, one of which had a pine growing over it three feet in diameter.

New York Times, August 10, 1880
Two Very Tall Skeletons
From the *Harrisburg (Penn.) Telegraph*

The following was copied verbatim from a note made in his pocket almanac by the late Judge Atlee: "On the 24th of May, 1798, being at Hanover (York County, Penn.,) in company with Chief Justice McKean, Judge Bryan, Mr. Burd, and others, on our way to Franklin, and, taking a view of the town, in company with Mr. McAlister, and several other respectable inhabitants, we went to Mr.Neese's tan yard, where we were shown a place near the currying house from whence (in digging to sink a tan-vat) some years ago were taken two skeletons of human bodies. They lay close beside each other and measured 11 feet 8 inches in length: the bones were entire, but on being taken up and exposed to the air they presently crumbled and fell to pieces. Mr. McAllister and some others mentioned that they and many others had seen them, and Mr.; McAllister, whom is a tall man, about 6 feet 4 inches high, mentioned that the principal bone of the leg of one of them, being placed by the side of the leg, reached from the ankle a considerable way up his thigh, pointing a small distance below the hip bone.

History of Erie County, Illustrated, 1884

Many indications have been found in the county proving conclusively that it was once peopled by a different race from the Indians who were found here when it was first visited by white men. When the link of the Erie and Pittsburgh Railroad form the Lake Shore road to the dock at Erie was in process of construction, the laborers dug into a great mass of bones at the crossing of the public road which runs by the rolling mill. From the promiscuous way in which they were thrown together, it is surmised that a terrible battle must have taken place in the vicinity at some day so far distant that not even a tradition of the event has been preserved. The skulls were flattened, and the foreheads were seldom more than an inch in width. The bodies were in a sitting posture, and there were no traces that garments, weapons or ornaments had been buried with them. On account of the superstitious notions that prevailed among the workmen, none of the skeletons were preserved, the entire collection as far as was exposed being thrown into the embankment further down the road. At a later date, when the roadway of the

Philadelphia and Erie road, where it passes through the Warfel farm, was being widened, another deposit of bones was dug up and summarily disposed of as before. Among the skeletons was one of a giant, side by side with a smaller one probably that of his wife. The arm and leg bones of this Native American Goliath were about one-half longer than those of the tallest man among the laborers; the skull was immensely large; the lower jawbone easily slipped over the face and whiskers of a full-face man, and the teeth were in a perfect state of preservation.

Another skeleton was dug up in Conneaut Township some years ago, which was quite remarkable in its dimensions. As in the other instance, a comparison was made with the largest man in the neighborhood, and the jawbone readily covered his face, while the lower bone of the leg was nearly a foot longer than the one with which it was measured, indicating that the man must have been eight to ten feet in height. The bones of a flat head were turned up in the same township some two years ago with a skull of unusual size. Relics of a former time have been gathered in that section by the pailful.

History of Erie County, Pennsylvania, Illustrated, 1884

An ancient graveyard was discovered in 1820, on the land known as the Drs. Carter and Dickinson places in Erie, which created quite a sensation at the time. Dr. Albert Thayer dug up some of the bones, and all indicated a race of beings of immense size.

Otzinachson, A History of the West Branch Valley of the Susquehanna, 1889

That Indians frequented this stream in considerable numbers there is no doubt, as they left abundant numbers traces of their occupation behind them, both in ruined huts and graves. As late as 1873, at the village of Sterling Run, while Mr. Earl was excavating for a cellar, seventeen Indian skeletons were disclosed. All except two were of ordinary grown stature, while one measured over seven and one-half feet from the cranium to the heel bones. The bones had all remained undisturbed. They lay with their feet toward each other in a three-quarter circle, that is, some with their heads to the east, and then northeasterly to the north, and then northwesterly to the west. There had been a fire at the centre, between their feet, as ashes and coals were found there. The skeletons, except one smaller than the rest, were all as regularly arranged as they would be naturally in a sleeping camp and similar dimensions; many of the bones were in a good state of preservation, particularly the teeth and jaw bones, and some of the leg bones and skulls. The stalwart skeleton held a stoneware or clay pipe between his teeth as naturally as if in the act of smoking.

History of McKean, Elk, Cameron and Potter Counties, Pennsylvania, 1890

The Freaty Indians, whose old country they entered, were comparatively modern settlers. There were men here before them, who lived in the age of giant nature. On the Fisher farm, near Bradford, in the Tuna Valley Flats, there were relics of a large race exhumed years ago. It appears an aged tree was felled and uprooted to make way for improvements, and beneath were found large skulls, any of which could encase the head of any modern man; while thigh-bones and shin bones were several inches longer than those of the present people.

History of Erie County Pennsylvania from its Settlement, 1894

There are also remains of an Indian fort between Gerard and Springfield. From a grave in this vicinity, some years ago, a thigh bone was exhumed which measured four inches longer than that of a man with which it was compared, who was six feet and two inches in height.

In Scoalers woods, east of Erie, is an Indian burying ground. Mr. Fredrick Zimmerman described a very large skeleton, which was found there; with it were two copper bowls perforated at the edges and laced together with a buckskin thong, which fell to dust soon after being exposed to the air. The bowls, which would contain about a pint each, were found filled with beads.

Smithsonian Institutions Bureau of Ethnology, 1898-99

On the upper terrace, within the corporate limits of Monongahela City, are situated the garden and greenhouse of Mr. I. S. Crall. Two ravines on the east and west sides open directly south into Pigeon Creek, and their erosion has lowered the ground until it is surrounded by higher land on every side except along the bluff next to the creek. The further side of the creek being bounded by a high hill, the view from the level land between the ravines is shut off in every direction, except through a narrow pass looking up the river, thus the tract is surrounded on every side by hills close at hand, ranging from 40 to 250 feet above its level. In excavating for foundation walls and other purposes, Mr. Crall has, at different times, unearthed skeletons, some of large size; the ground is strewn with mussel shells, flint chips, etc.

On the eastern side of this level, near the break of the ravine, and close to a never-failing spring, stands the largest mound above the one at McKees rocks, measuring 9 feet in height by 60 feet in diameter… At the center a hole measuring 3 feet across the top and 2 feet into the original soil. In this were fragments of human bones too soft to be preserved. They indicated an adult of large size. The gray clay was unbroken over this hole. Directly over this, above the clay and resting upon it, were portions of another large skeleton, with which was found part of an unburned clay tube or pipe.

Twentieth Century History of Erie County, Pennsylvania, Vol. I, 1909

Among the best known is that found in Wayne Township a short distance from Coury, which consist of a circular embankment of earth surrounded by a trench from which the earth had been dug, the whole enclosing about three acres… Smaller that the Wayne mound or circle is that of the John Pomeroy place on Conneaut Creek, near Albion. It encloses an area of a little less than an acre, and the embankment of this was three feet high and six feet broad at the base… On the same farm there is an interesting mound a hundred feet long and fifty feet wide by twenty-five feet high. There are stories of finding the skeletons of giants in one of the Conneaut Township mounds.

Rock Valley, Iowas Bee, January 7, 1921
UNEARTH SKELETON OF GIANT
Bones of Supposed Mound Builder
Those of Man Eight or Nine Feet High

Dr. W. J. Holland, curator of the Carnegie museum, Pittsburgh, and his assistant, Dr,. Peterson, a few days ago opened up a mound of the ancient race that inhabited this section and secured the skeleton of a man who when in the flesh was between eight and nine feet in height, says a Greensburg (Pa) dispatch to the Philadelphia Inquirer.

This mound, which was originally about 100 feet long and more than 12 feet high, has been somewhat worn down by time. It is on the J. B. Secrist farm in South Huntington township. This farm has been in the Secrist name for more than a century.

The most interesting feature in the recent excavation was the mummified torso of the human body, which the experts figured was laid to rest at least 400 years ago. Portions of the bones dug up and the bones in the legs, Prof. Peterson declares, are those of a person between eight and nine feet in height. The scientist figures that the skeleton was the framework of a person of the prehistoric race that inhabited this section before the American Indians.

The torso and the portions of the big skeleton were shipped to the Carnegie museum. Drs Holland and Peterson supervised the explorations on the Secrist mound with the greatest of care. The curators believe the man whose skeleton they secured belonged to the mound builder class.

Ohio

Firelands Pioneers, **1858,**
Vermillion Township, Erie County, Ohio

There are quite a number of mounds in the township, where the bones, and sometimes the whole skeleton of the human race have been found. The bones and skeletons found are very large, and some of the inhabitants think they much of belonged to a race of beings much larger in size than the Indians found here by the settlers.

Ohio Democrat, **(New Philadelphia, Ohio) January 14, 1870**
The Cardiff Giant Outdone
Alleged Discovery of the Skeleton of a Giant in the Oil Regions

The Oil City Times of Friday is responsible for the following:

On Tuesday morning last, while Mr. William Thompson, assisted by Robert R. Smith was engaged in making an excavation near the house of the former, about half a mile north of West Hickory, preparatory to erecting a derrick, they exhumed an enormous helmet of iron, which was corroded with rust. Further digging brought to light a sword, which measured nine feet in length. Curiosity incited them to enlarge the hole, and after some little time they discovered the bones of two enormous feet. Following up the "lead" they had unexpectedly struck, in a few hours time they had unearthed a well preserved skeleton of an enormous giant, belonging to a species of the human family which probably inhabited this and other parts of the world at that time of which the Bible speaks, when it says: "And there were giants in those days." The helmet is said to be the shape of those found among the ruins of Nineveh. The bones of the skeleton are remarkably white. The teeth all in their places, and all of extraordinary size. These relics have been taken to Tionest, where they are visited by large numbers of people daily. When his giant ship was in the flesh he must have stood eight feet in his stockings. These remarkable relics will be forwarded to New York early next week. The joints of the skeleton are now being glued together. These remains were found about twelve feet below the surface of the mound which had been thrown up probably centuries ago, and which was not more than three feet above the level of the ground around it. Here is another nut for antiquarians to crack.

History of Huron and Erie Counties, Ohio, 1879

Near these forts were mounds or hillocks, which were found to contain human bones, promiscuously thrown together, as if a large number of bodies had been buried at one time. The skull bones, when found entire, were shown to be larger upon average, than those of the present race, and all exhibited marks that would indicate that life had been taken in deadly combat.

History of Lorain County, Ohio, 1879

Their mounds are a proof of their existence, for their character and the place and mode of their erection attest the handiwork of intelligent beings, while the bones, weapons of warfare, stone implements and arrow heads which have been discovered and are still found buried in these earthworks, furnish a still stronger proof of the existence of a pre-historic people. The skeleton remains of human beings of almost gigantic proportions were exhumed from their ancient cemeteries by the first settlers. The Indians, disclaiming them as kindred, could give no information in regard to them."

History of Ashland County, Ohio, 1880

"About thirty-five years since, while engaged in cutting a bluff, on the bank of the creek, east of the residence of the late Patrick Murray, for the purpose of improving the trail-road alluded to, a number of human skeletons were unearthed, among which was one supposed to have been over seven feet high, when erect. The bones were in a good state of preservation. This giant must have loomed up among his aboriginal kinsmen like a Colossus."

History of Ashland County, Ohio, 1880

Two mounds were found in the north part of Perry Township, about one mile from the fort. They were about thirty feet apart, and occupied level ground near a brook. The larger one was about five feet high, and twenty-five feet in diameter at the base. The smaller one was probably twelve feet in diameter at the base and three and a half feet high. William Hamilton extirpated the larger one in digging a cellar; and about four feet below the natural surface found a triangular wooden post, and three human skeletons, one of unusual size, embedded in the sand.

American Antiquarian, Vol., 3, 1880

"A skeleton which is reported to have been of enormous dimensions" was found in a clay coffin, with a sandstone slab containing hieroglyphs during mound explorations by Dr. Everhart near Zanesville, Ohio.

A mound near Toledo, Ohio held 20 skeletons, seated and facing east with jaws and teeth "twice as large as those of present day people" and beside each was a large bowl with "curiously wrought hieroglyphic figures."

The History of Brown County Ohio, 1883

Mastodonic remains are occasionally unearthed, and, from time to time, discoveries of the remains of Indian settlements are indicated by the appearance of gigantic skeletons, with the high cheek bones, powerful jaws and massive frames peculiar of the red man, who left these as the only record with which to form a clew to the history of past ages.

History of Erie County, Ohio, 1889

"On the highest points and some distance back from the creek banks, in fields of light, sandy soil and clay sub-soil, are found circular deposits of extremely black earth varying in depth from one to three feet, in which are found skeletons of a 'race'-not Indians. The skull is well developed, being full in forehead, broad, with good height above the ears, and in all respects, different from the Indians. The skeletons of adults are above average size and some of them gigantic. The writer, together with Dr. Charles Stroud and Mr. T. L. Williams, have dug up a number in different localities, and always, with one exception, with the same results."

History of the Villages and Townships of Erie County, Ohio, Berlin Township, 1890

This creek has a branch called the West Branch. The two branches have had at different times over two dozen saw mills built along their course through the township. The Chapelle empties into the lake in the township of Vermilion. There is a mound on the farm of Henry Hoak, in the eastern part of the township, which covers one-eighth of an acre, with large trees growing on it; and in digging a cellar, some time ago, for a new house, near one which was built in the first settlement of the township, a large human skeleton was found, in a sitting posture. Others have been found at the same place, also many arrowheads, stone axes, and other relics, evincing that the spot built upon must have been a mound.

American Antiquarian, Vol. 13, 1890

Bones in a Gravel Bed-- Some workmen in Auglaize County, Ohio, recently came across some human bones in a bed of gravel. Mr. Charles Jones, a well-known and wealthy landowner of Spencerville, Allen County, says of the discovery:

"There was a remarkable discovery of prehistoric remains in our section the other day. The instance came under my own observation. Last week I had occasion to visit the farm of I. Hemley, about two miles west of Kossuth, just across the border in Auglaize County. Some workmen were engaged in digging a well, and had descended to a depth of 32 feet, when they struck a gravel drift, from which they exhumed a skull, 38 inches in circumference. Further down the other bones were found. There can be no doubt as to the kind of remains. The thigh bone measured three feet two inches long. All the bones were in an excellent state of preservation, and were probably those of a prehistoric warrior who was killed in battle, as the skull seemed to have been crushed with a blunt instrument. The whole skeleton measured eight feet eleven and one-half inches in height, and when clothed in flesh must have been a tremendously powerful man. A huge stone ax weighing twenty-seven pounds and a flint spear head of seventeen pounds weight were found with the bones, and were, no doubt, swayed by the giant with the greatest ease. A copper medallion, engraved with several strange characters, was also found with the bones. This is a startling discovery. The scientific value of the discovery is also considerable, and may lead to some interesting developments.

Newark Daily Advocate, (Newark, Ohio) July 22, 1895

Workman at Wooster O., unearthed the skeleton of a giant in an old graveyard

Sketches and Stories of the Lake Erie Islands, **1898**

A large quantity of human bones was discovered in a fissure in the limestone near the United States Coast Guard lighthouse. A crude tomb of black stone slabs, of a formation not known on the island, was found many years ago beneath the roots of a huge stump. Eight skeletons were found, one measuring over seven feet in height.

Fort Wayne News, **(Fort Wayne, Indiana,) April 20, 1898**

Toledo Ohio, April 30-Workman in the company of the Fergusan Construction Company excavating for the new Toledo and Ottawa Beach railroad, a little beyond the city limits unearthed three skeletons, evidently relics of some great race, as they were about seven feet in length. Just where the ears should be on the head are singular bony protuberances which curl forward. The finds were made in solid yellow clay about eight feet below the surface.

The cut is through a large mound not half of which has yet been torn up. Several stone tomahawks of larger size have been picked up in the locality.

Newark Daily Advocate, **(Newark, Ohio) August 14, 1902**
Giant Skeleton
Found in Bed of Sand in Northwestern Ohio-Man was Eight Feet High

Bowling Green, O., Aug. 14-While excavating for sand for building on the Charles Whirmer farm, Wm. Jones unearthed the skeleton of a gigantic man. It is in a fair state of preservation and will be preserved, as it is thought that it may have some scientific value.

The skeleton was found in a sitting posture, and when the bones were placed in a horizontal position they indicated that the height of the man in life must have been over eight feet. The head is of enormous size, being 12 inches in diameter.

It is believed that it is the skeleton of a member of a prehistoric race of giants. Further excavations will be made to see if other graves cannot be found.

History of the County of Williams, Ohio, **1905**

On fractioned section 12, about 2 miles north of Montpelier, two large mounds which were six on seven feet in height and fifty or sixty feet in diameter... taking there from two skeletons, one very large and the other of ordinary size.

History of the Maumee River Basin, **1905**
Defiance County, Ohio,

A mound was found on the high south bank of the Maumee River, a few rods west of the middle north and south line of section twenty and seven of Defiance Township... This mound was about four feet above the surrounding land, about thirty feet in diameter. Brice, who gave the writer this information, opened this mound in the year 1824. A small quantity of bony fragments were found which readily crumbled between the fingers on being handled. Human teeth were found, some of which were of large size.

History of Fulton County, Ohio, **1905**

Of the works examined in this county, those most worthy of mention are situated on the farm of the late Hon. D. W. H. Howard, in section 9, Pike Township. These mounds were explored during the summer of 1892.

These mounds are in a group of twelve in number, of which eleven are located and clearly identical and the site of the twelfth is plainly indicated... The mound mentioned as being located in the public road is, as stated above, entirely obliterated, but in an early day Col. Howard found in its center a circle of stones about four feet diameter, containing within the circle about a bushel of charcoal and ashes. The stones are what are known as "nigger heads."

Nearly all of these mounds were opened and examined by judge Handy, and the report of two of them we will give in the judge's own language. Of one he calls Mound No. 7 he writes, "Sandy soil, light yellow sand: about eighteen inches from surface found longest thigh bones yet discovered. No trace of fire-no disturbance of soil here to fore-bones crumbled on exposure-highest of the mounds-found near center skeleton with his head to the north, lying on his back and limbs extended-near hem found skeleton No. 2, with head to the east and lying on his face. Both being large men.

About another mound in this group it was written: "The part of the skull above the nasal bones was well preserved, and compared with the skull of an Indian found intrusively buried in neighboring mound, was a distinctly different type of man.

History of Sandusky Ohio, **1909**

Underneath the roots, and seven or eight feet from the surface of the ground were several large flat stones covering the skeletons of a number of Indians of varying stature, surrounded by wood, ashes, and charcoal. The adult skeletons indicated very large beings.

History of Sandusky Ohio, **1909**

Near the residence of Mr. Williams and not far form it, was found a mound about fifty feet in diameter, which much have been a very ancient construction. Mr. Williams said that about the year 1820 he assisted in cutting down a white oak tree which stood on the very summit of the mound, for the purpose of capturing a swarm of bees which had long been in the tree, and that this tree was then near three feet in diameter, and the elevation of the mound was eight feet above the general level of the surrounding land-the mound was afterward opened by Mr. John Shannon, of this county, and his brother, about the year 1840. The mound he said attracted considerable observation and much speculation among the observers as to what it was raised for, and what might be in it. The stump of the oak had then so far decayed that it was removed without much difficulty. On removing the earth from a considerable space and a little below the general level of the surface around the mound, they found the teeth of a human being in good preservation. Upon further carefully removing the earth they found, marked in a different colored earth from that surrounding it, the figure of a man of giant size, plainly to be seen.

History of the Villages and Townships of Erie County, Ohio, Berlin Township

This creek has a branch called the West Branch. The two branches have had at different times over two dozen saw mills built along their course through the township. The Chapelle empties into the lake in the township of Vermilion. There is a mound on the farm of Henry Hoak, in the eastern part of the township, which covers one-eighth of an acre, with large trees growing on it; and in digging a cellar, some time ago, for a new house, near one which was built in the first settlement of the township, a large human skeleton was found, in a sitting posture. Others have been found at the same place, also many arrowheads, stone axes, and other relics, evincing that the spot built upon must have been a mound.

A Twentieth Century History of Hardin County, Ohio, 1916

The most important evidences that the Mound Builders once occupied this region are the mounds that they left here and there throughout the county. One of the most important of these lies in Hale Township between Mt. Victory and Ridgeway, and from which hundreds of loads of gravel have been taken to build pikes. This mound covered an area of about one and one-half acres, and the first settlers of that part of the county said it was covered with a very heavy growth of timber when they came. At first it was thought the Indians had it for a burying ground, but the Indians knew nothing of the bodies there buried. In 1856 when a railroad was built through the southern part of the county connecting Cleveland with Cincinnati, this mound was ruthlessly torn to pieces for the purpose of furnishing ballast for the track. While the excavation was going on more than three hundred skeletons were dug up, most of which were dumped with the gravel on the railroad track. A few of the bones were saved, some of them being gigantic in size. Many of the bodies had been buried in sitting posture, and all about them were evidences of fire and the remains of various article. As the Indians knew nothing of the fact of who was buried there, it is safe to say the bodies must have been placed there many years before the white man knew anything about the country."

Historical Collections of Ohio, Howe V. I., Pt, 1.,
Ashtabula County

There were mounds situated in the eastern part of the village of Conneaut and an extensive burying ground near the Presbyterian Church, which appear to have had no connection with the burying places of the Indians. Among the human bones found in the mounds were some belonging to men of gigantic structure. Some of the skulls were of sufficient capacity to admit the head of an ordinary man, and jaw bones that might have been fitted over the face with equal facility; the other bones were proportionately large. The burying ground referred to contained about four acres, and with the exception of a slight angle in conformity with the natural contour of the ground was in the form of an oblong square. It appeared to have been accurately surveyed into lots running from north to south, and exhibited all the order and propriety of arrangement deemed necessary to constitute Christian burial. On the first examination of the ground by the settlers they found it covered with the ordinary forest trees, with an opening near the center containing a single butternut. The graves were distinguished by slight depressions disposed in straight rows and were estimated to number from two to three thousand. On examination in 1800, they were found to contain human bones, invariably blackened by time, which on exposure to the air soon crumbled to dust. Traces of ancient cultivation observed by the first settlers on the lands of the vicinity, although covered with forest, exhibited signs of having once been thrown up into squares and terraces, and laid out into gardens.

A Study of the Glacial Kame Culture,in Michigan, Ohio, and Indiana, 1948
Wilbur M. Cunningham
Zimmerman Site

On the farm of Arthur Zimmerman, in McDonald Township, about three miles north of Belle Center near the Logan County line in Hardin County, Ohio, is a huge gravel kame. It is the highest [point in the neighborhood, and from the top of it one can see for a distance of twelve to fifteen miles.

In the summer of 1931 when gravel was being hauled from this pit, 148 human skeletons were discovered from six feet to twenty-two feet below the surface; badly decayed human bones were uncovered about three and one-half feet from the surface, but no complete skeletons were buried at depths less than six feet. Mr. Zimmerman stated that in addition to the 148 skeletons counted, an unknown number was removed in his absence.

Some of the skeletons were buried face down. Others appear to have been buried "standing up," and still others were in a sitting position. With the skeleton in the deepest grave were 148 shell disk beads. In the deeper burials, in which copper was present, the skeletal remains were not well preserved.

According to the story, two skeletons of giant size, one male and one female, were found. Notwithstanding the reportedly large skeleton, the skull of the female was no larger than that of a child.

Indiana

Ft. Wayne, Daily Gazette, **July 26, 1872**
An Account of Fossil Remains Recently Discovered by the Opening of Mounds Near Laporte.

Those who feel an interest in the Neanderthal skull which was illustrated and described in *Harper's Weekly* a short time since, as well as those who have given the subject of races a more extensive research, may be interested to know that a race formerly inhabited Indiana and the adjacent country whose crania exhibit much the same peculiarity of structure, and a full knowledge of whose advancement in the arts of civilization might cause our phrenologist to modify, to some extent, their theories.

At Union Mills, in Laporte county, is a remarkable group of mounds, fifteen in number, or rather six double mounds, with three standing alone, all within a circle of about one-fourth of a mile in diameter, built mostly on the brow of a high tableland, overlooking the valley. This was probably the site of their village, a delightful spot, and with Mill Creek running through it, and the Kankakee Marsh on one side, and Lake Michigan not far distant to the other, well calculated to support a large population, who probably subsisted mainly by fishing, hunting and to some extent, by agricultural pursuits.

On Tuesday last, I exhumed from one of these mounds fragments of two skeletons, one a man and the other in probability, a woman.

They had been placed in a sitting posture on the original surface of the ground, and a mound of earth raised over them, which, after the lapse of many centuries, perhaps was yet eight feet high, with a base about fifty or sixty feet. That they were buried in a sitting posture is evident from the fact that all the bones and the skulls had fallen in one heap, except the leg bones, which extended from the heap in opposite directions. The skulls and bones were crushed by the superincumbent earth, and much decayed, so that a large portion of them was as soft and crumbled as easily as the surrounding earth.

Enough was preserved, however to prove that they belonged to a race entirely different from the modern Indians, and approaching nearly to the connecting link between mankind and his less favored brother-the ape.

The late Col. Foster says: "the Neanderthal skull affords the nearest approach hitherto observed to the confines of that gulf which separates man from the anthropoid types." But the LaPorte skulls if we can judge correctly between the actual and the pictured skulls, will help to bridge the gulf, for while the supercilliary ridge is less prominent in one, than in the Neanderthal skull, and is almost wanting in the other, their eyes looked out from beneath a frontal plate little more elevated than the skull of a turtle, and bearing some resemblance to it.

No tools or weapons were found in this mound, but a few flakes of flint lay with the bones.

In another of the mounds, which was cut through in constructing the Peninsular railway, besides the skeleton it contained, were an earthenware jar and a bowl, both of which have been broken and lost. Near these, which probably contained food and drink for the departed on his unknown journey, were placed his pipe, which is a beautifully carved stone, two copper needles, one pointed on both ends, and a number of flint spear and arrow heads, so that he should not lack means to defend himself, procure food on his way to the "happy hunting grounds. "

Smithsonian Annual Report, 1874
Dekalb County, Indiana

We next went to the farm of Henry Gonzer in Fairfield Township, there a mound once overlooked a small lake, which is gradually filling from the wash of the surrounding hills. The mound is now nearly obliterated by cultivation. We were informed by Mr. Gonzer that it was opened about twenty years ago, when the skeleton was found the thigh-bone of which was as long as his leg, and the skull as large as a half bushel measure. We dug a little below the surface, and found a few bones, among which was a broken thigh bone of ordinary size.

The Gonzer mound was obliterated by the plow, but this circular, Iroquois earthwork is still visible. From *The Nephilim Chronicles, A Travel Guide to the Ancient Ruins in the Ohio Valley,* 2010.

Several mounds of considerable size have been found in various parts of the county. Six of these mounds were found within the present limits of Marion, but only one remains, being just back of Buchanon and Son Marble Shop on Third Street.

The first frame courthouse was built on a mound, which stood just east of the present courthouse. This was about sixty feet in diameter and ten feet in height, which was among the largest found in the county, the average diameter being from ten to fifteen feet. The mound in the courtyard furnished the material out of which the brick was made for the present courthouse.

Excavations into these mounds, show that they are composed of alternate layers of gravel and sand. On a level with or just below the surface of the surrounding ground, the skeletons of human being in many instances have been exhumed. These seemed to have been buried in a sitting posture and the stature of some must have been seven feet. The bones when exposed seem much decayed, crumbling on the slightest touch. Articles of pottery ware, stone axes, pipes and various implements have been found, and some interesting collections have been formed out of these antique relics.

Northern Indianian, **April, 1881**
Kosciusko County, Indiana

On Tuesday and Wednesday of this week Mr. O.P. Jaquest had a number of hands employed in removing dirt from a strip of ground belonging to him lying between the C.W. & A.M., R.R. and the Goshen wagon road, in the northeast part of this city.

While so engaged on Tuesday they found four human skeletons that had evidently been buried in a trench, their bodies in a recumbent altitude, and after they had been covered with about one foot of dirt, another body had been buried on top of that in a sitting posture.

A short time after another trench was uncovered in which 13 bodies had been interred, they have evidently been laid in regular. They were all ages and both sexes.

On Wednesday two more bodies were found lying near each other some having a piece of mica (or isinglass) over his face. A piece about six inches long and four inches wide was secured intact and is now in Dr. Moro's possession.

A flint arrowhead and a stone about three inches long, one and a half inches wide and nearly a half inch thick with a hole bored through the center were also found with the same skeleton.

There is evidence of fire in the trenches, and two small pieces of what has evidently been a human skull burned to a substance resembling charcoal has been found. It seems that cremation was practiced long ago.

Who were these people? The arrowhead and peculiar shaped stone were common among the Indians and are frequently found in Indian graves; but the heads of these skeletons are remarkable from the contrast between them and those of the Indians or White, or indeed of any known race.

A skull of an old man shows some little evidence of intellectual powers; the forehead rising nearly one half inch above the eyebrows, but is very narrow transversely. The back part of the head and the width between the ears is immense.

The skull of a young woman shows absolutely no forehead at all. When alive a straight stick lay flat on her head would have touched her eyebrows and the crown of her head.

The skull of the man that the sheet of mica over his face is about halfway, as far as intelligence is concerned between the two.

A peculiar thing about the piece of mica is that it appears to have been set in an iron frame. As the entire circumference is covered with a thick coat of iron rust. Professor Moro made an analysis of the substance in order to be certain in regards to it, and it is unmistakably iron.

Another remarkable thing about the skull of these skeletons is their wonderful thickness. The adult skulls all of them at least one fourth of an inch in thickness, some of these more than that. A thighbone has been got out that shows the owner of it was at least six feet tall.

The place where they were buried was marked by three small mounds standing close together and for years the project has been frequently discussed digging into them and discussing their contents.

Histories of Lagrange and Noble Counties Indiana, **1882**
Lagrange County

A number of years ago, two mounds were opened in Section 13, Milford Township. A quantity of crumbling human bones was taken from one of them, among them being a skull quite well preserved. Some of the teeth were almost as sound as their ever were, and the under jaw, a massive one, was especially well preserved. In the other mound was found a layer of ashes and charcoal, extending over two or three square yards of ground. This was undoubtedly a mound where sacrifices were offered to the deity of the Mound Builders, and where burial rites with fire were performed.

History of the Counties of Whitley and Noble County Indiana, **1882**
Noble County

This much has been given on the authority of Schoolcraft, Wilson, Pidgeon, Smucker, Foster and the American Encyclopedia, to prepare the way for the classification and detailed description of the ancient earth and stone works in this county. No effort has been made in past years to gather together the pre-historic history of Noble County. No importance or value has been attached to disclosures of skeletons, the majority of citizens through out the county regarding them as belonging to the Indians, and consequently, the mounds which have been opened in years past in different parts of the county were not carefully examined, and no doubt much interesting, and perhaps valuable information has been hopelessly lost...

On Section 2, Elkhart Township, on what is called Sanford's Point, there are several mounds, one of which was opened some eight or ten years ago by the neighbors, who expected to unearth some valuable trinkets. Quite a number of bones were found, and these were scattered around on the surface of the ground, where they were left. No trinkets were found, an inferior maxillary bone found is said to have been remarkably large and sound.

Indiana Department of Geology and Natural History, **1883**
Grant County, Indiana

The largest of mounds found in Grant County is that one found two miles south and one mile west of Upland, in Jefferson Township. About forty years ago the mound was five or six rods on diameter and about fifty feet high. At that time it was covered with all kinds of timber.
After people began settling near the mound they began clearing away the timber. The dirt was carried about a quarter of a mile. There is a basin near, rather deep, and at that time it was covered with trees, the same as the same as the mound. The supposition of the old settlers was that the dirt in the mound was carried from where the basin is now.
The owner of the mound gave many people permission to dig into it. One day two men were given permission to dig. They dug a trench north and south about four feet deep. After, digging they found a part of a skeleton of a man, the thighbone, ball and socket joint, and many small bones. When the small bones were exposed to the air they immediately crumbled. The ball and socket and thighbones

were taken to a physician in Upland and he estimated the bones were of a man at least nine feet tall and weighing not less than three hundred pounds and the man was not fleshy.

Six mounds were once visible near the last mound described, but have all been leveled by a gravel company, save this last one located across the road. *The Nephilim Chronicles, A Travel Guide to the Ancient Ruins in the Ohio Valley,* **2010.**

History of Randolph County, Indiana, **1885**
Jay County

In a ditch dug by Joseph Stevens, in the northeast part of Green township, nearly south of Powers Station (Jay County), to drain a pond, great numbers of human bones were taken out, many being of unusual size. The jawbones were full of teeth.

There was found also what seemed to be a shriveled hand, like the hand of a little child.

Indiana Geological Report 16th, Annual Report, **1885**
Pulaski County, Indiana

In Indian Creek Township at a point opposite Pulaski Mills, in the "bottom" or alluvium of the Tippecanoe, is a large mound about one hundred feet in diameter at the base, and which was, before being plowed over, fully twelve feet high. Many years ago an excavation was made in this mound by a minister then sojourning in the neighborhood, with the result of unearthing several crumbling human skeletons. The bones were reported to have been very large and strong, but yielded to the action of the air and crumbled to dust.

Davenport Morning Tribune, **February 5, 1889**
Many Skeletons of an Extinct Indian Race Unearthed in Indiana

Whitlock, Ind., Feb. 4.- A huge graves pit was opened here recently. Soon after the excavating began a skeleton was found, and as the pit widened other skeletons were until at last thirty graves had been opened and many skeletons brought to light, evidently the remains of an Indian tribe-the Shawnees, probably, who had the villages in this region. One skeleton was found beneath a large stump, and yesterday another was found twelve feet under ground. The graves appear in regular order, and the occupants were buried in a sitting posture. In one grave three skeletons, supposed to be those of a woman and two children, were found. Yesterday the largest specimen was unearthed, the body of a person who in life must have been a giant. A peculiarity of the skeletons is that of the teeth are nearly all in a perfect state of preservation. In one grave beside the human skeletons was that of a dog, a copper spear-head, and earthen pot, and numerous beads, proving that some important personage had been put to rest there. The city of the dead is undoubtedly 150 years old.

Democrat, **November 24, 1892**
BURIAL PLACE OF GIANTS
Skeletons of an Ancient Race Unearthed in Indiana
Many Traditions Brought Out by the Discovery-Evidence of an Extinct Tribe of Very Large Americans

A rich archaeological find was recently unearthed two miles west of Crawfordsville in a gravel pit along the high bluffs of Sugar creek. Thus far twenty-five skeletons of Brodingnagian stature have been exhumed, and the unburying of these mammoth bones is still going on. This necropolis of long ago is filled with exited hunters of curios and scientific students from Wabash college almost continually, and as soon as removed from the gravel their rattling bones carried away to become parts of departments of archaeology, which are being established all over the city.

The last skeleton taken from the burial ground was a gigantic one, measuring seven feet in length. The femur alone would prove that the skeleton was that of a giant, and the pelvic bones twice as large as those of an ordinary man. The grinning skull of the giant had a perfect set of teeth, not one cracked or decayed, and with an enamel as beautiful as polished marble. The bones were perfect in every detail, notwithstanding the fact that they must have interred here for centuries. The entire absence of vegetable matter in the soil and the perfect drainage would account for the preservation of the bony structure.

Of the whole number of skeletons thus far found only two indicate immature development, the remainder representing the framework of a race of men evidently extinct for centuries. This is certainly the first discovery of skeletons in which the characteristic development of giants has been observed. It is thought by local scientist that these bones belong to a tribe of aborigines, but this theory cannot be fully established by the material structure of the skeleton.

Although no implements or ornaments were found buried with the bones, yet in close proximity many instruments of warfare and domestic utensils were found. They are mostly composed of stone, though some are composed of copper and a few of shell and bone. The stone implements are flint spears and arrow heads, and appear to be wrought with exceeding great skill. Pottery is found in great abundance. For many years specimens of these pots have been unearthed in this region, especially along the banks of the creek.

None of these skeletons was found in a separate grave, they being for the most part piled together in one conglomerate mass. Ten were found in one place in close contact, facing the setting sun, and arranged in a sitting posture. Many of the bones found farther down the bank, and in a soil in which there was more vegetable matter, crumbled to a dust as soon as exposed to the atmosphere, and the symmetry of a single bone could not be distinguished.

Many traditions have been brought out since the discovery. One old settler has called to mind the fact that fifty years ago a tree was uprooted on this same spot, exposing three skeletons of gigantic dimensions, and as they were beneath the trees, it must have sprung up long after the bodies were buried.

Gen. Lew Wallace says he remembers the sections of a stranger, who several years ago spent many months digging along the banks of Sugar creek in search of a gold spoon supposed to have been buried long ago when this part of the country was inhabited by savage tribes, and the owner of the land on which these remains were found calls to mind a tradition often related by his grandfather that a Spanish treasure had been buried here in the long, long ago, when the country was a wilderness and Chicago a barren waste of impenetrable swamps.

The excavations are being continued, and it is thought that rich developments are in prospect, for there is not a foot of the soil removed that does not contain some relic or grinning skull.

Alvards History of Noble County, Indiana, 1902
"A Big Indian"

Of the interesting collection on Dr. Egles possession the most prominent were the entire skull, dorsal and lumber vertebra, pelvic bones, dorsal and lumbar vertebra, pelvic bones, and left femur and forearm of a skeleton exhumed from one of the pre-historic mounds of Noble County, located on the farm of Jeremiah Noel, section 1, Elkhart township. Some measurements were taken, which are given below, with the common names of the measured parts: skull, from base to nose over the top of head to base of occiput, 11.5 inches: around the skull, from the middle of forehead, 15.75 inches; over the top, from ear to ear 11.5 inches; around the back of skull, from ear to ear 10.75 inches. Thighbone, 18 inches long, large and showing by the size of the muscular attachments great solidity and power of muscle. Forearm, 12 inches, large and strong. This skull, in size and proportions, was superior to those of many whites; and the pelvis, backbone and thighbone, all indicated that the form, when clothed in flesh and animated by the living spirit, must have been a noble specimen of manhood. The cranial developments showed capability of a high degree of intellectual culture...

The skeleton just described was found in a large mound on Noels farm, as above stated, with parts of twenty-seven others, by explorers in the interest of Balty & Co., publishers of a History of Lagrange and Noble Counties. In describing the excavation of this mound and others in the same vicinity, the principle writer of that history notes the posture of the skeletons as identical with known modes of Indian burials; and in alluding to the fact of a "remarkably large and sound maxillary bone", indicating comparatively recent burial, adds, "The reader must remember that these are the bones of the Mound Builders, not Indians

History of Park and Vermillion Counties, Indiana, 1913
Vermillion County

In March, 1880 while a company of gravel road workers were excavating gravel from the bank on the ridge at the southwest corner of the Newport Fairgrounds, five human skeletons were found... In the gravel bank along the railroad, at the southeast corner of the Fairground, another skeleton was found. No implements of war were found with the bones but ashes were perceivable. A collection of a dozen skeletons shows by measurements of the thigh bones found that the warriors, including a few women, averaged over six feet and two inches in height...the trochanters forming the attachment of muscles show that they were not only a race of giant stature, but also of more than giant strength.

Indianapolis News, **November 5, 1921**

Portland, Ind., November 5 - with the passing of Twin hills, the object of many sight-seeing trips in Jay County, and which are now being levels for the valuable gravel that they contain many interesting discoveries are being made.

The bones of these skeletons were found recently and those who know something about the human race say that the skulls look as though they might be from an early race of Indians or more remote races that was known to the early settlers.

Indiana Burial Place

It is the opinion of some who have given the subject some study that the Twin Hills, situated just northwest of this city, was the burial place of Chief Godfrey and his Indian followers in this vicinity. In the last few weeks more than twenty-five skeletons have been unearthed. Many of the skeletons have been found buried in sitting postures facing each other, and there is evidence of fire, which many believe indicates this race gave burned offerings to their gods. Charred bones have been found between the graves. It is believed these bones are of animals.

Some are of the opinion that the skeletons are those of the mound builders. The skeletons taken from the hills seem to differ from most of the skeletons that are being found in this part of the country. No ornaments are buried with them, much as been found buried with other skeletons believed to be those of the Indians. In m any other graves have been found stone pipes and tomahawks.

Exhibits Skull

O.O. Clayton, city engineer, has been exhibiting a skull in the streets of the city which is of queer shape. The skull was large but not as large as many that have been found in the hills lately, he said. The front part of the head sloped back almost straight. The teeth were in good condition, considering the time they have been buried. Many of the skeletons and bones found now are on display in the office of the county surveyor.

Indiana History Buletin Vol., III, **Oct. 1925**

The South Bend Tribune of October 4, contains an announcement of the discovery of important prehistoric remains in St. Joseph County. This announcement is confirmed by a letter from John D. Hibberd, Secretary of the Northern Indiana Historical Society. the circumstances are as follows:

Carl Litchfield of Teegarden, and Jesse Litchfield, who lives just north of Teegarden, recently excavated a mound on the farm of,Grove Vosburg some three miles north of Walkerton. The mound is reputed to be of great antiquity and this seems to be confirmed by the memory the owner of the farm has of an oaktree a yard in diameter formerly growing on top which fell down abut twenty years ago. The mound was at one time about 25 feet high but in recent years its height has been decreased. At a depth of about 12 feet, the Litchfields found eight skeletons buried in an arrangement somewhat like the spokes of a wheel with their heads toward the center. In the skull of one of the skeletons said to be of large size, a fine flint arrow was embedded. With this same skeleton several places of copper were found. The excavation also brought to light a number of other articles, bands beads, etc., and two pipe bowls, one smooth, the other elaborately carved.

History of Just 100 Years, Vol., I
Laporte County, Indiana 1938

   Near the junction of the Kankakee with the Little Kankakee on the farm of Wm. Flannigan...Directly under the apex of the principle mound, on a plane about one to one and one-half feet about the same level, was found three skeletons in a semi-reclining attitude and facing northeasterly. The central one was that of man, fully matured, of much more than the ordinary Anglo-Saxon's size and proportions of the present day. At his right was one that was markedly feminine, much smaller and younger, and at the left, one scarcely discernable, but from the bone outlines and their structure, also young and not so large as the central figure.

Michigan

Gazetteer of Michigan, **1839**

   In Bruce township, in the county of Macomb, on the north fork of the Clinton are several (fortifications) The latter consist mostly of an irregular embankment, with a ditch on the outside, and including from two to ten acres, with entrances, which were evidently gateways, and a mound on the inside opposite each entrance. In the vicinity there are a number of mounds.
Several small mounds have been found on a bluff of the Clinton River, eight miles from Lake St. Clair. In sinking the cellar of a building for a missionary, sixteen baskets full of human bones were found, of a remarkable size.

Pioneer Society of Michigan
***The Rabbit River Mounds and Circles* by H.D. Post 1878**

   We visited another burial mound near the Rabbit river, on the northeast quarter of section one, town three north, range fifteen west, Manlins, which measured twenty feet in diameter and three feet high, but found that it had already been dug open. The remains of a very large skeleton were near the surface.

History of Cass County, **1882**

   The largest mounds in the county are those upon the farm of Joseph Walter. Three beautiful and regular mounds occur here, situated in a line from east to west. A short distance south of them is a well defined ditch which forms a perfect horseshoe, measuring about one hundred and sixty feet in length by one hundred feet in width. There is no trace of embankment in connection with the excavations. For what purpose the horseshoe-shaped enclosure was made by the ancient people can, of course, only be conjectured. There is no probability, however, that it was designed, as many suppose it to have been, for work of defense.
One of the three large mounds, which have been mentioned, was excavated in September 1878, by Dr. E.J. Bonine, of Niles, who operated under the auspices of the Smithsonian Institution. It was a mound about thirteen feet high (originally it must of greater altitude), and the diameter of its base was about fifty feet. On the summit of the mound, within the memory of the settlers, stood a burr oak tree four feet in diameter, and probably three hundred years old. A shaft was sunk by the excavators into the center of the mound. was found to be composed throughout of the same soil as that of the surrounding

plain--a rich black loam. Almost invariably the human remains found under the mounds rest upon the natural surfaces of the earth, the mounds simply being heaped over them, but in this case the interment was several feet below the original level. Several skeletons were found, being those of men, women and children, a number of fragments of pottery, a curious bone or ivory ornament, bearing some resemblance to a walrus tooth, several amulets pierced with holes, through which thongs had doubtless once been placed to attach them to the person, several bone implements and five copper hatchets of fine edge and good formation. Portions of the skeletons were in a good state of preservation. The femur, or thigh bone, of one of the males, which Dr. Bonine has now in his possession, is of great size and indicates that its owner must have been at least seven feet in height.

History of Bay County, 1883
There are also four fortifications on the Rifle River, in Township Twenty-two north. They comprise from three to six acres each, containing several mounds of large size. They are also situated on the bluffs. The walls can yet be traced, and are from three to four feet high and from eight to ten feet wide, with large trees growing on them. A friend of mine opened one of these mounds and took from it a skeleton of a larger size than an ordinary person. He says he also saw several large mounds on the Au Sable River

History of Bay County, 1883
One of the highest elevations in Bay County is the mound or ridge at the east approach to the Lafayette avenue bridge. In 1905 we find on it the massive buildings of the Bay City Brewing Company, a hotel, livery stable, the venerable old McCormick homestead, and, on the northern spur, the palatial home of Ex-mayor George D Jackson. The elevation comprises about two acres.... In excavating for the massive brewery, Indian skeletons were found four to five feet below the surface, while five feet deeper down were found skeletons of another and apparently an older race, buried with oddly-formed burned pottery and quaint stone and copper implements. Some of these implements showed that this strange prehistoric people had the art of hardening copper, and of working in metals...In grading 22nd street, through the north end of this mound, three skeletons of very large stature were found at a depth of 11 feet, with large earthen pots placed at there head of each sarcophagus

History of Genesee County, Michigan, 1886
Another instance in the east part of the county, where a number of skeletons (also of large size) were found buried in a circle directly beneath the stump of a gigantic pine-tree of the oldest growth; but in both cases the finding of the bones was wholly accidental, as there was no mound or other surface-mark to indicate the places of burial.

Memorials of a Half Century, 1886
When mounds are opened in most cases, it is impossible to determine from the reports whether the skeletons found belong to original or intrusive burials. According to some accounts the skeletons indicate a race of very inferior size; according to others, they show a race of giants. The elasticity of these ancient relics, to suite the zeal of the narrator, is truly wonderful. On one occasion I accompanied an old pioneer and worthy Judge to visit several mounds in Western Michigan. My guide gravely

informed me that, twenty years before, he had dug from one of these mounds a skeleton which, when laid out upon the turf measured eleven feet, eight and three quarter inches, and a skull of which fitted entirely over the judicial head!

Memorials of a Half Century, 1886

By far the finest group of mounds that has come to my knowledge occurs on the banks of the Grand River, three miles south of Grand Rapids. They were still perfect when the writer had the satisfaction of seeing them in 1874.

The largest of these mounds has a diameter of 100 feet, and a height of 15 feet or more above the general surface. Close by are two others of nearly equal size, all very regular in shape and conical. They are in a line about 100 feet apart, and 500 feet from the river. Around them cluster seventeen smaller tumuli, without regular arrangement, and varying in height from eight to two feet. All within an acre of two and a half acres.

This group occupies the first terrace, which is overflowed in high water to the foot of the mounds. It lies in the shadow of the ancient, untrimmed forest, consisting principally of sugar maples. Trees were growing on the mounds of two to three feet diameter, and there were evidences of still older ones which have perished.

Seven of these tumuli were opened during the year preceding my visit, by Captain Coffinbury and others, and among them one of the largest. This was found to be wholly composed of the richest portions of the surrounding alluvial soil, differing in this respect from the others, which were composed of the gravel of the uplands. No relics were disclosed, except a copper awl. Patches of ochreous earth were met with, a bushel in a place, as though dumped from a basket. The absence of skeletons in this tumulus, and the red earth, together with ashes, mingled with commuted bone, would imply that this mound was appropriated to such bodies only as were cremated.

Of the smaller mounds, six were opened. In all skeletons were found, generally one only in each, and all were so decayed that it was impossible to preserve them. They were of ordinary size, except one, which is pronounced gigantic, the proportions "indicating a stature of seven feet" All were in a sitting posture, and faced to different points.

The early Iroquoian mound group is still visible near Grand Rapids, Michigan. They are left overgrown and being systematically destroyed by university excavations. *From The Nephilim Chronicles, A Travel Guide to the Ancient Ruins in the Ohio Valley*, **2010**

History of Genesee County, Michigan 1886

Another instance in the east part of the county, where a number of skeletons (also of large size) were found buried in a circle directly beneath the stump of a gigantic pine-tree of the oldest growth; but in both cases the finding of the bones was wholly accidental, as there was no mound or other surface-mark to indicate the places of burial.

Salt Lake Tribune, September 14, 1894

ELEVEN FEET TALL

Skeletons of Prehistoric Giants Unearthed in Michigan Mounds

A Carson City (Mich.) correspondent of the Detroit News writes that the remains of a forgotten race were recently dug up from the mounds on the south side of Crystal lake, Montcalm county. One contained five skeletons and the other three. In the first mound was an earthen tablet, five inches long, four wide and a half an inch thick. It was divided into four quarters. On one of them were inscribed curious characters. The skeletons were arranged in the same relative positions, so far as the mound was concerned.

In the other mound there was a casket of earthenware, ten and a half inches long and three and a half inches wide. The cover bore various inscriptions. The characters found upon the tablet were also prominent upon the casket. Upon opening the casket a copper coin about the size of a two-cent piece was revealed, together with several stone types, with the inscription or marks upon the tablet and casket had evidently been made.

There were also two pipes, one of stone and the other of pottery, and apparently of the same material as the casket. Other pieces of pottery were found, but so badly broken as to furnish no clew as to what they might have been used for.

Some of the bones of the skeletons were well preserved, showing that the dead men must have been persons of huge proportions. The lower jaw is immense. An ordinary jawbone fits inside with ease. By measurement, the distance from the top of the skull to the upper end of the thigh bone of the largest skeleton was five feet five inches . A doctor, who was present stated that the man must have been at least eleven feet high.

One of these mounds was partly covered by a pine stump, three feet six inches in diameter, and the ground showed no signs of ever having been disturbed. The digging had to be done amongst the roots which had a large spread.

Much speculation is rife as to who these giants of a prehistoric race may have been.

The American Antiquarian, 1905

Burial mounds in Michigan

Near this place the skeleton of a man was found which was encased in a certain kind of clay, unlike any clay ever found in this country, which clay had been burned after it was adjusted to the to the subject, some of the charcoal still remaining. The person was supposed to have been more than six feet in height, having very large bones, a very broad under- jaw, the front of the head receded so much as to leave no forehead. A burned clay vase or urn, of about three feet in height was found standing upright, into, into which the whole skeleton of a man had been compressed, the top of the urn being covered with burned clay. Resting against the outside of the urn, was a similar skeleton, supposed to be that of a female. Pipes were found with the skeletons which had been molded from some plastic material and glazed and burned, on which were skillfully portrayed the likeness of different human faces, and other fancied or real objects. Although each pipe has several faces exactly alike, yet they are entirely unlike those on the other pipes. The whole group of faces represents the broad, round, oval, oblong and conical shaped faces, not caricatures, but careful, skillful, and truthful representations of nature. Stone tubes, and stones which had been wrought into many curious and beautiful forms, many being perforated, all requiring great skill and patience, were found in the mounds with the above described skeletons. How long those bones had reposed in their air tight burned clay enclosures, in the bowels of these mounds, or to what particular race of human beings they belonged, are questions which are too profound and grave to induce an opinion from me. But my long acquaintance with the language, manners, and customs of the Indian tribes of Michigan, enables me to say that they never disposed of their dead as indicated above. Nor did they make such highly ornamented pottery as there found. Nor did they posses the skill to make such marvelous things, especially the ornamented pipes. Moreover, if it were possible, it is probable that they would have represented the human faces and other objects as seen by them rather than with the features which these relics contain. And furthermore, I believe, Indians were never arrayed in such a snug, neat fitting single garment, reaching from the neck to the feet, and showing such a natural and graceful formed bust, as was displayed by the female figure on one of these pipes. It is very obvious to me that the wild, savage Indians, never invented nor manufactured these implements, but adopted and used them whenever they found them.

Wisconsin

Janesville Gazette, **July 6, 1860**
DISCOVERY OF LARGE HUMAN SKELETONS
   Buck, of Driesbach City, six miles north of LaCrosse, sends the following account of the discovery of large human skeletons to the Winona Republican:
   A. L. Jenks, of this place, in prospecting in one of those mounds that are so common in this western country; discovered at the depth of five or six feet, the remains of seven or eight people of very large size.
   One thigh bone measured three feet in length. The under jaw was one inch wider than that of any man in this city. He also found clam shells, pieces of ivory or bone rings, pieces of kettles made of earth and coarse sand. There were at the neck of one of these skeletons, teeth two inches long by one-half to three-fourths of an inch in diameter, with holes drilled into the sides, and the end polished, with a crease around it. Also an arrow five inches long by one and a half wide, stuck through the back bone, and one about eight inches long stuck into the left breast. Also the blade of copper hatchet, one and one-half inch wide and two inches long. This hatchet was found stuck in the skull of the same skeleton. The mound is some two hundred feet above the surface of the Mississippi and is composed of clay, immediately above the remains, two feet thick; then comes a layer of loam, then another layer of clay six inches thick, all closely packed that it was with difficulty that it could be penetrated. There are some four or five different layers of earth above the remains. There is no such clay found elsewhere in this vicinity.

12th Annual Report of the Bureau of Ethnology to the Secretary of the Smithsonian Institution
1890-1891
Sheboygan County

There are some scattering mounds on the hills bordering the Sheboygan marshes on the north. These are usually isolated, simple conical tumuli, though some are in irregular groups on elevated situations.

The only one opened (the rest had been previously explored) was situated on a sandy ridge half a mile north of the marsh and 100 feet above it. It was about 50 feet in diameter at the base and 5 feet high. After passing through 18 inches of surface soil the central mass was struck, which appeared to be composed of earth mingled with fire beds, charcoal, ashes, and loose stones. Near the center of this mass, at the bottom of the mound, a large human skeleton in a sitting posture was discovered, apparently holding between its hands and knees a large clay vessel, unfortunately in fragments. These were covered over by large an irregular layer of flat boulders.

Daily Northwestern, **July 13, 1891**
RELICS OF THE CAVE
The Skeletons Unearthed at Clifton
POSSIBLY THOSE OF THE MOUUND BUILDERS
Remains Discovered on the East Shore of Lake Winnebago Supposed to Belong to a Pre-Historic Race-The Evidence Presented-An Antiquarian Expected to Make an Investigation of Calumet's Big Mounds

The recent discovery of skeletons at Clifton has attracted a great deal of interest throughout the sate. In the town of Harrison. Calumet county, on the east side of Lake Winnebago, is on a piece of land that contains as many as fifty mounds-nearly all being perfect images of men with outstretched arms, but of gigantic proportions. A number of ridges run through this property, off which are these mounds. A portion of the property is owned by A. W. Miller of Milwaukee [...]

The length of the mound from which this skeleton was taken is 12 feet and the width of the outstretched arm 5 feet [...]

The skeleton has a tremendous chin, a high forehead, an extraordinarily large humorous bone, the elbow joint [...] strong in proportion. A well posted gentleman says that the skeleton is different in every way from that of an Indians. The Indians have prominent cheek bones: these are not. The Indian lived from the result of the hunt" everything about the head of this skeleton shows its possessor to have been a vegetarian. The remaining teeth are solidly set in. It is probable that the man, of which only these crumbled bones remain, was nearly 7 feet tall. It is worthy of study

New York Times, **August 9, 1891**
THE WISCONSIN MOUNDS
Elaborate Systems of Defensive Works-Madison the Centre Of An Ancient Race
Their Burial Places And Their Weapons

MADISON, Wis., Aug 9-The largest prehistoric work in this state heretofore described, and of which the Smithsonian Institution has published a complete report, is Fort Aztalan, near Lake Mills, so named from the pyramidal mounds found there, which greatly resemble those found in Mexico. But without doubt the most stupendous and elaborate system of defensive works in the State are found in the vicinity of this city. The celebrated mounds of Ohio and Indiana can bear no comparison either in size, design, or the skill displayed in their construction with these gigantic and mysterious monuments of earth-erected we know not by whom, and for what purposes we can only conjecture. That the unknown race was semi-civilized is certain, as art of a high type flourished among them. Carving in stone, especially was brought to a high degree of perfection. The art of weaving and dying cloth was known and practiced, the colar used being invariably red.

Madison was in ancient days the centre of a teeming population numbering not less than 200,000 souls. It is situated on the northern end of a chain of five lakes, between Lakes Mendota and Monona, and extending south to Lake Wingra. It is built on a chain of hills which slope gently down to the water's edge or end of high bluffs. This was the mound builders paradise in bygone ages, and the region has lost none of its natural beauty.

On the land of Gearge Catterson, seven miles south of Madison, is a prehistoric foret. It occupies the summit and southeast side of a huge hill overlooking Lake Kegonsa. It is bounded on the east by a marsh and the cliffs of the lake on the south. It is undoubtedly a strong position for defense. The fort is square in shape. Its four outer walls are each 400 feet in length, and from the center of each side high walls, 800 feet long, stretch out. Inside the fort, about ten feet from the first of breastworks, extends a second parallel to the others. In this line gates were left in the corners, and these were protected by round mounds, the tops of which show evidence of fire, for a few inches below the surface are found quantities of charcoal. In the centre are three mounds in a direct line, connected with each other by a thin bank of earth. The tops of these mounds are sunken showing that they served the purpose of " caches" being hollow, but in the lapse of ages the tops have caved in.

Scattered about inside the second line are six rows of earthworks about twenty feet long. A group of seventeen burial mounds occupied the northeast corner of the fort, arranged in the shape of a turtle. Two of these were opened and interesting finds made. In the first mound opened a layer of forest mold six inches in thickness was first removed; then seven feet of yellow clay was penetrated and a thick bed of ashes and charcoal, in which were scattered arrow heads of flint, and pottery prettily ornamented in various patterns was brought to light. Below this was a foot of clay so hardened by the fiores as to turn the edge of the spade. Beneath this was a rudely made coffin off large flat stones probably brought from the lake.

Upon being opened this coffin was found to contain a large sized skeleton in a sitting posture, the earth within the coffin having held it in shape. The hardened clay above prevented the least moisture from entering thus preserving the bones in fairly good condition. At the side of the body was found a curiously-carved pipe in shape resembling a human head with peculiar characters rudely cut on the sides. Near the right hand was an axe of banded slate in the form of an ancient double edged battle axe, a number of arrow heads and a gorget of slate.

In opening a new road through the hill a landslide occurred exposing a hollow place about six feet square, which contained the skeleton of a person who must have been a giant in his day. Beneath the hand was an axe of synenite, finished with great skill and very nicely polished and grooved, which weighed five pounds.

The Centralia Enterprise and Tribune, **June 10, 1899**
Giant Indian Bones
Discovery of an Extraordinary Skeleton Near Fond du Lac
Fond du Lac, Wis., June 6-(Special)

An Indian skeleton was dug up on the farm of Matt and Joseph Leon, one mile south of St. Cloud, Sunday. There is nothing strange in finding an Indian skeleton, but this one was a giant in size, his frame measuring seven feet. He must have been a man of note among his people, for he was buried in a large mound, sixteen handsome arrows surrounding his body. The skull was brought to this city and is on exhibition in one of the Main street windows. Near the Huber gravel pit skeletons by the hundreds have been dug up for the past several years. Most of the bodies are in a sitting position, with their face towards the east, to face the rising sun.

Two Red Ochre skulls uncovered in Wisconsin. Skulls have archaic features of a prominent brow-ridges, sloping forehead, and a projecting upper jaw. *Wisconsin Archaeologist,* **Vol 45.**

The New North, **(Rhinelander) July 23, 1908**
GIANT SKELETON FOUND
Massive Human Bones and Indiana Relics Unearthed Near Pelican Lake.

That human beings of enormous size inhabited this section od the country ages ago was proven last Sunday, when the massive skeleton of an Indian was unearthed near Pelican Lake.

The interesting discovery was made by Geo. Patton and L. H. Eaton, two Chicago tourist, who are spending the summer there. For several days the men noticed a mound on their travels through the woods and at last led by curiosity decided to excavate it. Procuring spades they fell to work and after digging down to a depth of about four feet were surprised to find the bones of a large human foot protruding through the earth. Digging further they gradually uncovered the perfect frame of a giant. The skeleton was nearly eight feet in height and the arms extended several inches below the hips. Buried with the bones were numerous stone weapons and trinkets. Among these were a curious stone hatchet, a copper knife, several strange copper rings and a necklace made of the tusks of some prehistoric animal. The skeleton is no doubt that of an Indian who was one of a tribe of giants who roamed this part of the state over one thousand years ago.

Last year near Monico, there was unearthed the bones of a human arm three feet in length. This former discovery goes to show that the Pelican Lake giant was not alone on earth. The Chicago men will present the skeleton to some geological museum.

The New York Times, **May 4, 1912**
STRANGE SKELETONS FOUND
Indications That Tribe Hithero Unknown Once Lived in Wisconsin.

MADISON, Wis., May 3.- The discovery of several skeletons of human beings while excavating a mound at Lake Delvan indicates that heretofore unknown race of men once inhabited Southern Wisconsin. Information of the discovery was brought to Madison today by Maurice Morrisey, of

Delavan, who came here to attend a meeting of the Republican State Central committee. Curator Charles E. Brown of the State Historical Museum will investigate the discoveries within a few days.

Upon opening one large mound at Lake Lawn farm, eighteen skeletons were discovered by the Phillips Brothers. The heads, presumably those of men, are much larger than the heads of any race which inhabit America today. From directly over the eye sockets, the head slopes straight back and the nasal bones protrude far above the cheek bones. The jaw bones are long and pointed, bearing a minute resemblance to the head of a monkey. The teeth in front of the jaw are regular molars.

There was also found in the mounds the skeletons, presumably of women, which had smaller heads, but were similar in facial characteristics. The skeletons were embedded in charcoal and covered over with layers of baked clay to shed water from the sepulcher.

The Sons of God and the Nephilim

The sons of God are mentioned in several times in the *Book of Job*. In one quote Satan is amongst the sons. Job 1:6 "Now there was a day when the *sons of God* came to present themselves before the Lord and Satan came also among them to present himself before the Lord." The *sons* are mentioned again in the Book of Job; revealing that the *sons* were with God from the beginning. It also shows the importance of math and geometry in the formation of the earth by God. It was this *secret* knowledge of math and geometry, that the *sons* would later share with their giant offspring. Job 38:4 "Where were you when I laid the foundations of the earth? Tell me, if you have understanding. (5) Who determined its measurements? Surely you know! Or who stretched the line upon it? (6) To what were its foundations fastened? Or who laid its cornerstone, (7) when the morning stars sang together, and all the *sons of God* shouted for joy."

One of the most mysterious quotes in the Bible is found in Genesis 6:4 that mentions the *sons of God* and their offspring of giants or Nephilim. Origins of the word "Nephilim" in the Greek version of the *Old Testament,* called the *Septuagint,* renders the Hebrew term "Nephilim" as "giganates." The ancient writings indicate that the Nephilim were physically quite large; they were "giants." The word, which literally means "earth-born" or the "sons of Gaia," that refers to a savage race destroyed by the Gods of Greek mythology.

Genesis 6:1-4 *"*Now it came to pass, when men began to multiply on the face of the earth, and daughter were born to them, (2) That the *sons of God* saw the daughters of men, that they were beautiful; and they took wives for themselves of all whom they chose. (3)And the Lord said, "My Spirit shall not strive with man forever, for he is indeed flesh; yet his days shall be one hundred and twenty years. (4) There were giants on the earth in those days, and also afterward, when the sons of God came in to the daughters of men and they bore children to them. Those were the mighty men who were of old, men of renown."

Key phrase is "of old," meaning that in the distant past there were a race of giants. The angels mating with women and creating a race of giants is further elaborated in the *Book of Enoch.* This is one of the many original texts of the Bible that over the years has been removed. Enoch was the son of

Jared and the great grandfather of Noah. He is given "angel" status in *Genesis 5:24* ; "And Enoch

walked with God: and he was not; for God took him." This transformation happens when Enoch is 365

years old.

The Book of Enoch.. Chapter 6: "And the angels, the children of heaven, saw and lusted after them, and said to one another: 'Come, let us choose us wives from among the children of men and beget us children." And Semjaza, who was their leader, said unto them: "I fear ye will not indeed agree to do this deed, and I alone shall have to pay the penalty of a great sin.' And they all answered him and said: ' Let us swear an oath, and all bind ourselves by mutual imprecations not to abandon this plan but to do this thing.' Then sware they all together and bound themselves by mutual imprecations upon it. And they were in all two hundred; who descended in the days of Jared on the summit of Mount Hermon, and they called it Mount Hermon, because they had sworn and bound themselves by mutual imprecations upon it. And these are the names of their leaders: Samlazaz, their leader, Araklba, Rameel, Kokablel, Tamlel, Ramlel, Danel, Ezeqeel, Baraqijal, Asael, Armaros, Batarel, Ananel, Zaqlel, Samapeel, Satarel, Turel, Jomjael, Sariel." Chapter 7: And all the others together with them took themselves wives, and each chose for himself one, and they began to go in unto them and to defile themselves with them, and they taught them charms and enchantments and the cutting of roots, and made them acquainted with plants. And they became pregnant, and they bare great giants."

As the grandchildren of God, the Nephilim had a covenant with God, in the time of the flood that is

never specifically mentioned, but is alluded to. The covenant is revealed after the flood, when the

giants were not destroyed. Their survival both before and after the flood is found in the statement;

"There were giants on the earth in those days, and also afterward." Meaning, before and after the flood.

There are also Rabbinical and Eastern traditions that claim that Og, the Amorite king and giant,

survived the flood by wading next to the ark, with another version stating that Og survived by Moses

feeding him through a hole in the roof of the ark.

The Nephilim or giants are referred to within the Bible as Rephiam, Nephilim, with tribal names of

Amorites, Canaanites, Emmin, Zanzummin, Amalekites, Horim and the Anak. The Emmin that

translates to "terrible men" were an accounted tribe of giants that lived east of the Dead Sea. The

Anakim are defined in " *A Dictionary of the Bible*" as "A race of giants, descendants of Arba (Josh.

15:13, 21:11), dwelling in the southern part of Canaan, and particularly at Hebron, which from their progenitor received the name of "city of Arba." Besides the general designation Anakim, they are variously called Anak (Num. 13: 33), descendants of Anak (Num. 13: 22) and sons of Anak (Deut. 1: 28). These designations serve to show that we must regard Anak as the name of a race rather than that of an individual, and this is confirmed by what is said of Arba, their progenitor, that he "was a great man among the Anakim" (Josh. 15:15).

The most renown of the giants were the Amorites, whose two kings Sihon and Og, who ruled the rich lands that are described as being bounded by Jabok on the north, the Arnon on the south, Jordan on the west, and the "wilderness" on the east (Judges 11: 21, 22), this land was noted as the "land of the Amorites." Their possessions are distinctly stated to have extended to the foot of Mt. Herman. (Joshua 13:11) Mt. Herman is where the *sons of God* swore the oath to mate with mortal women.

2000 B.C the Amorites are living with, an allied to Abraham, who was living in Hebron, under the "oak grove." The tribes of giants had been subjugated by Chedorlaomer , and rebelled. Genesis 14: 5 "And in the fourteenth year came Chedorlaomer, and the kings were with him, and smote the Rephaims in Ashteroth in Karnaim, and the Zuzims in Ham, and the Emmins in Shaveh Kiriathaim, (6) And the Horites in their mount. [...] and smote all the country of the Amelekites, and also the Amorites, that dwelt in Hazezon-tamar."

Abraham's alliance with the Amorites is revealed in (14:13) That states he dwelt in the plain of Mamre the Amorite, and these were confederate with Abram. A little known association of Abraham with the Sons of God and the Nephilim, was uncovered in *Praeparatio Evangelica, The Old Testament Pseudepigrapha* , Charles H. Charlesworth. " In anonymous works, we find that Abraham traced his ancestry to the giants. These dwelt in the land of Babylonia. Because of their impiety, they were destroyed by the gods. One of them, Belos, escaped death and settled in Babylon. He built a tower and lived in it; the tower was called Belos, after its builder. After Abraham had learned astrology, he went to Phoenicia and taught it to the Phoenicians."

In Genesis 14:15, God promises Abraham a child to produce the multitudes that would possess the "promised land." However, the possession of the "promised land" is delayed by the covenant God had with his grandchildren, the Amorites. Genesis 15:16 "But in the fourth generation they shall come hither again: for the *iniquity* of the Amorites is not yet full."

The Hebrews can not take possession of the "promised land" because "the inequity of the Amorites is not yet full." This has to be interpreted as the "giants," who were the offspring of the *sons of God* had a covenant, that up to this time had not been broken. The mysterious origins of the Amorites is shown in Ezekiel, chapter 28, in the judgement of the prince of Tyre. Tyre is a Phoenician/Amorite city of trade on the western shores of the Medeteranean. In the *Book of Ezekiel* are versus that place the Amorites within the Garden of Eden, where they were "created." Is the Bible saying that Amorites who were of the Cro-Magnon species the first men that inhabited the Garden of Eden? Ezekiel 28:13:

"Thou hast been in Eden, the garden of God. The workmanship of thy tabrets and of thy pipes was prepared in the day that thou was created." Thou art the anointed cherub that covereth; and I have set thee so: thou wast upon the holy mountain of God; thou hast walked up and down in the midst of the stones of fire. Thou wast perfect in thy ways from the day that thou was created, till *iniquity* was found in thee."

Iniquity, is defined as a wicked, unjust or unrighteous act. The iniquities the Amorites committed is explained further in *The Book of Enoch,* that tells of godlessness the Nephilim-Amorites were engaged in. The quote acts as a checklist in identifying the Amorites; as the introducer of metals or weapons, temples constructed to track the sun and the moon, evidence of the use of mathematics in the temples construction, astrology, the "signs of the earth," which is the worship of the Earth Mother, and the identification of sacred springs, hilltops, mountains, boulders, groves, trees and rivers. The knowledge passed from the *sons of God* to the Nephilim is as relevant and observable in ancient land of the Amorites, as it is in the British Isles and the Ohio Valley.

Enoch 8:1-3: "And Azazel taught men to make swords, and knives, and shields, and breastplates, and made known to them the metals of the earth and the art of working them, and bracelets, and ornaments, and the use of antimony, and the beautifying of the eyelids and all kinds of costly stones, and all (2) colouring tinctures. And there arose much godlessness, and they committed fornication, and they were led astray, and became corrupt in their ways. Semjaza taught enchantments, and root cuttings, Armoras the resolving of enchantments, Baraqijal taught astrology, Kokabel the constellations, Ezwqeel the knowledge of the clouds, Araqiel the signs of the earth, Shamsiel the signs of the sun, and Sariel the course of the moon.

Azazel "made known," to the Nephilim metals, and how to fashion them into swords and knives. This is historically consistent with the beginning of the Bronze Age around 2700 B.C. The Amorites were mining and trading copper and tin and innovating improved types of knives and swords. It is also around this time that marks the beginning of the Megalithic building of stone circles, that were used to track the motions of the sun and moon and to pay homage to those deities.

The *secrets* taught to the Nephilim by the 8 *sons of God* are important in identifying the Amorites in the Levant, the British Isles or in the Ohio Valley. In each location, the Amorites introduced metals and built open-air stone and earthen temples to worship the Sun and the Earth. The temples were constructed using complex mathematics and were used as solar celestial observatories. They placed their dead in burial mounds, where their gigantic skeletons have been uncovered, revealing that they were an *archaic* type of human, closely related to the Upper Paleolithic, Cro-Magnon.

H. P. Blavatsky, *Isis Unveiled,* reveals that the legend of the *sons of God* mating with the daughters of man is not isolated to the Torah or the Bible but is widespread in the stories of many ancient texts.

Isisi Unveiled, A Master Key to the Mysteries of Ancient and Modern Science and Theology, 1877, "The fables of the mythopœic ages will be found to have but allegorized the the greatest truths of geology and anthropology. It is in these ridiculously expressed fables that science will have to look for her "missing links."

Otherwise, whence such strange "coincidences" in the respective histories of nations and peoples so widely thrown apart? Whence that identity of primitive conceptions which, fables and legends though they are termed now, contain in them nevertheless the kernel of historical facts, of a truth thickly overgrown with the husks of popular embellishment, but still a truth? Compare only the verse of *Genesis vi:* "And it came to pass, when *men began to multiply* on the face of the earth, and daughters were born unto them, that the sons of God saw the daughters of men that they were fair; and they took them wives of all which they chose...There were *giants in the earth in those days,"* ect., with this part of the Hindu cosmogony, in the *Vedas,* which speaks of the descent of the Brahmans. The first Brahman complains of being alone among all his brethren without a wife. Notwithstanding that the Eternal advises him to devote his days soley to the study of the Sacred Knowledge (*Veda),* the first-born of mankind insists. Provoked at such ingratitude, the eternal gave Brahman a wife of the race of the *Daints,* or *giants,* from whom all the Brahmans maternally descend. This the entire Hindu priesthood is descended, on the one hand, from the *superior* spirits (the sons of God), and from *Daintany,* a daughter of the earthly giants, the primitive men. "And they bare children to them; the same became mighty men which were of old; men of renown" (Mythologie des Indous)

The same is found in the Scandinavian cosmogonical fragment. In the *Eddas* is given the description to Gangler by Har, one of the three informants (Har, Jafuhar, and Tredi) of the first man, called Bur, "the father of Bor, who took for a wife Besla, a daughter of the giant Bolthara, of the race of the *primitive giants."* The full and interesting narrative may be found in the *Prose Edda,* sects. 4-8, in Mallett's *Northern Antiquities.*

The same groundwork underlies the Grecian fables about Titans; and may be found in the legend of the Mexicans-four successive races of *Popol-Vuh.* It constitutes one of the many ends to be found in the entangled and seemingly inextricable skein of mankind, viewed as a psychological phenomenon. To say that it sprang up, and grew and developed throughout the countless ages, without either cause or at least firm basis to rest upon, but merely as an empty fancy, would be to utter as great an absurdity as the theological doctrine that the universe sprang into creation out of nothing."

The Amorites

Archaeology and the Bible, George Barton, 1916 "The Amorites belonged to the Indo-European type, the nose was straight and regular, the forehead high, the cheek bones somewhat prominent. We find that this was precisely the character of the face of the Mentone and Cro-Magnon skulls. The Amorites at one time occupied the mountainous regions of Asia Minor from the south-west of the Caspian to the Aegean Sea. Here they flourished, and for many centuries horde after horde of these people descended into the rich and fertile plains of Syria and Mesopotamia."

The Amorites, were the accounted giants of the Bible who possessed the land of Canaan before the conquest of the Israelites. The Amorites brought with them a new tribal organization, pottery, and burial customs in the Intermediate Bronze Age. In early Babylonian inscriptions the "land of the Amorites" consisted of all the "western" lands that included Syria and Canaan.

The artist who ornamented the Egyptian monuments has there illustrated the various people of Western Asia conquered by Thothmes III, and among them were the Amorites who, we know from Babylonian records, were a powerful race inhabiting the country north and east of the Dead Sea, as early as 3800 BC.

Amorite depicted on Egyptian bas-relief, From *The Early History of the Hebrews,* 1901

Archaeology and the Bible, George Barton, 1916 "We are accustomed to call this Semitic people Amorites, and it probable that this is right. About 2800 B.C, under a great king named Sargon, a city of Babylonia called Uru, or Amurru, and Agade conquered all of Babylonia. The dynasty founded by Sargon was Semitic and ruled Babylonia for 197 years. Even before Sargon conquered Babylonia, Lugalzaggisi, King of Erech, had penetrated to the Medeteranean coast. Sargon and two of his successors, Naram-Sin and Shargali-sharri, carried their conquest to the Medeteranean lands. A seal of the last mentioned king was found in Cyprus. It is probable that the coming of the Amorites began in the north with the conquests of these kings.

To the east of Lebanon the Princeton expedition found stone structures similar to Babylonian *Ziggurats,* which they attribute to the Amorites, and hold to indicate the prevalence of Babylonian influence in this region.

It is probable that the Amorites slowly worked southward, occupying different cities as they went. Mr. Macalister estimated that they reached Gezer about 2500 B.C. It is not, therefore, unreasonable, though they may have arrived there a century earlier than that. This was the beginning of that long intercourse with Babylonian language and script for the purpose of expressing written thought in Palestine long after the Egyptians had conquered the country. The intercourse was the more natural because the Semites who came to Palestine were of the same race as those who were dominant in Babylonia.

Meantime, the Egyptians had begun to take notice of Palestine, an officer of Pepi I of the sixth Egyptian dynasty, relates that he crossed the sea in ships to the back of the height of the ridge north of the "sand-dwellers and punished the inhabitants." This refers to the coast of Palestine in the neighborhood of the Phillistine cities of Gezer. This time was between 2600 and 2750 B.C. Egypt was at this time only anxious to make her own borders secure; she had no desire to occupy this Asiatic land.

Again, between 2300 and 2200 B.C., a fresh migration of Semites, apparently also of the Amorite branch, invaded Babylonia and in time made the city of Babylon the head of a great empire. This race furnished the first dynasty of Babylon, which ruled from 2210 to 1924 B. C. Its greatest king, Hammurabi, who gave to Babylon a code of laws in the vernacular language, conquered the "west land," which means the Medeteranean coast.

It was probably under his successor, Shamsu-iluna, but certainly under one of the kings of this period. In this same period there lived in Babylonia an Abraham, the records of some of whose business documents have come down to us. We also find three men who bore the names Yagubilu (Jacobel) and Yashubilu (Josephel) and one who was simply called Yagub, or Jacob. Palestinian evidence from a later time leads us to believe that men bearing all these names migrated during this period to Palestine and gave their names to cities which they built or occupied.

Egyptians also came to Palestine during this period. There was apparently considerable trade with Egypt at this time. Men from Palestine often went there for this purpose. Trade with Egypt is also shown to have existed by the discovery of Egyptian scarabs of this time of the Middle Kingdom in the excavation of Gezer, Jericho, Tannach and Meggiddo. As Egypt was nearer and commerce with it easier, its art affected the art of Palestine during this period more than did the art of Babylon, although the people were akin to the Babylonians. In the reign of Sesostris III, 1887-1849 B.C., the Egyptian king sent an expedition into Palestine, and captured a place called in Egyptian Sekmen, which is though by some to be a misspelling of Shechem. This expedition probably stimulated Egyptian influence in the country, though the Egyptians established no permanent control of the land at this time.

Between 1800 and 1750 B.C a migration occurred which greatly disturbed all western Asia. There moved into Babylonia from the east a people called Kassites. They conquered Babylonia and established a dynasty which reigned for 576 years. Coincident with this movement into Babylonia there was a migration across the whole of Asia to the westward which caused an invasion of Egypt and the establishment of the Hykos dynasties there. As pointed out previously, it is possible that this movement, in so far as the leadership of the invasion of Egypt was concerned, was Hittite. In any event, however, many Semites were involved in it, as the Semitic names in the Egyptian Delta at this time prove. It is customary to assume that it was in connection with this migration that the Canaanites came into Palestine. We thus feel sure that there was an increase of population and when next our written sources reveal to us the location of the nations, the Canaanites were dwelling in Phoenicia. The Egyptian scribes of a later time called the entire western part of Syria and Palestine "The Canaan." Probably, therefore, the Canaanites settled along the sea coast. We, therefore, infer that they came into this region at this time. With the coming of an increased population, the Amorites appear to have been in part subjugated and absorbed, and in part forced into narrower limits."

In *Phoenician Origin of the Britons & Scots,* 1925, Waddell observed, " that the old ruling race of Asia Minor and Syria-Phoenicia, from immemorial time, were the great imperial, highly civilized, ancient people generally known as Hittites, but who called themselves *"Khatti"* or *"Catti,"* which is the self-same title which the early Briton kings of the pre-Roman period called themselves and their race."

The Amorites had been subjugated by the Hittites prior to the Egyptian invasion and may have been the impetus for them leaving the Medeteranean for destinations in North Africa, Britain or North America. A letter from the Hittite King, Mursilis, written about 1300 B.C. to the king of the Acheans, Ahijawa, complained about one of the Amorite residents, who agreed to be a vassal of the Hittite King, only to wage war on him. The name of this Amorite King was Tawagalawas. King Mursilis was angered that he could not locate Tawagalawas, and presumed he and his people had left by sea.

An important aspect of the Amorites and their trading prowess was in the supply of materials to arm the first large armies in history. They were engaged in trade with both the Egyptians to the south and the Hittites to the north. In the Late Neolithic (3,500-2700 B.C). copper was first used for personal ornaments and some spear heads. This was followed by adding tin to copper that used to make bronze, called the Early Bronze Age (2700-2000 B.C) While this early date is given for the use of bronze, it isn't in widespread use until about 2000 B.C., and later in other western Medeteranean areas. In many places, the use of copper remained in widespread use until the MBA.

The God Kings and Titans, Bailey, 1973 " In the Bronze Age, the life and death struggles of competing states, Egyptian, Syrian, Hittite, Hurrian, Cretan, Babylonian, Assyrian, Elamite and the Indus peoples, among many others, demanded that they should have sufficient supplies of copper and tin to make bronze weapons for their armies and bronze tools for their workers. They also built places with walls of bronze. There was a fortune to be made supplying it; disaster overtook a state without it"

The Amorites were masters of the sea dedicated to the mining of copper and tin to make bronze. Their sea captains and trading companies became rich. By 1900 B.C. their influence spread west across the Medeteranean,where they built great cities in Carthage and Crete, where grand palaces were built at Knosses, Phaistos and Mallia. The cities in Crete had running water and a sewage system that wouldn't be duplicated for another 2000 years.

Weapons from from the EBA that carry over to the MBA were daggers, spears and javelins. Daggers: the short strait untanged dagger, with rivit holes in the base for attaching the the haft, common in the EBA remained in use in the MBA, however a new type that was longer with a pronounced mid-rib and a tang in which several rivets were drivin to improve the attachment were new to the IBA

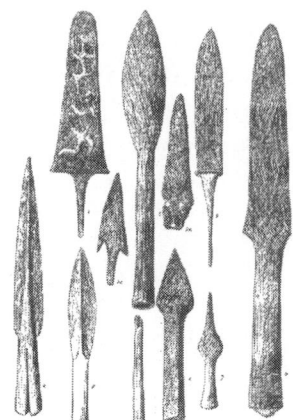

Weapons uncovered from the ancient city of Gezer. Weapons include tanged and socketed spears and knife with rivit holes. A practice that was used from the Early to the Intermediate Bronze Age. *Archaeology and the Bible, George Barton,* **1916,**

The Art of Warfare in Biblical Lands, Yigael Yadin, 1963 " In addition to the curved or sickle swords, the MBA also produced a series of short strait swords, some what like daggers. They were designed no doubt for defense in hand to hand combat. Unlike the strait narrow sword of the EBA, the blades become broader during MBII, taking the shape of a pointed leaf. They were designed primarily for stabbing and the blade was therefore strengthened by a central spine or rib."

Spear-head from Israel with long tang for hafting, with a central spine or rib for additional strength. *The Art of Warfare in Biblical Lands,* **Yigael Yadin, 1963**

"At the end of third millennium 2,000 BC, the armorors were still grappling with the problem of finding an effective method of attaching the spearhead and javelin-head to a wooden staff, and had come up with nothing better than the tang. This also marked the type of spear and javelin in use at the beginning of the second millennium, and not until a later stage in the first half of the second millennium [1500BC] was the socketed type to be developed and more commonly used."

The socket, developed about 1500 B.C. was a revolutionary new method of hafting a spear to the wooden staff. The socket was also employed in the use of the hoe in agriculture, greatly improving the the productiveness of farming.

The Amorites were trading throughout the Medeteranean and had also secured and forged trading routes that stretched from Japan to the Indus Valley to Northern Europe. The Amorites who travelled by land into Europe would be known as the Beaker People; bringing with them the knowledge of the metals and leaving their indelible mark upon the soil of Europe in the form of burial mounds.

The early Amorites buried their dead within burial mounds which contained megalithic dolmans within their interior that acted as a sarcophagus. *Archaeology and the Bible, George Barton,* 1916, "Megalithic tombs were constructed in the Land of Israel during phases of the Chalcolithic [The use of stone and copper] and early Bronze periods. In the late third millennium, two types of the megalithic tombs, extant already in the Early Bronze Age, became widespread: stone or earthen tumuli and dolmens. The base of the tumulus was bounded with one or more rings of small stones."

Dolman burial chamber located in Jordan, outside of Ammon. When constructed, the dolman would have been covered by an earthen mound. From the *Palestine Exploration Fund,* 1911.

123

Many of the burial mounds of the Amorites have long ago been effaced from the landscape of Israel, Jordan, Palestine and Syria. But in the Persian Gulf, on an island once known as Tylos, Dilmun and currently called Bahrain, over ten thousand burial mounds still exist. *The Persian Gulf,* Lt. Col. Sir Arnold Wilson, 1928. "Of the mounds so far investigated, the entrance faces west; the building is two storied, of carefully hewn blocks of stone, the lower story being more lofty than the upper. On both sides of a passage or corridor, leading to the east, are niches or chambers which were designed to hold cists, stacked one above the other.

"The plan on which the tombs are built agrees in striking fashion with those known of the Phoenicians; this was noticed by Strabo, who says that "the islands of Tyrus and Aradus have temples resembling those of the Phoenicians." The use of the double chamber or sepulchre has a Phoenician parallel, for there are examples of two-storied tombs in the cemetery of Amrit in Phoenicia, in Sardinia, and at Carthage. The similarity of the place-names, Tylus-Tyrus and Aradus in the Persian Gulf, and Sur and Arvad on the Phoenician coast, is also noteworthy."

Burial mounds on the island of Tylos in the Persian Gulf. The mounds contained a two storied sarcophagus, similar to the mound at Moundsville, West Virginia.

Archaeology and the Bible, George Barton, 1916, " Fields containing thousands of dolmans may be found in Transjordan and the Golan. The Late Stone Age or Neolithic men in Palestine much more is known. This knowledge comes in part from the numerous cromlechs, menhirs, dolmans, and "gilgals" which are scattered over eastern Palestine. A cromlech is a heap of stones roughly resembling a pyramid, a menhir is a group of unhewn stones so set in the earth as to stand upright like columns; a dolman consists of a large unhewn stone which rests on two others which separate it from the earth, and a "gilgal" is a group of menhirs set in a circle. On the west of the Jordan megalithic monuments were probably once numerous, since traces of them still survive in Galillee and Judea, but the later divergent civilizations have removed most of them. In the time of Amos, one of the "gilgals" was used by the Hebrews as a place of worship, of which the prophet did not approve.

These monuments are the remains of men of the stone age who dwelt here before the dawn of history. They were probably erected by some of those peoples whom the Hebrews called Rephaim or "shades," people who, having lived long before, were dead at the time of Hebrew occupation."

Phoenician Origin of the Britons & Scots, 1925 "Rude stone avenues and remains,compared by De Saulcey to Celtic dolmens, still exist among the hills of Moab. (Dead Sea, 1835, p. 546). Mr. Stanley (pg. 272) describes a circle of rough upright stones, a few miles to the north of Tyre, of which people have tradition, reminding us of similar tales at home, that they are "men turned into stone for scoffing at 'Nabi Zur.'"

Remains of a megalithic stone tower located east of the Jordan River in Israel. From the *Palestine Exploration Fund*, 1911. Two similar stone towers are on each side of the Ohio River at Moundville, West Virginia.

Syria-Phoenicia is as yet little explored, " a circle of rough upright stones" is reported to stand a few miles north of Tyre itself, and several "Stone Circles" have been reported by Conder, Oliphant and other in South Syria as well as in Hittite Palestine, and especially to the east of Jordan; and Mcalister has unearthed at Gaza, ect., rows of Megaliths, with the "cup marked rocks in their neighborhood." But we have seen, that the later restricted Roman province of "Phoenicia" itself formed only a part of the Eastern Phoenician empire, while the Persian Gulf are which the earlier Phoenicians occupied before coming to the Levant, Stone Circles like Stonehenge, dolmans and other megaliths are reported along with "*Catti*" names.

A series of stone circles surrounds a central stone cairn on the Golan Heights. It has been called the Circle of the Nephilim and Og's Circle.

Between the Persian Gulf and the Red Sea, in the district of Kasin, are reported three huge rude Stone Circles, which are described as being "like Stonehenge" and, like it, composed of gigantic trilithons about 15 ft. high; and several huge Stone Circles in the neighborhood of Mt. Sinai, some of them measuring 100 ft. in diameter. On the old caravan route from the Cilican coast via "Jonah's Pillar" to Persia (or Iran of the ancient Sun-worshippers), several megaliths are incidentally reported by travellers. Near Tabriz, to the east of lake Van, are "several circles" of gigantic stones ascribed to the giants "Caous" (Cassi?) of the Kainan dynasty. In Parthia, at Deh Ayeh, near Darabgerb, is a large circle. On the N.W. frontier of India, on the route from Persia near Peshawar, is a large circle of unhewn megaliths about 11 ft. high, and resembling the great Kenwick Circle in Cumberland. And amongst the many megaliths along the Mediterranean coast of Africa, so frequented by the Phoenicians, are several Stone Circles in Tripoli and Gaet-uli hills with trilithons, like Stonehenge. The Tri-lithons are located in Tripoli, North Africa.

Tri-lithons located in Tripoli, North Africa. From *Remains of the Prehistoric Age in England*, 1904

Amorite Symbolism

Simple numerals were written by early Sumerians by strokes, such as / for 1, // for 2, /// for 3 and so on up to 9- a system which has survived in the Roman numerals up to IIII, and on the dials of modern clocks and watches. When engraved on stones, these lower numeral stokes were at first formed by easier process of drilling by the jewelled drill worked by a bowstring fiddle, thus forming circular holes **O,** to so-called "cups."

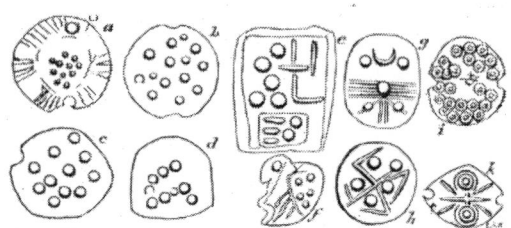

Hitto Sumerian Cup-marks on seals and amulets from Gaza pictured in *Phoenician Origin of the Britons & Scots,* 1925. Easily recognized is the horned moon in "g." The swastika in "h." Two circles in opposition in "k" was symbolic of the movements of the sun between the winter and summer solstice.

The occult values attached to certain numbers by the Sumerians, through ideas associated with particular numbers, was the origin of the mystical use of numbers in the ancient religions of the East and Greece referred to by Herodutus and other writers, as current amongst the adepts in the mysteries of the Magians, Pythagoras, Eleusis, and later amongst the Gnostics, and surviving in some measure in religion to the present day. This use of numbers ro reveal secret meanings is also known as Gematria. Thus "one" as "unity" and "First," was secondarily defined by the Sumerians as "complete" and "perfect," and thus represented "God, of heaven and earth." When formed by a circle or "cup mark," it represented the Sun and Sun-god, who are represented by a circle with a central dot in Egyptian hieroglyphs. ◎

0   =1 or 10 "The One" Sun-god or Earth, Heaven and Sun

00   =2 or 20 Represents the visible Day and Night or "ressurecting" sun

000   = 3 or 30 Moon or moon-god or Death or Fate

0000   =4 or 40 Mother Goddess Ma-a (Maya or May) and numerically the four quarters or the cardinal points.

The origin of the earliest form of the True Cross, was the crossing of the twin tinder sticks used for the producing by their friction the sacred fire, symbolizing the Sun.. The Cross was thus used as the symbol of Divine Victory of the Sun on the earliest Sumerian sacred seals from about 4000 B.C., and continued to be used by the Amorite Phoenicians, Hittites, Trojans, Goths and Ancient Britons.

Sumerian sign for Sun-god *Bil* [Baal] or Fire-god with word value *Bar*, also *Pir* or Fire and defined as Flame from *Phoenician Origin of the Britons & Scots, 1925*

Phoenician Origin of the Britons & Scots, 1925   "This simple equilateral form of the Sun Cross of Divine Victory, was sometimes ornamented by the Catti (or Hittites) and Sumerians by doubling its borders, so as to superimpose one or more crosses inside each other, as in the "Cassi" Cross and by decorating with jewels or fruits and broadening its free ends to form what is called "The Maltese" Cross, which is found on the ancient Sumerian sacred seals and as amulets on the necklaces of the priests-kings in Babylonia, ect. And it is a variety of this amulet or necklace form, with a handle at the top, or pierced with a hole above for stringing on a necklace or rosary, which has hithero been called 'The Phoenician" or "Egyptian" *Crux anstasa, o*r "key of life-to come"

Maltese cross found on Trojan, Egyptian and Phoenician pottery and amulets, from *Phoenician Origin of the Britons & Scots, 1925*

"Another common form of this simple Sun Cross is the *Swastika,* which we have carved, in the center of the Phoenician votive pillar to Bel at Newton. This is formed from the simple "St George's Cross" by adding to its free ends a bent foot, pointing in the direction of the Sun's apparent movement across the heavens, ie., towards the right hand and thus forming the *"Swastika"* or what I call the "Revolving Cross." This discloses for the first time the real origin and meaning of the Swastika Cross and its feet, and its talismanic usage for good luck. This Swastika form of the Sun Cross occurs on early Hittite and Sumerian seals and sculptures."

The swastika on the left was carved into a neolithic cave (called Og's Cave) inhabited by the Horites at Khubet El-Ain, Israel. Swastika on right is commonly found symbol with Sumerians, Amorites, Cretans, Megalithic Britons and the Allegewi Hopewell mound builders in the Ohio Valley.

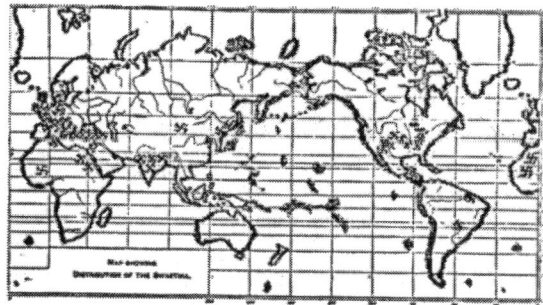

Map showing the distribution of the Swastika, that reveals the Amorites trade routes. The swastika symbol is found most frequently in the Medeteranean and north into the British Isles and Scandinavia. Swastikas, megaliths and stone circles are also found in the Indus Valley of India, Korea and Japan. The occurrence of swastikas in the Ohio Valley is consistent with the open-air circular sun temples and legends of giants that are found in all of these regions.

The simple equal-limbed cross was also sometimes figured inside the circle of the Sun's disc and sometimes intermediate rays were added between the arms to form a halo of glory. This now discloses the *Catti* or Hittite origin of the "Wheeled" Crosses of pre-Christian Britain

Called a circular type swastika that is depicted with 8 cogs (Og).

The symbolism is important in understanding the religion of the Amorites that consisted of a host of gods and and goddesses, with the two primary deities being the Sky or Sun Father and the Earth or Moon Mother. The Sun god is known by number of titles in different pagan cultures including Baal, Og, Adonis, Bel, Bachus, Moloch, Jehovah. The Earth or Lunar Mother was known as Ashtoreth, Astarte, Ishtar, Isis, Astoreth, Venus and Eve.

The stone circles constructed in the Levant is evidence of the importance of the Sun-god. More obscure was the worship and importance of the Earth Mother. The Amorite King Og ruled over sixty cities with the chief cities being Edrei and Ashtaroth-Karnaim. Joshia 13:4 "And the coast of Og king of Bashan, which was of the remnant of the giants, that dwelt at Ashtaroth and at Edrei." Ashteroth Karnain, known as "Ashtaroth of the two horns or peaks," was a place of great antiquity and the abode of the Rephaim at the time of the incursion of Chedorlaomer (Gen 6:5). Ashtaroth or Astaroth, was a city on the east side of the Jordan, in Bashan, within the kingdom of Og. Its name, so called from being the seat of the worship of the goddess of the same name.

Ashtaroth was the principle female divinity of the Phoenicians. The worship of Ashtaroth seems to have extended wherever Phoenician colonies were founded. She symbolized the productive power while Og symbolic of the generative power; and it would be natural to conclude that as the sun is the great symbol of the latter, and identified with Og. The moon and the Earth are the symbols of the former and can be identified with Ashtaroth. Ashtaroth called by the Roman Astarte was identified by many ancient writers with the goddess Venus or Aphrodite as well as with the planet of that name.

The name of the idles for Ashtaroth was called Asherah, which translates to "grove." Kings II 23:13: "And the high places that were before Jerusalem, which were on the right hand of the mount of corruption, which Solomon the king of Israel had builded for Ashtoreth..(14) And broke in pieces the images and cut down the groves."

Earth Mother/Goddess, Peter Knight writes, " The concept of the "Mother Earth" or "Mother

Nature" has roots that go back thousands of years, to prehistoric cultures devoted to the worship of an omnipresent feminine deity. In ancient times the Earth Mother or Goddess was a primary deity, the spiritual driving force that inspired the building of so many temples and megalithic sites worldwide."

The groves that were worshipped is an example of animism that is part of Earth Mother worship, where trees, rocks, springs and rivers were thought to be endowed with spirits. Menhirs and stone circles are in some instances made up of un-worked stones. Whole stones are tied to the Earth Mother. An example of the importance of the stones in their natural state is evident when God commanded Moses to build an alter. "Deuteronomy 27:5,6; "And there shalt thou build an alter unto the Lord thy shalt not lift up any iron tool upon them. Thou shalt build the alter of the Lord thy God of whole stones:" The practice of anointing the stones was practiced by the Phoenicians and the early Hebrews. Genesis 28:18 " And Jacob...took the stone, and set it up for a pillar, and poured oil upon the top of it."

Paganism was practiced by, and described by the early Hebrews. No doubt from their proximity to, and intermarrying with the last remnants of the Amorites. Ezekiel 16:3 "And say, Thus saith the Lord God unto Jerusalem; Thy birth and thy nativity is the land of Canaan; thy father was an Amorite, and thy mother an Hittite." The Amorites and Hittites lived in Jerusalem prior to the Israelites conquest, where their place names still exist today.

Phoenician Origins of the Britons and Scots, 1925. The presence of Gentile Sun-priests in the temple on Mt. Moriah at Jerusalem is explained by the fact that; besides the name "Moriah"-which is recognized as meaning "mount of the *Morias* or Amorites"-that temple, long before the occupation of Jerusalem by David and its rebuilding by Solomon, was a famous ancient *Sun*-temple of the Hittites or Amorites."

Ezekiel has a vision shown to him by God of the abominations of the Amorites and their temple. This vison of the temple comes on the sixth year, in the sixth month, in the fifth day of the month. This surely was symbolic of 666 which was the number in gematria for the Sun. In Ezekiel 8:14, he is

132

describing the temple, and the sun worshippers, "Then he brought me to the door of the gate of the Lord's house which was towards the north; and behold, there sat women weeping for Tammuz." Tammuz is identified as Adonis the Amorite sun-god. It was the practice of the Amorite women to yearly mourn Adonis at the summer solstice sunrise, when the sun was at its furthest point in the northern sky and the day was the longest. In Ezekiel 8:16 he describes men in the temple "and their faces toward the east; and they worshipped the sun toward the east." This practice was allowed by Solomon who built the temple and reserved the outside courtyard for the worship of the sun.

The landscape mythology of the Amorites acted as a vehicle to join the powers of the Earth and the Sky. *Theory of Geometry, Geomacy and Sacred Landscape,* Cerrig, 2004 "Amongst the earliest sites of this kins were sacred springs where the stuff of life, enlivend by the earth energies, bubbled from the mother earth. There would have been trees that by their placement and age showed the union of heaven and earth and so echoed the great World Tree which in many mythologies support the universe."

The union of Heaven and Earth is called the *Sacred Marriage* or the *Holy Union of Opposites* and is the cornerstone of all of the sacred sites that are found associated with the Amorites. To know the sacred landscape that existed before the Hebrews conquest of Israel, is to know the archetype of the burial mounds and circular temples dedicated to the Sun and Earth that are found in the British Isles and in the Ohio Valley. From the early Biblical lands we have the accounted giants, that have been described as Cro-Magnon, state of the art weapons, burial mounds surrounded by rings, circular open-air sun temples, that included avenues or sacred vias, sacred stones, springs and groves dedicated to the Earth Mother. Symbolism included cup stones, swastikas and crosses. The practice of gematria, implies a knowledge of complex mathematics.

Amorites in the British Isles

Phoenician Orgins of Britons and Scots 1925 "The approximate date for the initial erection of these rude Stone Circles and other early megaliths in Britain appears to have been many centuries and even a millennium or more before the arrival of Brutus about 1100 B.C., or about 2800 B.C., or earlier. This is evident from the geographic and geological correlation of these monuments to the prehistoric tin and copper mine working, flint factories and neolithic villages. These relationships make it clear that these monuments were erected by the earlier branch of the sea-trading Phoenicians, who were exclusively engaged in mining for the bronze trade in the East, and using that metal in Britain sparingly themselves, and not engaged to any considerable extent, if at all, as agricultural colonist, such as were Brutus and his later Brito-Phoenicians, who used bronze more freely, as attested by their tombs, bronze sickles, ect. Whilst the numerous "Barat," "Catti" and "Cassi" place-names on so many of their sites and the "Catt-Stanes" testify that their erectors were "Catti" or "Cassi" Barats or Brito-Phoenicians, as were the Amorites. The physical type of the builders of these Stone Circles and Megaliths is obviously that represented by the skeletons of tallish Nordic type found with Iberian or Pictish type found in the long barrow burial burial mounds, chambered cairn and stone cist of the late Stone and Early Bronze Ages in the neighborhood of these circles."

Dolman in the southern England. From *Remains of the Prehistoric Age in England,* 1904

The British Isles A Resume of Skeletal History, Coons 1939 "By far the most important Neolithic movement into Great Britain, and into Ireland as well, came by sea from the Eastern Medeteranean lands, using Spain as a halting point on the way. It was this invasion which passed up the Irish Channel to western and northern Scotland, and around Denmark and Sweden. The settlers who came by sea were the Megalithic People, who belonged to a clearly differentiated variety of tall, extremely long headed."

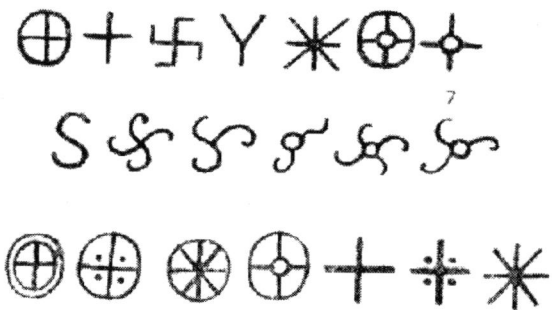

Sun and Fire Symbols from Bronze Age Denmark. *Archaic England,* **Bayley, 1920**

"It was long ago observed that the distribution of these prehistoric megaliths or "great stones" over part of the world followed mainly the coast lines, thus presuming that their erectors were a seafaring people. Moreover, a these monuments are most numerous in the East, it is generally agreed that this cult in Britain, Brittany, Scandinavia, Spain and the Mediterranean basin was derived from the East.

   The distribution of the megalithic monuments in different parts of the world would suggest that their builders were engaged in exploiting the mineral wealth of the various countries. These megaliths all the world over are located in the immediate neighborhood of ancient mine workings for tin, copper, lead and gold or in the area of the amber trade."

Phoenician Origin of the Britons & Scots, 1925. "The great "prehistoric" Stone Circles of gigantic unhewn boulders, dolmens (or "table stones") and monoliths, sometimes called *Catt S*tones still standing on weird majesty over many parts of the British Isles , also now appear to attest their

Phoenician origin."

It has been estimated that there were somewhere between five and ten thousand megaliths in the British Isles that stretched from south England to Ireland, Scotland and as far north as the Orkney Islands and the Shetlands. In *Megalithic Sites in Britain,* Thom writes, " It is remarkable that 1000 years before the earliest mathematicians of classical Greece, people in these islands not only had a practical knowledge of geometry and were capable of setting out elaborate geometrical designs but could also set out ellipses based on Pythagorean triangels. We need not be surprised to find that their calender was a highly developed arrangement involving an exact knowledge of the length of the year or that they had set up many stations for observing the eighteen year cycle of the revolution of the lunar nodes."

Evidence of the widespread prevalence of "sun-worship" amongst the ancient Catti Barats or Britions who erected the prehistoric Stone Circles in Britain is shown in their "Cup-markings" which are sometimes found carved upon the stones of these circles, in funeral barrows [mounds], upon some standing stones, dolmans and stone cist coffins, and on rocks near Ancient Briton settlements over a great part of the British Isles, and in Scandinavia and other parts of Europe and the Levant, associated with megalith culture.

Sun Cross and Cup-markings which are are found carved upon the stones within circles, upon standing stones, dolmans and stone -cist coffins. They are found over the greater extent of the British Isles, Scandinavia and other parts of Europe and the Levant, associated with the Amorites and megalithic culture. Standing stone or menhir with a sun cross and cup stones, located in England from *Phoenician Orgins of Britons and Scots* 1925

These Stone Circles have been supposed to have been used to track the movements of the sun has been inferred from the existence of special entrances at the cardinal points, and also from the elaborate avenues attached to some of them, and supposed to have been used for ritualistic processions; and also suggested by the apparent later use of some of them by the Druids as temples. They were undoubtedly considered sacred, as seen in the frequency of ancient burials in their neighborhood. This is especially evident at Stonehenge where the great numbers of tombs of the Bronze Age in the neighborhood of that monument.

The space in which these circles are found are usually deliminated by an earthwork called a "henge." A henge is described by Wikipedia encyclopedia as, " A roughly circular or oval-shaped flat area over 20m in diameter which is enclosed and delimited by a boundary earthwork that usually comprises a ditch with an external bank. Access to the interior is obtained by way of one, two, or four entrances through the earthwork. Internal components may include portal settings, timber circles, post rings, stone circles, four stone settings, monoliths, standing post, pits, coves, post alignments, burials, central mounds, and stake holes."

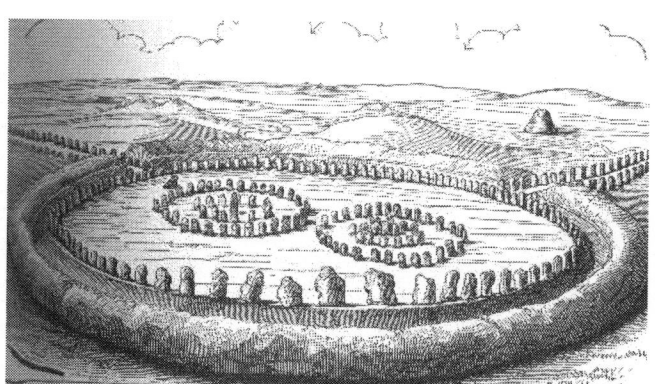

Avebury Henge with interior stone circle and stone sacred vias.

Phoenician Origin of the Britons & Scots, 1925. " The mysterious race who erected these cyclopean monuments, wholly forgotten and unknown, now appears from the new evidence to have been the earlier wave of immigrant mining merchant Phoenician Barats, or "Catti" Phoenicians of the

Mur, Mer or Martu clan-"The Amorite Giants" of the Old Testament tradition. And from whom it would seem that Albion obtained its earliest name (according to the First Welsh Triad) of Clas Myrd-in (or Merddin)" or "Diggings of the Myrd."

"This early Phoenician title of Muru, Mer, Martu or Maratu, meaning "Of the Western Sea (or Sea of the setting sun)", which now seems obviously the Phoenician of the name "Mauret-ania" or "Morocco" with its teeming megaliths, and of "Mor-bihan" (or "Little Mor) in Brittany, with its sun cult megaliths, is also found in several of the old mining and trading centers of the earlier Phoenicians in Brition associated with Stone Circles and megaliths and mostly on the coast, eg., Mori-dum, port of Romans in Devon, and several More-dun."

"Amorites of Syria Phoenicia-Palestine are called "giants" by the Hebrew in the Old Testament. They are moreover, also called them the sons of (Beni-anak) Now "Anak" in Akkadian is the name for "Tin" and Tarnish, which, as Tarz or Tarsus, we have seen was the chief port of the Amorite Phoenicians, and was actually visited and conquered by Sargon I., is thus celebrated in the Old Testament in connection with Tyre of the Phoenicians. "Tarnish was thy merchant by reason of of the multitude of all kinds of riches; with silver, iron, tin, and lead, they traded in thy fairs."(Ezek, 27, 12)

It would thus appear that the tin which was imported into ancient Palestine, and which entered into the bronze that decorated Solomon's temple, and formed sacred vessels in that sanctuary, was presumably obtained in most part, if not altogether, from the Phoenician Tin-mines of Ancient Britain."

There is also the possibility that some of the bronze was made from Lake Superior copper, that was being mined at the same time.

Prehistoric menhir in Wigtownshire with Hitto Phoenician Sun Cross, swastika and cup marks. *Phoenician Orgins of Britons and Scots* **1925**

Remains of the Prehistoric Age in England, 1904, "Extensive settlements and many ruined chambers are found throughout western and northern Britain. The open flint-bearing chalk uplands of Wiltshire and the similarly unforrested Cotteswold Hills were the scene of extensive settlements. The chambers of this area were built in "long barrows," an elongated mound of a form occasionally found in Brittany or Amorica, thus divulging the Amorite connection. The multiple chambers of many of the English and Welsh "passage graves" and the frequency of cremation also suggests that Brittany was the dominant intermediary. Megalithic chambers are lacking in the long barrows of southern parts of Wiltshire and Dorset. Local supplies of suitable stone are lacking in these more southerly districts and the culture was apparently not sufficiently vigorous to induce the transportation of the megalithic blocks over long distances, as was done in Iberia and in Britain itself in the later, Bronze Age monument of Stonehenge.

  The burned timber found in these barrows suggests that here the megalithic chamber was reproduced in wood and consumed by fire after the completion of the internments."

Long Barrow at West Kennett, England, surrounded by standing stones. From *Remains of the Prehistoric Age in England,* 1904

They Came by Sea, 1945 "The physical type of the builders of these stone circles and megaliths is obviously that represented by the skeletons of the tall Nordic type (with some others of the smaller river-bed and mixed Iberian or Pictish type) the long barrow mounds, chambered Cairns and stone cist of the Late Stone Ages in the neighborhood of these circles. And it was presumably early pioneer stragglers of this same Nordic stock at the end of the Old Stone age who are who are represented by the "Red Man" of Paviland cave, in the Gower peninsula of Wales, of the mammoth age, and the "Keiss chief" in the stone cist. Both of these interred with rude stone weapons, and are interred with rude stone weapons, and are of the superior and artistic Cro-Magnon type of early men, which seems to have been the proto-Nordic or proto-Aryan. These early Nordic people, who buried near the Circles, were generally found in their tombs laid on their right side, and their face usually facing eastward to the rising Sun, thus evidencing their solar religion and belief in resurrection."

Amorite, Beaker People

Henge with central mound located near Stonehenge in Wiltshire, England. From *Ancient History of Wiltshire,* **1812**

According to the *Archaeology of the Bible* by George Barton, The Aryans (Amorites) separated into an eastern and western branch; the latter we can trace from Asia to Europe. They founded colonies in the Delta of the Danube, and spread westward along the Danube into Germany and the valleys of the Rhine, the Loire and Seine, and still pushing westward established themselves into Brittany and Great Britain.

The Bell Beaker people took advantage of, and improved transportation routes by sea and rivers to mine and trade mineral resources. They introduced the use of copper and bronze weapons as they expanded into new lands. Artifacts suggest an Iberian source for the earliest copper, followed by central European mines being exploited. Distinct to this people was their cord impressed, tulip shaped pottery known as the Beaker.

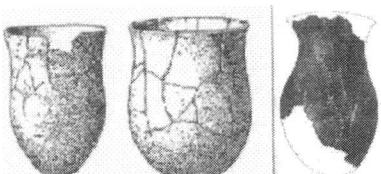

Beaker type pottery with a tulip shape that was copied from Egyptian designs. Two pots on the left are from Michlesburg, Germany and the on the right from Bridgeport, Michigan.

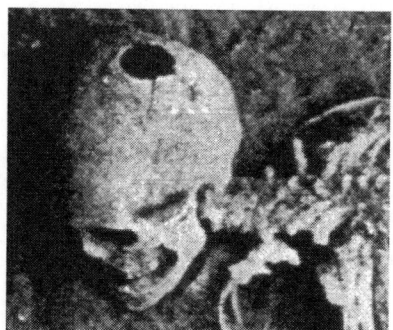

Bell Beaker (Dinaric) traders's grave from the cemetery of Beradz on the upper Vistula, southern Poland. From *The Prehistory of Eastern Europe*, Gimbutas, 1956. Note the extremely high forehead and the trephaned skull.

Stephen Coons, *The British Isles, a Resume of Skeletal History,* " With or shortly before the introduction of metal, the British Isles were invaded from both sides by fresh settlers. From the west came a triple combination of Boreby brachycephalics, Corded people, and Eastern Mediterranean Dinaric, under the hybrid auspices of the Zoned Beaker Culture."

Stephen Coons writes in *The Bronze Age in Britain,* "The Bronze Age people of England as represented by this Beaker series, were clearly heterogeneous. The three ancestral elements which met in the Rhinelands may be distinguished easily. All three were tall [...] The Corded element, however was the tallest, and the Borreby element the shortest. On a whole, the heavy-boned, rugged quality of the Borreby type seems to have influenced the bodily build of the total group."

The accepted date for this invasion of the Beaker People into the British Isles is about 2700 B.C. Stephen Coons also writes, *"*The Beaker people did not exterminate the builders of the long barrows, who continued for a while to build their characteristic earth-covered vaults, in some which Beaker pots have actually been found." They incorporated many of the religious customs of the earlier settlers, most notably was the worship of the sun within the earthen henges.

The most easily recognized cultural elements that manifest in the British Isles consists of their unique bural mounds and the continuation of building circular, open air solar and celestial observatories known as henges. The henges include sacred avenues or processionals of stone, much like those that were described in the Levant. The earthworks were built using using mathematics, geometry and utilizing sacred numbers in their construction. Their purpose was to track the movements of the sun, moon and the host of heaven. They were places where the people would pay homage to the Sun god and the Earth Mother.

The most impressive of these henges, besides Stonehenge is Avebury, sometimes written as Abury. The iconic design of this temple is explained in *Here Be Dragons: The Strange Enigma of Serpent Mounds,* Phillip Gardner "In Egyptian hieroglyphs we can see the symbol of the snake going over the solar disc, merging head erect. Overlaid onto Avebury it is the same image! Adding to this, that the snake is often depicted with the ancient Egyptian Ankh symbol dangling from its emergent neck-the Ankh being the symbol of new life-the great cycle of Avebury simply has to be the 'solar disk' and the pathway of the snake-thus illustrating in a painfully labor-intensive way, the ritualistic path of the serpent worshipper new life."

"Of course even as Ave Burym the Ave reverts back to the root of 'Eve" which we know means 'female serpent' The pathway of Avebury passes through a large circular Temple of the Sun emerging and then winding again and ending with an oddly, not quite circular head-directly in line with the "Snakes Head Hill" (Hackpen.). The central circle is symbolic of the Sun, which is the male principle in the creative process and is symbolized elsewhere as a bull or lion. Once the serpent has passed through or around the sun circle it is recharged for a new life. The archaeology of the area shows that people used to walk outside the pathway of the serpent, leaving the pathway for the priests."

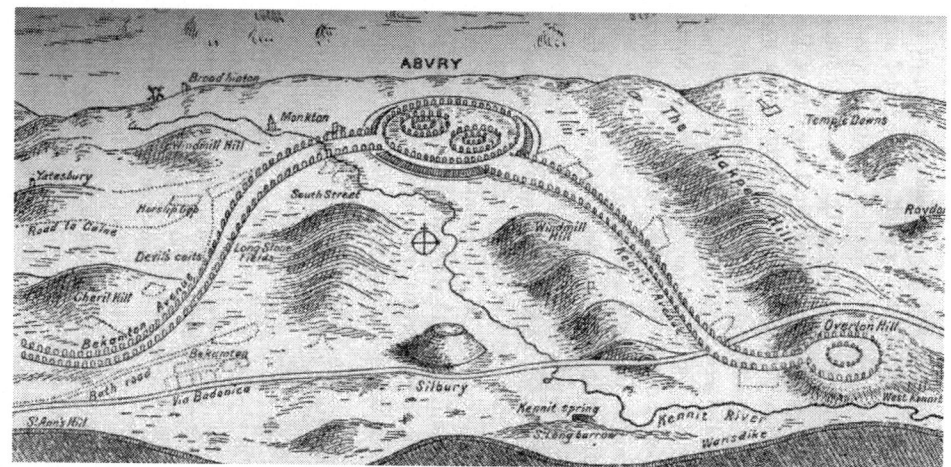

Avebury henge with stone circles and avenues in the form of a serpent, by Stuckeley, 1740. Avebury is Britons largest henge with a diameter of 1250 feet. The serpent is a common eastern Medeteranean symbol for the sun, and is also symbolic of the female or Earth Mother. The Phoenicians adored the animal as beneficent genius, with superior power and wisdom.

Stonehenge also contains this sublime message of the symbolic representations of the *Sacred Marriage* or *Holy Union of Opposites* of the Sun Father and Earth Mother. *Marriage of Earth,* Dee Finney, "Stonehenge people worked with symbols and integrated symbolic meanings into the stones, scholars have identified enough meanings, using parallels from early agrarian cultures elsewhere, to reconstruct an outline of the lost mythology. The formulation of these concepts into a coherent and unified theory is due to physics professor Dr. Terrance Meaden, an archaeophysicist.

To commence Stonehenge, Avebury and other stone monuments of Western Europe were built over 4 millennia ago in an era when Neolithic farmers believed in an Earth Mother and a Sky Father. With this knowledge, scholars in ancient religions, proficient in the interpretation of archaic art forms see that the ordered stones of Stonehenge could constitute an open-sky temple implicitly dedicated to the worship of the Earth Mother. This is because thee monument is heavy with feminine symbolism. Above all, the concentric circles and the U-settings appear to represent the womb of the Earth Mother while the trilithon arch in the outer circle is her vulva."

146

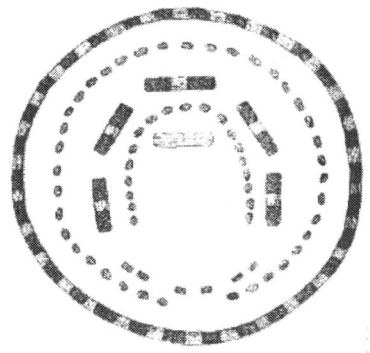

Overhead view of Stonehenge's, "concentric circles and the U-settings that appear to represent the womb of the Earth Mother while the trilithon arch in the outer circle is her vulva."

"Next the axis of the monument is directed at the rising sun on midsummer's day.[Summer Solstice] It is only on midsummer morning that the rising sun penetrated the middle arch of the womb to illuminate the internal Goddess Stone with its radiant energy. Watchers would see the stone sparkling in the reflected light of the Sky God, serving here in his role of the Sun. This constituted a dramatic spectacle in which the actual Marriage and Consummation of the Gods was witnessed."

Heel stone at Stonehenge that is aligned with the earthen gate of the henge on the summer solstice sunrise.

"Hence the inferred Creation Myth of the Stonehenge and Avebury Peoples is that Earth Mother and Sky Father came together to beget the world and that the midsummer spectacle was the anniversary and dramatic re-enactment of the primordial event." *The Stonehenge Solution,* 1992 Professor G. T. Meaden, Oxford physicist and antiquarian, describes how the consummation of the celestial marriage was ritually acted out by the interplay of light and shade among the standing stones.

Conical burial mounds are surrounded by a ditch, earthwork or stones with a single causeway to the circular platform on which the mound sits. This is similar to the burial mounds described in the Levant that were erected by the Amorites. In some mounds, burials were placed in a sun burst or spoked postion. Symbolism of the Amorites is also present that included cup stones, crosses and swastikas representing the movement of the sun.

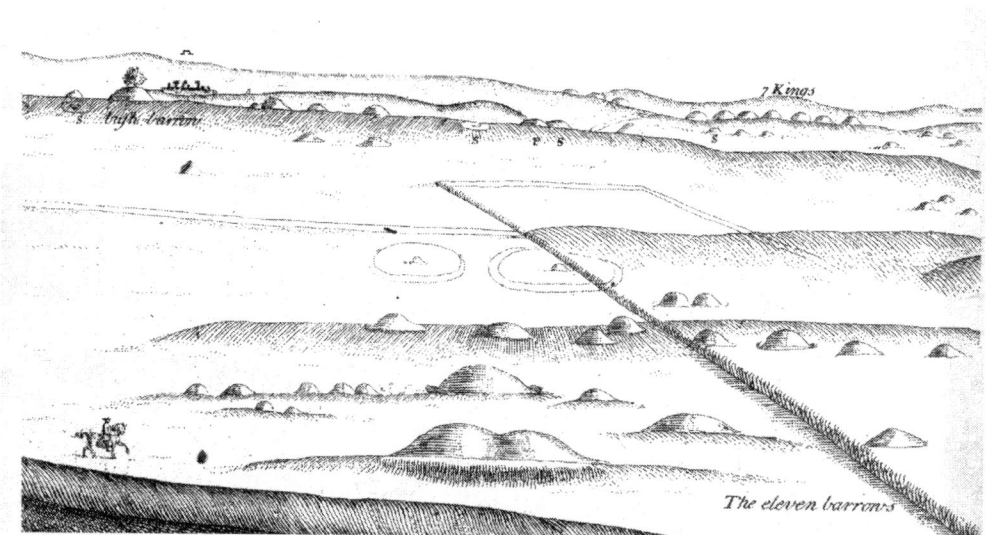

"Eleven Barrows" by John Stukeley, 1740, showing different mound types that includes a conjoined mound in the foreground and in the background mounds surrounded by earthworks.

Cultural similarities in the form of burial mounds and symbolism have led many researchers to conclude that the Corded people were a northern branch of the Amorites. The majority of the Corded people were long headed, as were the Amorites, opposed to the Dinarics who were round headed. In

England, three times as many round headed people were found in the mounds opposed to the long heads, with as many somewhere in the middle. This would imply that the Corded and Dinarics were intermarrying and had not yet fused into one people.

By examining the three cultural elements of the Beaker People in England, consisting of the Dinaric, the Corded People (Scythians) and Boreby Cro-Magnon, all under the auspices of the Beaker People we can find all of the cultural elements that would remanifest in the Ohio Valley and labeled as the Adena or the Allegewi.

Skeletal remains of the Beaker people are very tall with skull features that are still considered identical to those of the Upper Paleolithic Cro-Magnon. These features include protruding brow ridge, massive jaws and thick skull walls. The size of the skeletal remains is sufficient evidence to conclude that they were related to Cro-Magnon.

The Bronze Age in Britain, "They form a rare group in the world with a cranial length of 184 mm and a index over 80. This peculiarity they share with the few known Brachycephalic crania of the Upper Palaeolithic. Again reminiscent of the Upper Palaeolithic skulls in the ruggedness of muscular marking, the prominence of the brow-ridge and occipital lines and the depth and breadth of the mandible." The only other population group with this type of skull were the Allegewi mound builders of the Ohio Valley.

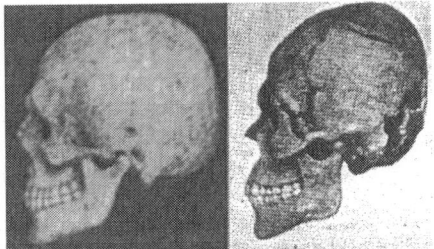

The skulls of the Dinaric are described as being tall short heads, a high bridge of the nose and the back of the head appearing "cut away" at the back. Facial prognathism or a jutting upper jaw is also present. On the left is a Dinaric skull from Germany, on the right is a Dinaric skull from Ohio.

The Dinaric spread through conquest out of the Caucasus into central Germany to Northern France. From France, the Dinarics advanced into the British Isles. Another group of seafaring Dinarics are found throughout the Medeteranean. There is evidence that the Dinarics were in the Levant at the time of the Amorites. Several of the Dinaric skulls were found in Palestine and Israel, that at first were believed to be Peruvian skulls, however identical skulls were found and it was realized that these unique head shapes represented a different type of people. One of these skull was found in Damascus, within the realm of the Amorites and Og.

One of these Dinaric skulls was discovered by Prof. Retzius, who described it in the *Proceeding of the Royal Academy of Science*, 1902 "adducing arguments to strengthen that supposition. A Peruvian skull which had been brought to Europe as a curiosity during the reign of Charles V. and afterwards thrown aside. His communication appeared in Muller's "Archive fur Anatomie" The opinion of the learned traveller was, however, subsequently reversed by the discovery at Atzgerdorf, near Vienna, of another simular cranium. More recently others have come to light at the Vollage of St. Roman in Savoy, and in the Valley of the Doub near Mandense. Dr. Fitzinger has probably investigated this subject with more thoroughness than any other writer, and has shown in his articles in the "Transactions of the Imperial Academy of Vienna", that this custom was native to the Scythian region in the vicinity of the Moetian Moor, and prevailed in the Caucasus and along the shores of the Black and Caspian seas and the Bosphorus. One of these deformed skulls was discovered in 1856 by J. Hudson Barclay, in a large cavern near the Damascus Gate at Jerusalem. The skull was of unusually large size and decayed, but the skull, which was pretty well preserved, was brought to this country and is preserved in the collection of the Academy of Natural Sciences of Pennsylvania."

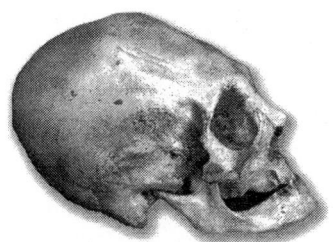

Peruvian skulls with flattened back of the head and a high forehead have been mistaken for Dinaric. Note that facial prognathism, or a jutting upper jaw is not present, nor is the mandible as large as what would be found in Upper Paleolithic skull types.

The Amorites or Dinarics had a trading stronghold on the island of Crete, with a people known as the Minoans. The island was destroyed in 1400 B.C. when it was enveloped by a tidal wave caused by the island of Thera's violent volcanic explosion. After 1400 B.C Crete no longer played a role in world history.

The Dinaric first appeared in the Medeteranean in last centuries of the third millennium B.C. Since the metal ages of the middle and northwestern Mediterranean were later than those farther east, the chronological aspect of this theory presents no contradictions.

Swastika crosses on dress of Phoenician Sun-priestess carrying sacred fire. From terra-cotta from a Phoenician tomb in Cyprus. The Earth Mother was worshipped in Crete. The Earth Mother was the goddess of the air and was symbolized by the dove, She was also the goddess of the underworld where she was represented by the Serpent.

Stephen Coons writes in *The Copper and Bronze Age in the Western Mediterranean"* The evidence of the racial composition of the Copper Age sailors who reached Italy and the Italian islands is simple and direct. The moderately tall long headed, mid narrow-nosed Megalithic people who were implanted, during the Late Neolithic, upon the small Mediterranean type which had preceded them, were followed, during the Aeneolithic by other, of the same kind, in the company of equally tall brachycephals. The latter resembled the same Dinaric head form in Cyprus, Crete, and the Aegean, and without doubt formed a westward extension of the same movement."

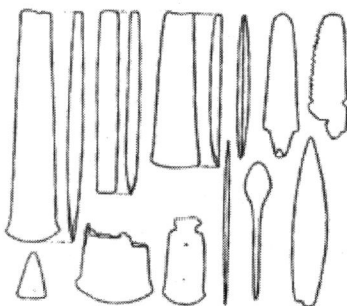

"Tanged daggers and flat copper celts from the seaport located in southeast Spain within Almeria and El Argar are of the forms that were being made in Egypt, Troy and Cyprus in the early part of the third millennium." from the *American Anthropologist,* January, 1930.

The American Antiquarian **June, 1911**
Mining in the Stone Age
   A most interesting glimpse into prehistoric mining in the Stone Age has recently been revealed upon the reopening of the Oural and Aram copper-cobalt mines in Spain.
   A few of the men seem to have been of great size, and all of them must have been of extraordinary muscular development. The heavier stone hammers which they used weighed as much as 20 or 22 pounds. In spite of their muscular development, the majority of the miners were evidently of extremely slim build, for some of the galleries are literally polished by the rubbing of bodies, and in these galleries, penetrating through the solid rock.

There are similar legends of the Maritime Dinarics who landed on the Irish shores and the Amorites, who were living in Israel and Palestine. The *Babylonian Talmud* states that some of the Amorites were endowed with a double row of teeth. The Irish also have traditions of an early people called the

Fomorians, (Muru or Amorites) who were pirates from the Medeteranean coasts of North Africa. They are described as a ferocious tribe of giant demons who scourged the Irish coasts. There burial mounds and remains are also found in Scotland and north to the Orkney Islands. Some of these giants were also said to be endowed with a double row of teeth.

Also found within the burial mounds of the British Isles are the Corded Ware culture. The Corded people like the Dinaric were known to introduce the art of metal working. Artifacts within burial mounds characteristically includes copper daggers and archer's writs guards. The Corded Ware culture represented some of the largest skeletons found within the British barrows. Their remains can be found over a wide expanse of northern Europe, including Germany, Poland, Lithuanaia and the Ukraine. They had also moved north into the area of the Hunters and Fishers as early as 2900 B.C. in the present countries of the Netherlands, Denmark, Sweden and Finland.

Mounds associated with the Corded People almost mirror the mounds that are found in the Ohio Valley that are associated with the Allegewi. These included conical mounds of earth or stone, the presence of an alter within the mound, wooden chambers within the mound and the mound being surrounded by a stone fence, moat or earthwork with an entryway to the mound.

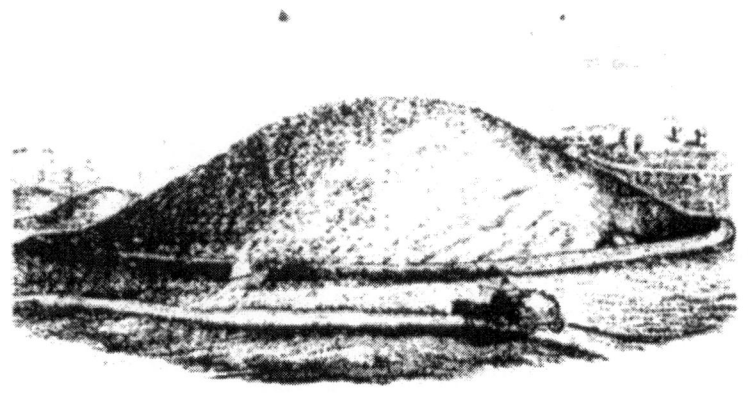

Burial mounds surrounded by an earthwork in the Ukraine, dating to 400 B.C.

The influence of the Corded People within the Scandinavian countries led to the transition of glacial kame burials to those in burial mounds. The earliest mounds were of the megalithic type that included passage graves. The megalithic tombs were the similar to the charnel houses in the Ohio valley, in that they were storehouses for dead that could viewed or visited. They show a similar ancestral worship where the remains of the dead were venerated.

After the Megalithic period burial mounds in Denmark are found with vertical stratigraphy; with the oldest burials in sub floor pits, the second grave above this, with a third level above those. This type of burial being most reminiscent of the Allegewi in the Ohio Valley, whose mounds grew in size as interments were added. This type of burial is also evident in early Iroquois burials within mounds where the charnel house was burned and an layer of dirt added to cover the remains; followed by additional burials on top of those.

The Corded People influenced the type of mounds constructed in the southern England and are archetypes of the mounds that are found in the Ohio Valley. Their preferred burial method was in rectangular pits that were lined with stones with the inclusion of red ochre. Sometimes these stones were arranged like spokes of a wheel within the burial mounds. Their mounds also had a circular ditch or earthwork that surrounded the mound; with a gateway or causeway. Corded mounds have also been found that had a circular pavement of stones that surrounded the mounds.

Drawing showing different mounds types in the British Isles. The long barrow (A) are the oldest and were contemporary with the megalithic peoples. (B) is a henge with no apparent gateway. The conical mounds attributed to the Dinaric and Corded Ware under the auspices of the Beaker People have a ditch or earthwork that surrounds them. One of the mounds in the background is two mounds that are joined, with surrounding ditch and earthwork.

The Corded people are also are associated with a hill fort people. Enclosures are on hilltops and promontories. This monumental architecture consisted of earthen walls and ditches. The ditches are crossed by earthen bridge like causeways. The largest causeway represents the entrance. What all these hill top enclosures share is their large size.

They are most numerous in northern Germany and in England. German legends of the Corded People confirm that they were a race of race of giants as reported in the *American Antiquarian, Mitteilungen der Schlesischen Gesellschaft fur Volkskunde,* 1903 by Dr. H. Seger. Dr. H Seger contributes a brief article on "Die Denkmaler der Vorzeit im Volksglauben" in which he discusses folk-lore and folk-belief concerning prehistoric stone graves, which the folks know as "giants graves," "giants ovens," "Huns graves," ect.. Mounds of the Bronze Age called "Huns graves" are coupled with the ramparts and other fortifications, sometimes called "Tarter walls" or "Tarter forts."

New York Times, **October 3, 1892**
A Race of Giants in Old Gaul

In the year 1890 some human bones of enormous size, double the ordinary in fact, were found in the tumulus of Castelnau (Herault,) [Germany] and have been carefully examined by Prof. Kleiner, who, while admitting that the bones are those of a very tall race, nevertheless finds them abnormal in dimensions and apparently of morbid growth. They undoubtedly reopen the question of the "giants" of antiquity, but do not furnish sufficient evidence to decide it.

The third group that made up the Beaker People was Borreby type. Borreby skulls have been described as being highly developed in the forehead, a pronounced brow ridge with a receding forehead. In describing Bell Beaker skulls from Middle Germany they were said to have the greatest similarity with the Danish Borreby form and the Swedish brachycranial Neolithic skull from Karleby. Thes similar skulls were described as having, round skulls with steep a forehead, steep occiput, powerful browridges and a high face.

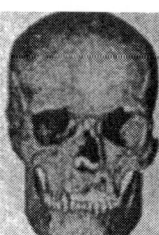

Borreby Cro-Magnon

Origins of the Allegewi mound builders are a combination of cultural and physical traits from the Dinaric, Corded People and Borreby Cro-Magnon. These traits include: Conical mounds surrounded by an earthwork, ditch or ring of stones, henges used as solar and celestial observatories, evidence of Earth Mother worship, hill top enclosures, and most persuasive evidence being the unique Dinaric type skulls and the large size of the skeletons. The uniqueness of physical and cultural traits appear to be overwhelming evidence that a connection exist between the Amorites, the Beaker People and the Allegewi mound builders of the Ohio Valley. This hypothesis does not contradict current archaeological theories that profess that they do not know where the Adena (Allegewi) came from, nor where they went.

Giant's Remains in the British Isles

Burial mound near Stonehenge encircled by a ditch and earthen wall

The skeletal remains found in burial mounds in England, Scotland and Ireland represent the Dinaric, Boreby Cro-Magnon and the Corded People. The passage graves found within the long mounds, utilized as ongoing receptacles for the dead were replaced by conical mounds, with single internments and cremations of the dead. The use of flagstones in the construction of a coffin is most common. The Amorite and Corded People, influence in mound construction is seen with many of the mounds being surrounded by a ditch, berm or both.

Their round barrows, containing flint weapons, barbed and tanged arrow-heads, copper axes and daggers. It is believed that the remains of males in the mounds were warriors. Large females are also found within the mounds, that may descended from the infamous Amazonian women of the Corded People. Most burials were accompanied by a beaker-type pot, seemingly to hold a drink for the deceased to drink on his journey.

Skull types are predominately round headed or brachycephalic, as found with the Dinarics, with some admixture of the long headed Corded People. All of the skulls exhibiting archaic type features that include a prominant brow ridge, a receding forehead, massive jaws and thick skull walls. The large size of the skeletons along with the archaic skull features show that they are of Cro-Magnon species.

Crania Britannica, 1865

At the Green Gate Hill Barrow the skull was described as being brachycephalic, with a well-expanded forehead it well formed cranium. The cheek bones are prominent and the face broad from the direction of the malar bones. The abrupt prominences of the nose rising from the deep depression below the well marked frontal sinus, is a characteristic feature of this skull; the nasal index, however,is not to high. Another example of the ancient Brachycephalics of England is to be seen in the museum above referred to No. 297 described as having a deep nasal notch, receding external margins of the orbits, with prominent cheek bones and well formed jaws.

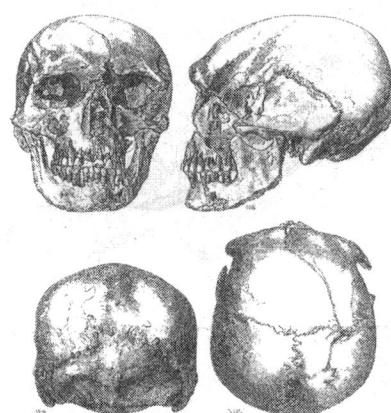

Archaic type, brachycephalic type skull found within the Green Gate Hill mound. The projecting brow ridge, sloped forehead and massive jaw, is indicative of Upper Palaeolithic, Cro-Magnon populations.

Our Early Ancestors, an Introductory Study of Mesolithic, Neolithic and Copper Age Cultures in Europe and Adjacent Regions, by M.C. Burkett, M.A., F.S.A., F.G. 1876

The invaders differed somewhat from the former inhabitants of the land. The Neolithic folk seem to have been of moderate stature, long headed, oval faced, narrow nosed, with small features. They were not at all powerfully built race. The new-comers on the other hand- according to Abercromby- were characterised by short square skull showing a great development of the supercilliary ridges and eyebrows. The cheek-bones, nose and chin were prominent and the powerful lower jaw was supplied with large teeth. They were a tall, strongly built race and must have presented- at any rate as far as men were concerned- a fierce, brutal appearance. The dead were buried in round barrows, inhumation being practiced. They knew about the use of copper and introduced into England the beaker type pot

Remains of the Prehistoric Age in England, 1904

Here again there is no lack of skulls and skeletons, and the descriptions of them are many. In the barrows of this period we find two classes of skulls, long and broad. The former may be those of the earlier people, the latter those of a race which had invaded the country. Or the collection may be explained without supposing the arrival of a different race, but these are points into which it is impossible to enter here. Suffice it to say that the skulls regarded as typical of this period are brachycephalic, of large size and with well-formed brow. There are salient ridges above the eyes, but these are not the monsterous projections of the Neanderthal type. One gains the idea that the cast of countenance of the possessors of these skulls must have been much more fierce and commanding than that of the milder race which preceded them.

The Way of the Sea, 1929

A further point arises here that is at least interesting, even if it be but a coincidence. It was apparently about the time that Hissarlik II was reaching the height of its greatness with the rebuilding of its walls, roughly dated about 2200 B.C., that the beaker became important on the European loess. It declined soon after Hissarlik was destroyed, as Frankfort thinks, by Anatolian, probably Hittite invaders. We must add that Hissarlik II seems to have had widespread interest, presumably of a commercial nature, in central Europe and throughout the Mediterranean.

In our last volume the story of important parts of eastern Europe was shown to hinge upon the conquest of cultivable areas by steppe warriors, the men of the stone battle-axe. It may well be that some of these warriors retained to a certain extent their ancient mobility and power of organization, a power which, we cannot but think, became of commercial rather than of military importance. In conjunction with local influences here and ther they may have developed the beaker culture, which seems to us to belong primarily to the loess regions of central Europe. The skeletons found in association with objects belonging to the beaker culture, especially on the loess, include long-headed men, who might well be related to the warriors of the battle-axe. In west-central Europe and on the west, on the other hand, and notably in Britain, are found broad- headed men of very strong build with powerful brow-ridges, the origin of whom Keith thought years ago could be traced back to Polish Galicia.

Historical, Topographical, and Descriptive View of the County of Northumberland, 1825

The ancient Britons were remarkable for the large stature of their bodies; their eyes were generally blue, which was esteemed a great beauty; and their hair red or yellow, though in many various gradations. They were remarkably swift of foot, and excelled in running, swimming, wrestling, climbing, and all kinds of exercises in which either strength or agility were required. Accustomed to hardships and despising cold and hunger, in recreating they plunged into the morasses up to the neck, where they remained several days. They painted their bodies with a blue dye extracted from woad, and at an early age they were tatood in a manner the most ingenious and hideous; and in order to exhibit these frightful ornaments in the eyes of their enemies, they threw off their clothes in the day of battle. When advancing to the combat their looks were fierce and appalling, and their shouts loud, horrid, and frightful.

Caledonia: or a Historical and Topographical Account of North Britain, 1888

Barrows of a greater or a less size may be found in every district of North-Britain, in the most southern as well as the most northern. Near the abbey of Newbottle there was once a remarkable Barrow, composed of earth and of a conic figure, in height thirty feet, and in circumference at the base ninety feet; it was surrounded by a circle of stones, and on its top there grew a fir tree. When this barrow was removed there was found in it a stone coffin, near seven feet long, and proportionably broad and deep; and from it was taken human skulls. Several other Barrows, both in South and North-Britain, have been also surrounded with circles of stone. There is a Barrow in the parish of Kirkmabreck, in Wigtonshire, which is called Cairny-wanie, and which is merely the Cairn-uaine of the Scoto-Irish, or Green-Cairn of the Scoto-Saxon: when Cairny-wanie was opened there was found in it a stone coffin, comprehending a human skeleton that was greatly above the ordinary size, together with an urn containing some ashes and an earthen pitcher. There was a sepulchral tumulus at Elie, in Fife, which, when opened some years ago, was found to contain several human bones of remarkably large size. In the parish of Logie, in Forfarshire, there are several tumuli, two of which have been opened: in one of these there was found a coffin, formed of flag stones, and containing a human skeleton, the bones whereof were of and extraordinary size, were mostly entire, of a deep yellow colour, and were very brittle when touched: in the other tumulus there were found, about a foot from the surface, four human skeletons, the bones whereof were exceedingly large:and near these was discovered a beautiful black ring, like ebony, of a fine polish, and in perfect preservation

Caledonia: or a Historical and Topographical Account of North Britain, 1887

In a large oblong Cairn about a mile west from Ardoch, in Pertshire, there was found a stone coffin, containing a human skeleton seven feet long. From those facts, with regard to the large size of the skeletons, the tradition on this suggest should seem not to be quite groundless, as indeed Tacitus, when describing the Caledonians, appears to intimate.

Historical, Topographical, and Descriptive View of the County of Northumberland, 1825

At the bottom of the hill, where stands Humbleton Burn House, and close to the barn, the plough in 1811 struck against a large stone. On removing this impediment, a human skeleton was exposed to view, lying in a kistaen, formed of six large flag stones. The bones were in a high state of preservation, of a close texture, and remarkably large. From the specimens sent by the late Mr. Alexander Kerr, of Wooler, to the publishers, the skeleton must have been at least seven feet long. An urn was found beside the remains of the ancient chieftain; but the place was not examined with any attention. The cone of the tumuli seems to have been levelled by tillage.

History of the Highlands, 1884

Many of these tumuli have been subjected from time to time to the prying eyes of antiquaries; and, as their researches are curious, a short notice of them may be interesting to the general reader. With in several tumuli which were opened in the Isle of Skye there were discovered stone coffins with urns containing ashes and weapons. In a Barrow which was opened in the isle of Egg, there was found a large urn, containing human bones, and consisted of a large round stone, which had been hollowed, while its top was covered with a thin flag stone. In a large oblong cairn, about a mile west from Ardoch, in Pertshire, there was found a stone coffin, containing a human skeleton seven feet long.

History and Antiquities of the County of Somerset, 1791

In sinking a well in some part of this parish in the year 1670 there were found at the depth of thirteen feet, the remains of one of the Cangick giants, a people supposed to have formerly inhabited these parts. The top of his skull was said to be an inch thick, and one of his teeth three inches long above the roots, three inches and a quarter round, and after the root was broken off, weighed three ounces and a half.

Somersetshire Archaeological and Natural History Society, **1860**

Therefore there were two distinct races occupying the country at an early date, if not three. There was a marked difference in the camps; in some there was a threefold arrangement of earthworks, of which the innermost was the most strongly fortified. These, he considered were aboriginal encampments, of which Worle-hill was an example. He believed that the encampment on Worle hill was one of the oldest in Europe, and had reason to think that it was earlier than Dolberry. He then pointed out marks of a track way, on each side of which were hut circles. This track way, Mr. Warre explained, led to a village without the works, and which probably arose there in a simular way to those that had sprung up in the neighborhood of castles.

On their return the members and their friends dined at the Royal Hotel, after a few remarks by the President, Mr. Freeman gave a detailed account of the various objects of interest examined during the excursion that day.

Lord Talbot De Malahide desired information in regard to the flint knives. The Rev. F. Warre said that these knives were found mixed among the rubble of the hut circles on Worle-hill. He had found simular ones on the Quantock hills. At the suggestion of Mr. Dickinson, Mr. Warre gave an account of the remains he had found in the hut circles. There were at the top, six to eight inches of surface mould, after which he came to rubble from the hill, then two skeletons, bearing marks of extreme violence, and apparently of two different races--one a gigantic race, with skull presenting the most uncivilized appearance, the other smaller and more advanced.

History and Antiquities of the County of Somerset, 1791
Nortan-Comitis

Is a small town, situated against the side of a hill, about a mile southward from the village of Hinton, and nearly equidistant from Bath and From, the turnpike road betwixt those places running through the eastern part of it; which road, betwixt this town and the village of Wolverton, is mended with a singular kind of stone. In digging for stone in the north part of the parish, about the year 1752, some workman found, at the depth of nine feet beneath a rock, a large quantity of human bones of various sizes, with part of a jaw-bone and several teeth in it of prodigious size.

History, Topography and Directory of Westmorland, 1885

The ancient Britons have left behind them but few traces of their occupation. Near Rawthey Bridge may be seen the holes wherein, some fifty years ago, there stood the upright blocks forming a stone circle, supposed to have been connected in some way or other with the religious worship of the ancient Druids. The stones were removed in 1822 for the repair of the bridge. At Rasate are several tumuli, in which human bones have frequently been found. These barrows were explored about ten years ago by Professor Rolleston and others.

At Hard Rigg was found a cylinder urn, two feet long and one foot in diameter, composed of well-fired clay, and containing the burnt bones of a female. A bracelet had also been placed in it. At Rassett Pike there was found, 11 or 12 feet below the surface, the skeleton of a man who must have been at least seven feet in height. A kistaven opened at Sunbiggen contained the skeletons of two females, lying in a direction north and south. Several mounds, or giants graves, are still to be seen, but no skeletons have been found in them.

***Historical Survey of the County Of Cornwall,* 1817**

In the village of Men, near the Lands End, a farmer, in the year 1716, removing a flat stone seven feet long and six wide, discovered underneath it a cavity formed by stone, two feet long at each end, and on each side another stone twice as long. In the middle was an urn, full of black earth, and round it were some very large human bones irregularly dispersed. In some sepulchers have been found bones much larger than those of the human body.

***Historical Survey of the County Of Cornwall,* 1817**

In the month of March 1761, some tinners being employed on a new mine, in the neighbourhood of Tregoney, one of them struck his pick-axe on a large stone coffin, on the lid of which were some characters, but so much defaced, as to be unintelligible. On opening it was found the skeleton of a man of gigantic size, but on being touched, the whole of it mouldered into dust, except one tooth, which remained entire. This tooth measured two inches and an half in length, and was thick in proportion. The length of this coffin was eleven feet three inches, and the depth three feet nine inches.

Early 19th century excavation of a burial mound in England. From *Historical Survey of the County Of Cornwall,* 1817

***A Description of Caerarvinshire* by Edmund Hyde Hall 1952**

About a mile and a half from the church, close upon the road to Nefyn, stands a small tumulus surmounted with a stone of memorial of stele, called Pen y Maen Carnedd. The evidence, coupled with tradition, in favour of this place is supposed to rest upon the discovery of the bones of a large sized man in a grave said to be still dimly visible.

***A Description of Caerarvinshire,* 1952**

Upon Mynnyd yr Ystym, a secondary hill above the house of Bodwrda, is a British post with a double agger and ditch. The diameter of the outer circle is about three hundred yards, and that of the smaller or inner about one hundred and eighty. The gates or gangways open towards the east and west, and near the last stands a stone memorial. A little below the spot towards the west is seen a heap of stones known by the name of the Giants Grave, in which, if I am not misinformed, were found upon opening it some time ago a congeries of human bones.

The History of Galloway, 1851

Nature seems to have been peculiarly profuse in its bounties to the Celtic nations. Their persons were large, robust, and well-formed; and they excelled in running, wrestling, climbing, and swimming. Both history and tradition assert these facts; and from the writings of Tacitus they receive extensive corroboration. Strabo mentions the Britons as taller in stature than the Gauls, and as differing a little from them in the colour of their hair. "For proof of their tallness," he says, "I myself saw very youths taller, by half a foot, than the tallest men." Besides, in some of the sepulcher remains of the earliest inhabitants of North Britain, have been discovered human bones of a large size. In a cairn about a mile from Ardoch, in Perthshire, there was found a stone coffin, containing a human skeleton about seven feet long. Some years ago, upon opening a barrow in the parish of Kirkmabreck, in the Stewarty of Kirkcudbright, a stone coffin presented itself, in which was a human skeleton, much above the ordinary size. A sepulchral tumulis was opened at Elie, in Fifeshire, which exhibited some very large human bones. In the parish of Logie, in Forfarshire, two tumuli were opened, in which, was a skeleton of extraordinary dimensions; the bones were of a dark yellow colour, and very brittle: The other tumulus presented four skeletons possessing exceedingly large bones: a black ring was found near them, apparently made for the very thick wrist.

The Britons, and consequently the primitive inhabitants of Galloway, wore little, or no clothing. According to the testimony of Julius Ceaser, they painted themselves with woad, which imparted a bluish colour to the skin, and a hideous appearance to their persons. Herosain says, they dyed their skins in such a mannner as to represent the figures of beasts, and wore no clothes. Ovid calls them "virides Britannos:" Martial, "Pictos Britannos:" and Lucan, when speaking of them, uses the words "flavis Britannis."

History of the County of Fife, 1815

The mode of sepulture adopted by these aborigines, is also worthy of observation. During the existence of paganism they were in the practice of burning their dead; but after their conversion to Christianity, they appear to have relinquished this mode, and to have adopted inhumation. Evidence of both of these methods still reamains in different parts of Scotland, though the progress of agricultural improvement has recently rendered the appearance of this much less frequent than it formerly was. The Barrows, Cairns, Cistvaens, and Urns, which have been so often the subject of antiquarian examination and research, are the sepulchral remains of these early Celtic tribes. These remains have been found if Fife as in other parts of North Britain; though in all probability its early advance in agriculture removed many of them, previous to the time when they became objects of attention. In the parish of Cupar, a barrow was opened some years ago, in which were found several heads of battle-axes, formed ov very hard white colered stone, neatly shaped, carved and polished. In the parish of Kettle, on a hill called the Knock of Cleish, a cistvaen was also found containing human bones and trinkets, with the brass head of a spear. Near Elie there was a sepulchral tumulus which when opened was found to contain several human bones of large size.

Woodland Daily Democrat (Woodland California) March 13, 1890
A Gigantic Skeleton

The following paragraph from the Dublin Freeman's Journal of August 1812 seems to show that men of gigantic stature were not unknown in Ireland in prehistoric times. It is not a little surprising, considering our veneration for Irish antiquities, that no notice should be taken of the skeleton recently disinterred at Leixlip. This extraordinary monument of a gigantic human stature was found by two laborers in Leixlip churchyard on Friday, the [...] when making a kind of sewer, near Salmon leap, for conveying water, by Mr. Haigh's orders. It appears to have belonged to a man of not less than ten feet in height. It is believed to be the same mentioned by Keating-Phelim O'Tool, buried in Leixlip churchyard, near the Salmon leap, 1,272 years ago. In the same place was found to be a large finger ring of pure gold. There was no inscription of characters of any kind upon it, a circumstance to be lamented, as it might throw a clear light upon this interesting subject. Our correspondent saw one of the teeth, which was as large as an ordinary forefinger.

The History of Ireland, Commencing with its Earliest Period, 1843
The ruinous remains of a circular temple near Dundalk, formed a part, it is supposed of a great work like that of Stonehenge, being open, as we are told to the east, and composed of similar circles of stone within. One of the old English traditions respecting Stonehenge is, that the stones were transported thither from Ireland, having been brought to the latter country by giants from the extremities of Africa; and in the time of Giraldus Cambrensis there was still to be seen, as he tells, on the plain of Kildare, an immense monument of stones corresponding exactly in appearance and construction with that of Stonehenge.

Cymmroder Magazine of the Honourable Society of Cymmrodorion, 1888
Exploration of another class of ancient tombs has brought to light the remains of an equally defined race of widely different physical characteristics, which settled in remote antiquity upon the coast of west and north-west Europe, and penetrated into our isles at long subsequent period. It came here as a conqueror, and afterwards mixing with the people I have described. It was a tall race, large boned, and presumably of great muscular strength; with a round skull, beetling eyebrows, promininant cheekbones and massive jaws. Its affinities are a matter on which anthropologist are not agreed, and it is bur a bare surmise which identifies it with the material of the earlier Celtic or Goidelic immigration.

Portrait of The Isle of Man, 1958
The Clovin Stones or Giants Grave, Baldrin, Lonan, have always been a puzzle. The site is now surrounded by houses and bungalows, so all that can be written about it must be taken from past records. In the Swarbreck MS (1815) in Manx Museum it is recorded " Mr. Millburne informed us that about seven years since, he with two or three miners opened the mound to the depth of five feet, and discovered a human skull and some thigh bones, which from their uncommon size must have belonged to a person of gigantic stature. When surveyed in 1865 a plan was drawn, showing it to be an almost circular barrow, with two compartment gallery grave. On the N.E. edge of the barrow there are two pillars 6 to 7 feet high, which presumably formed the entrance approach to the chambers. The taller stone is split from top to bottom along a cleavage line of the stone apparently while it was in situ. But the split has given rise to many traditions, the one most often told being that on this stone King Orry tried the strength of his sword, and presumably the strength of his arm, for he split it at a single blow.

The History of Powys Fadog, 1885

The oblong tumuli or long barrows that are found in almost all parts of the kingdom are the burial places of those inhabitants of this island who lived in the Neolithic age. The bones found in these tumuli were those of a short dolichocephalic race- that is, a race whose skulls were long and narrow- and the implements buried with them were either of stone or flint. The skeletons resemble in all respects those found in the graves of Gibralter, an account of which has been given in a work entitled "Cave Hunting", by Professor Boyd Dawkins, who states that these remains are those of the ancient Iberians, who in ancient times crossed over from Spain and occupied Britain. From the sepulcher discoveries, it appears that the Neolithic tribes occupied the whole of Britain themselves for perhaps many ages. All the short and dark races, such as Silurians, whether long headed or round-skulls, are treated as descendants of a primitive non-Aryan stock including "broad-headed Welshman, and the broad-headed dark Frenchman, and connected by blood not only with the modern Basque, but with the ancient and little known Ligurian and Etruscan races.

Subsequently, however, the Neolithic tribes were invaded by men of a different race, whose remains we find buried in round barrows. From these remains we find that the invaders were a tall race of men, with short, round, or brachycephalic skulls, and that all their weapons were made of bronze. These bronze weapons are always found buried with them, whether we find them buried with the Neolithic race, or separately in their round tumuli.

The Scottish Gael, 1876

The strong and robust bodies of the Celtae, their comeliness and great strength, have been remarked by all ancient authors who have had occasion to notice them. These qualifications must have been produced by a sufficient supply of food, by their temperance, and by the freedom and activity of their lives: hunting, pasturage, agriculture, and athletic amusements, being almost their sole occupation, when not engaged in warfare.

Both Celts and Germans were remarkably tall. They surpassed all other men in stature; and the largest, who were called Barenses, inhabited the extreme and most cold parts. The lowest of the Germans were taller than the tallest Romans. Hieronymus says, Gaul always abounded in great and strong men, who were wont to ridicule other people on their diminutive size. The Senones were particularly remarkable, being terrible for their astonishing bigness and vast arms. The Insubres are described as more than human. The Britons appear to have exceeded even the Gauls in height. Tacitus remarks the large limbs of the Caledonians; and some prisoners that Caesar carried to Rome, were exhibited as curiosities for their prodigious size. Strabo indeed says, that he had seen British young men at Rome, who stood half-afoot above the tallest men; The Celts were, however generally admired for their fine figures, as we learn from Polybius, Arrian, and other. Tacitus notices the advantage, which this height gave the enemy on occasion of crossing a river: while the Romans were in risk of being swept away, the Germans could keep themselves easily above the water. These people were celebrated for their strength, their stature, and their sinewy bodies, the Romans being certainly of inferior size compared with the barbarians...

Tall as the Celtae generally were, the princes and chief men usually exceeded the common people, both in stature and strength; for beauty and stateliness of person were generally characteristics of nobility in early society, and naturally proceeded from the constitution of a rude community, where superior strength and warlike accomplishments are the only recommendations in a chief or leader, and as they intermarry with families enjoying similar advantages, the race does not degenerate.

Prehistoric Mining in North America

Extensive ancient mines have been discovered in the Great Lakes region that have never been recognized as the works of the later Europeans. The ruins are a testament of the great effort that was employed to extract copper, lead and oil. Historical accounts of canals extending many miles, are the remnants of a once busseling commerce in North America.

The most extensive mines are found in the Lake Superior region of Isles St. Royal. Massive amounts of copper was removed from this region, with no evidence of its final destination. Copper artifacts are found the extent of the Great Lakes region associated with the Copper People. The artifacts date from 2700 B.C – 1500 B.C., which corresponds to the Megalithic and subsequent Beaker Culture influence in the British Isles and northern Europe and also with the Early to Late Bronze Age. Tanged and socketed daggers and spears were manufactured by the Copper people, but the amount of artifacts found doesn't come close to accounting for all of the copper removed from Isle St. Royal.

The weapons found that are associated with the Copper Culture are identical to what was being manufactured for the Hittite and Egyptian armies. The earliest of these weapons were the tanged daggers, that were made of copper before transitioning to bronze. There have been a few historians that have reported finding bronze weapons, but this is a rarity, most were of copper. An explanation of why bronze weapons were not found in greater numbers would be consistent with Phoenician/Amorites foreign trade policy that required the latest in weapons technology would not be shared in areas where mineral resources were being exploited.

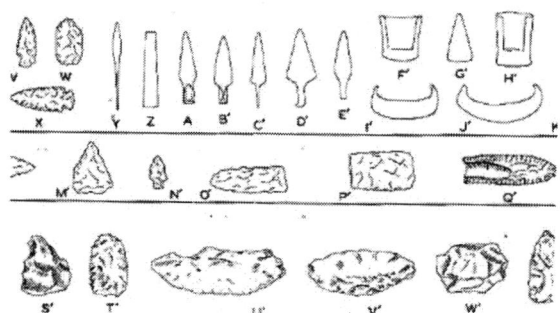

This is an implement sequence from the Middle to Late Archaic Periods from *Archaeology of the Northeastern United States,* MacNeish, 1952. The tanged and socketed daggers and spear heads shown were made from copper. How did weapons technology go instantaneously from Stone Age to Bronze Age?

Wisconsin surface find of a copper "tanged dagger" with a pronounced mid-rib for additional strength. It is more typical for a civilization or people to first produce metals into ornaments before making tools or weapons. Looking at these weapons in a global perspective in order to date when they were manufactured; the copper tanged dagger and spearheads were developed around 2,700 B.C in the Mid-East. By 2,000 B.C. the mid-rib becomes prevalent dagger and spear type. From *The Wisconsin Archaeologists, Vol. 48*

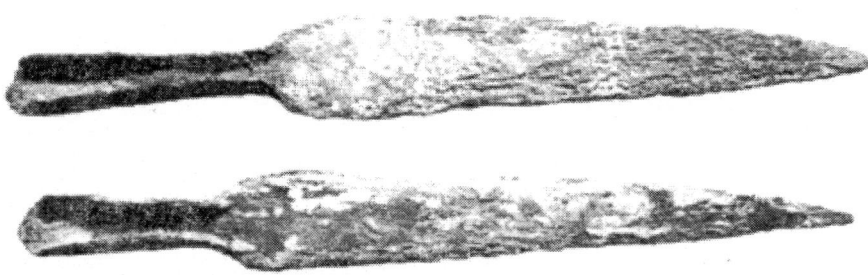

Spearheads from Wisconsin with a socket for hafting to a shaft. The socket was a revolutionary invention, not only for battle, but also for farming. The socket was in use in the Middle East around 1500 B.C. At about the same time mining ceased in the Lake Superior region. *The Wisconsin Archaeologists, Vol. 48*

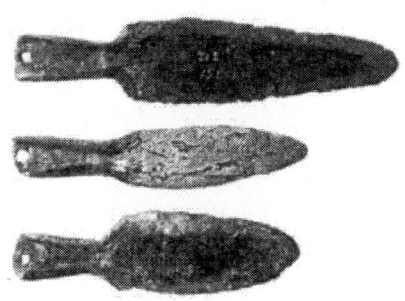

An improvement on the socketed spear heads was the addition of rivit holes for reinforcing the attachment to the spear. These spear heads were found in Wisconsin. *The Wisconsin Archaeologists, Vol. 48*

The Copper Culture is within the Lake Forest Tradition, that also includes another sub group or "phase" called the Brewerton. Their remains have been dated from 3,000-2,500B.C. William Ritchie formulated the Brewerton Phase of the Lake Forest Tradition in the 1930s. According to Ritchie, the distribution was chiefly in New York and southern Ontario. Associated with the Brewerton are winged bannerstones, polished gouges, adzes, celts, slate arrows and spears, plummets, bone awls gouges, mullers and shallow mortars. Many times, copper weapons are found within their burials.

The slate arrows, bone awls and plummets are similar to those associated with the Maritime Archaic. There is some evidence that the Brewerton may represent early Allegewi. Their head types (brachycephalic) are the same, as is their gigantic stature. A Brewerton point was found in the Allegewi, Daines mound in Athens County, Ohio.

Skeletal reamains of the Brewerton are slight, however one of these burials was discovered in New York that has similarities with the later Allegewi. These similarities being large skeletons that were extended, and in a spoked position within the burial vault. The copper weapons from the burial would date them much earlier than the accepted beginning of the Allegewi mounds in the Ohio valley, that date around 1000 B.C.

***History of New York, from "Prehistoric Man,"* 1877**

In 1856, Dr. Thomas Reynolds of Brockville exhibited to the Canadian Institute a collection of copper and other relics discovered in that neighborhood under singular circumstances; and possessing a special interest owing to the distance of the site from Lake Superior. They included a peculiarly-shaped chisel or gouge, six inches in length […], a rude spearhead, seven inches long […], and the small daggers or knives, […] all wrought by means of the hammer out of native copper which had been subjected to fire, as is proved by the silver remaining in detached crystals in the copper. They were found at the head of Les Galops Rapids, on the river St. Lawrence, about fifteen feet below the surface, along with twenty skeletons disposed in a circular space with their feet toward the center. Dr. Reynolds remarks of them: "Some of the skeletons were of gigantic proportions. The lower jaw of one is sufficiently large to surround the corresponding bone of an adult of our present generation." The condition of the bones furnished indisputable proof of their great antiquity. The skulls were so completely reduced to their earthly constituents that they were exceedingly brittle, and fell in pieces when removed and exposed to the atmosphere.

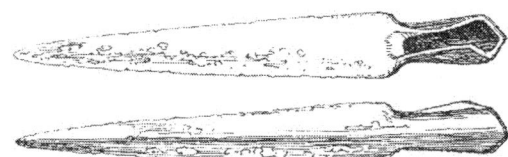

Copper socketed spear heads with mid-rib was found with the burials at Brockville.

The appearance of weapons technology that is parallel with those found in the Medeteranean and being disseminated by the Amorites and Beaker People into England is not serendipitous. Additional evidence of migrations to North America, were the discovery of mines that have no historic precedent. Two of the most sought after materials in the Bronze Age were copper and lead. The amount of copper being mined out of the Lake Superior region was summarized in, *"Science Frontier Online"*, originally published in *Ancient American*, September 1993.

"For some 1800 years, beginning abruptly about 3000 BC, some industrious peoples mined ore equivalent to 500,000 tons of copper from Michigan's Isle Royale and Kewenaw Peninsula. Who were these mysterious miners, and what happened to all that copper? It certainly hasn't been found in the relics of North American Indians. And where was the ore smelted? About all the unidentified miners left behind are some of the crude tools they used to pound out chunks of ore from their pit mines (5000 pit mines on Isle Royale alone). Outside of some cairns and slab rock ruins, there is little to help pin down these miners. Mainstream archaeologist attribute all these immense labors to a North American "Copper Culture"--certainly not to copper-hungry visitors from foreign shores. Admittedly, many copper artifacts have been dug up from North American mounds, but only a tiny fraction of the metal the Michigan mines must have yielded.

Curiously, North American Indian mounds have contained copper sheets made in the shape of an animal hide. Called "reels" their function, if any, is unknown. The reels do, however, resemble oddly shaped copper ingots common in European Bronze Age commerce. Their peculiar shape earned these ingots the name "oxhides." They have been found in Bronze Age shipwrecks, and are even said to be portrayed in wall paintings in Egyptian tombs. The standardized hide-like shape, with its four convenient handles, was useful in carrying and stacking the heavy ingots. Could the reels from the North American mounds have been copied from the oxhides?"

In the following report is a description of a copper mine, where a six ton piece of copper was being

removed before being abandoned. This is evidence of the large scale mining that was taking place, that

far exceeds the amount of copper weapons that have been found associated with the Copper Culture.

The technique of lighting fires to heat the copper and then pouring water over the heated metal to make

it crack is the same procedures used in the early Amorite mines in Europe.

Historical Collections of the State of Pennsylvania, 1843

From these considerations alone we are readily infer that the Mound Builders either engaged in mining or else trafficked with those nations who did so engage. In 1847, about one year before the ancient copper mines were discovered, it was pointed out that the probable source of the copper and silver was the region of Lake Superior.

The discovery of the ancient mines has set all the speculations to rest. Copper Mines: In the copper regions of Lake Superior have been found numerous excavations in the solid rock from which the copper has been extracted. Upon examination, it has been discovered that the whole extent of the copper-bearing region was resorted to by this ancient race. The ancient trenches and pits were found to be filled even with the surrounding country, and were not detected until many years after the region had been thrown open to actual exploration. Mining began effectively in 1845, and it was not until 1848 that S.O. Knapp, then the agent of the Minnesota Mining Company, made the discovery. In passing over a portion of the company's grounds, in the winter 1847-8, he observed a continuous depression in the soil which he conjectured was formed by the disintegration of a vein. Followed up the indications, he came to a cavern where he noticed evidences of artificial excavation. On clearing out the rubbish, afterwards, he found numerous stone hammers, and at the bottom was seen a vien with ragged projections which the ancient miners had not detected.

Two and a half miles east of the Ontonagon River (the center of the great copper region of Michigan) is the Minnesota mine illustrated. This shaft is situated in a wall of rock of compact trap. The excavation reached a depth of twenty six feet, which was filled up with clay and a matted mass of mouldering vegetable matter. At a depth of eighteen feet, among a mass of leaves, sticks and water, Mr. Knapp discovered a detached mass of copper weighing six tons. This mass had been raised about five feet, along the foot of the lode, on timbers by means of wedges, and was left upon a cob-work of logs. These logs were from six to eight inches in diameter, the ends of which plainly showed the marks of a cutting tool. The upper surface and edges of the mass of copper were beaten and pounded smooth, showing that the irregular protruding pieces had been broken off. Near it were found other masses. On the walls of the shaft were marks of fire. Besides charcoal there was found a stone sledge weighing thirty-six pounds, and a copper maul weighing twenty-five pounds. Stone mauls, ashes and charcoal have been found in all these mines. In further explanation of the engraving, the letter *b*, represents the original matter thrown out by the ancient miners; *a,* the angle of the shaft; *d,* three masses of copper.

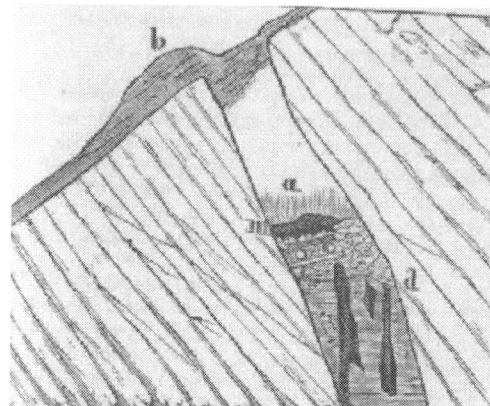

On the island known as Isle Royale, near the northern shore of Lake Superior, these ancient works of man are very extensive, and some of the pits are sixty feet in depth. On opening one of these pits of this island it was discovered that the mine had been worked through solid rock to the depth of nine feet. At the bottom was vein of pure copper eighteen inches thick. The works are scattered throughout the island, and are located on the richest veins. These miners were intelligent and experienced, for they not only showed rare powers of observation in locating the veins, but also displayed much knowledge in following them up when interrupted. The excavations are connected underground, and drains are cut into the rock to carry off the water. At one point the excavations extend for over two miles in a nearly continuous direction.

In these ancient mines have been found wooden shovels, used in scraping away the soil. Wooden bowls and troughs of cedar occur. From the splintered pieces of rock embedded in the rim of some of these bowls, it is inferred that they were used for bailing our the water; and as charcoal also occurs, it may be inferred that the rocks were heated and then water was dashed on in order to shatter and destroy the cohesion between the particles. Stone hammers, or mauls, and copper are frequently met with..
The wide distribution of the copper implements shows that an extensive business was carried on, and to penetrate to Lake Superior, from the valley of the Ohio, required a journal of a thousand miles, which must have performed during the summer. With them they must have carried their provisions, as there is no evidence of a settled life in that region, such as mounds, village plats, etc. The climate is too cold for the maturity of Indian corn, and hence it was necessary to go in well-organized companies.

Bronze Age traders of metals in the Medeteranean were seeking lead. There is no historical records

of historic Native Americans mining or using lead, which makes the following historical accounts even

more puzzling, unless we consider the possibility that metals were mined and sent to ready markets in

Europe and the Levant..

A remarkable, irregular trench, the vestiges of which can yet be seen, with occasional interruptions, runs from the upper lead mines to the neighborhood of the lower; it is at least six miles in length. It was found there by the earliest emigrants, and thirty years ago, stout trees grew on the banks of earth thrown out in excavating it. It was there, it is said, and ancient in its appearance, when Roberdeau erected or commanded the fort at the upper lead mines.

The American Antiquarian, **January 1889**
Ancient Mining in North America, by J. S. Newberry

The finding of various implements made of the more precious minerals would not in itself indicate that the Mound Builders engaged in mining. Copper was extensively used, and yet this material has been found in various localities. It has been found in pieces of several pounds weight in the valley of the Connecticut and near New Haven, where a mass was found in weighing ninety pounds. It is found in small pieces in New Jersey, Indiana and Illinois. Small pieces have been picked up in various localities in Ohio. It is probable that the copper found in Connecticut and New Jersey originates from the red sandstone formation, while in Indiana and Illinois it was deposited during the drift. In Ohio it has possibly been dropped by the hand of man. What is true of copper is not so of galena, obsidian, mica and silver.

Considerable quantities of galena have been found in the mounds of Ohio. Upon one of the alters within a mound in "Mound City" (three miles above Chillicothe) a quantity of galena was found, which had been exposed to the action of fire. It is frequent occurrence on the sacrificial alters, and met with in quantities of thirty pounds weight. "Plumb bobs" and net-sinkers are met with made out of this material, and yet no original deposits are known in the State of Ohio.

Obsidian, a peculiar glass-like stone of volcanic origin, is obtained from some of the mounds, but in very small quantities, and in the shape of arrow and spear-points and cutting implements. This mineral has not been met in situ north of Mexico and east of the Rocky Mountains.

Mica (commonly called isinglass) has been taken in large quantities from the mounds, and often ploughed up in the neighborhood of the enclosures. In these sepulchral mound in the center of the earth-work at Circleville, Ohio, there was taken out a sheet of mica three feet long, one foot and a half wide, and one inch and a half in thickness. In the year 1828, in one of the low mounds near Newark, Ohio, regular layers of mica plates, from eight to ten inches in length, four or five inches wide, and from half an inch to one inch in thickness, were found covering fourteen human skeletons in an advanced state of decomposition. From this mound there were taken about twenty bushels of mica. As mica is found in large quantities, and carefully laid away in the mounds, it is evident that it was regarded as of great value. It was used for mirrors, ornaments, and, as it has been found covering the skeleton, may have been looked upon as having supernatural properties. Mica is found in New Hampshire and North Carolina. In the former State it has been found from two to three feet in diameter; but there is no evidence that the Mound Builders penetrated that far east neither have any ancient mines been discovered there. Traces of wrought silver have been found, but they are exceedingly scarce, and constituted no technical importance among them.

The polished stone implements composed of a greenish slate of close grain have already received attention. This stone is not found in original deposits in the valley of the Mississippi, unless it be upon the rim of the basin. It belongs to the oldest sedimentary formation and occurs in considerable masses along the Atlantic coast, and has been observed from Rhode Island to Canada.

From these considerations alone we are readily infer that the Mound Builders either engaged in mining or else trafficked with those nations who did so engage. In 1847, about one year before the ancient copper mines were discovered, it was pointed out that the probable source of the copper and silver was the region of Lake Superior.

I have been much interested in reading the article on "Ancient Mining in America," by E.P. Appy, in March number of The Antiquarian, and I take the liberty of reporting to you some facts bearing on the subject with which he seems not to have been familiar.

The ancient copper mines on Lake Superior have been fully described by many writers. I have been much in that country and can testify to the accuracy of the descriptions of the ancient copper mines given by Whittlesy, Foster and others, as well as the review of the subject now presented by Mr. Appy. I will only add that so far as my observation has extended all the ancient workings on Lake Superior were abandoned many hundred years ago, for the heaps of debris that surround the pits made by the ancient miners were covered with forest trees which had obtained their maximum size, and I have heard of any of the old miners which did not show evidence of abandonment at least four hundred years ago.

The old mica mines of North Carolina and the quarries of serpentine in the Alleghenies, worked by the ancient inhabitants to procure materials for their pots, pipes, ect., show the same rude processes and I may add the same antiquity as the copper mines of Lake superior, for they, too, were overgrown by what seemed primeval forest when first visited by the whites.

To all the evidences of ancient mining industry in our country cited by Mr. Appy, I will add that some population of the Mississippi valley in ancient times worked our oil fields in many places, and at least on one case opened and extensively worked a vein of lead. This lead vein is situated on the Morgan farm, about six miles northeast of Lexington, Kentucky. Part of the area traversed by it has been long cultivated and the evidences of excavation have been thereby to some extent obliterated, but a part of the course of the vein runs through a tract of woodland which has never been touched by the axe. Here the ancient working is in the form of an open cut, six to ten feet wide, of unknown depth, and now nearly filled with rubbish. On either side of this trench the material thrown out forms ridges several feet in height, and these are everywhere overgrown by trees, many of which are as large as any found in the forests of that section.

In regard to the working of our oil fields in former times, I would say that I have found conclusive evidence that wells were sunk and oil collected on Oil Creek, near Titusville, Pennsylvania, In Mecca Ohio and at Enneskillen, Canada. In 1860 the first fountain well was opened by Brewer and Watson just below Titusville. I then resided in Cleveland, Ohio, and went to Titusville to examine the interesting geological phenomena presented by the newly-opened wells. In passing down the valley of Oil Creek, I noticed that the surface of the ground was pitted in a peculiar way; it was in places completely occupied by shallow depressions, ten to fifteen feet across and from one to three feet in depth. At first I thought they must have been produced by a wind-fall, in which the trees were all up-rooted, but I was familiar with the character of the depression made by the overturning of a large forest tree, and knew that the pit thus formed was oval, with a ridge on one side and none on the other. These pits were, however, quite symmetrical and were a puzzle to me. While I was talking with Mr. Brewer or Mr. Watson about them and asking questions to which I got no satisfactory answers, a man standing near told me if I would go with him to his well one hundred yards away the mystery would be solved. I did so, and found that he had begun the excavation of a well in one of these pits, and had sunk through the superficial material some twenty-five feet to the rock where he was to begin drilling. In sinking his pit followed down an old well, cribbed up with timber, and in it stood a primitive ladder, such as was so often found in the old copper mines of Lake Superior; a tree of moderate size, with many branches, had been felled and the limbs cut off a few inches from the trunk, thus forming a series of steps by which one could ascend or descend. The cribbing of the ancient well was rudely done with sticks six to eight inches in diameter, either split from a larger trunk or lengths cut from a smaller one. The sticks had been cut by a very dull instrument, undoubtedly a stone hatchet.

The method of gathering the oil practiced by the ancient inhabitants was evidently that followed in the Caspian region up to the time when the American method of drilling and pumping was introduced, viz.: a pit sunk in the earth, and the oil skimmed from the water.

What use was made of the oil we can only conjecture, possibly it was employed only medicinally, as the oil from the spring at Cuba, New York, was used by the Indians in that region; possibly for burning, as petroleum has been used from time immemorial in Persia, India and China. The large number of pits sunk in the valley of Oil Creek indicated, however, that the quantity taken out was large and that the oil served some important purpose among the ancient people. The pits described above were located in a dense hemlock forest in which many of the trees were three feet and more in diameter.

The amount of pits that were excavated would indicate that large amounts of oil was being removed.

It would seem probable that this oil was for the use of lamps. Several oil lamps have been found, that

are believed to belong to the Allegewi Hopewell. Lamps, like many every day items were not included

with burials in mounds. Archaeologists, while destroying hundreds of Allegewi Hopewell burial

mounds, have never excavated, nor looked for any Allegewi Hopewell village sites. Because this

important anthropological aspect has been completely ignored by university archaeologist, many

artifacts used in everyday life have never been found.

The following articles describe the oil pits found in Pennsylvania and New York.

In French Creek Valley (Pennsylvania) 1938

In writing of his visit to the pits near Titusville, William Reynalds quoted a part of an address deleivered in 1843 by William H. Davis: A short distance below the village of Titusville, and on the west side of Oil Creek, there are perhaps about two thousand pits, scattered over a level plain not exceeding five hundred acres. Some of them are very close together; as close as the vats in a tan yard, which they somewhat resemble, each being seven or eight feet long, four wide and six feet deep. These pits had nearly all bee filled; some of them entirely so by vegetable deposit, perhaps the accumulation of ages. The mounds raised at the sides of the pits by the excavation of the earth from it are distinctly visible. Close upon the margin of them on the very mounds of the earth excavated are trees whose size indicates an age of two or three hundred years.

The early settlers first discovered the pits from the regularity of size. They were induced to open them and found that each pit was walled with logs regularly cut and halved at the ends so that they could lie close together. It was found that the water rose in the bottom of these pits and in a few days would be covered with oil to the depth of a third or half inch.

Of the many pits that were cleaned out at this time and later, all were found to have been built and walled in like manner. Their probable purpose was for skimming what the first settlers knew as Seneca Oil. The Indians had no traditions about them and there are no records to show that they had been constructed by the French during their occupation.

History of Venango County, Pennsylvania, 1879

It is said large and deep pits exist in the region of Cherrytree, in this county, which were undoubtedly made long ago, by whom and for what purpose is not known, but probably to obtain some mineral. It is well known that when the English came here, they found deep pits along Oil Creek, in which large trees were growing, which appeared to have been used as reservoirs for the collection of petroleum, and fragments of notched logs, which had been used as ladders in obtaining the oil were found in them, which had been preserved from the entire decay by the saturation of the oil. Mr Henry, in his "History of Petroleum" states: " There is reason to believe that at some former period in the history of the American continent, the existence and uses of petroleum had been better understood than they were for some centuries before the recent artesian developments. The numerous pits, until recently, and perhaps even still to be seen cribbed with roughly hewn timber, but nearly hidden by the rubbish of ages, indicated a development comparatively extensive. Trees were found growing in the center of some of these pits, which we are told, on the evidence of the concentric circles in the wood, were shown to be the growth of centuries. Many circumstances concur in referring these excavations to a period of time, and to a race of people, who occupied the country prior to the advent of those aborigines, found here by our Latin or Saxon ancestors. They were, probably the work of that mysterious people who left the traces of their rude civilization in the copper mines about Lake Superior and the mounds of the Southwest. The mound builders were and offered sacrifices to the Sun God, and were, in these respects, like the people of Mexico at the time of the Spanish invasion. It is speculating too musch to say that they made annual or more frequent journeys here to obtain petroleum to maintain the perpetual fire used their worship and sacrifice?

History of Cattaraugas County, New York, 1879

Another early writer mentions that "numerous pits were found along Oil Creek and the Allegheny, cribbed with logs many years before discovery; and in the center of some of these pits trees were growing centuries old."

History of Crawford County Pennsylvania, 1885
Who Built the Oil Pits?

On an extensive plain near Oil Creek, there is a vast mound of stones, containing many hundred thousand cart loads. This pyramid has stood through so many ages that it is now covered with soil, and from its top rises a noble pine tree, the roots of which, running down the sides, fasten themselves in the earth below. The stones are, many of them, so large that two men can scarcely move them, and are unlike any in the neighborhood; nor are there quarries near, from which so large a quantity could be taken. The stones were perhaps, collected from the surface, and the mound one of many that have been raised by the ancient race which preceded the Indians, whom the Europeans have not known. These monuments are numerous further north and east, and in the south and west are far greater, more artificial and imposing.

When first visited by the whites in 1787, in the valley of French Creek, were old meadows destitute of trees, and covered with long, wild grass and herbage resembling the prairies; but by whom those lands were originally cleared will probably forever remain a matter of uncertainty.

The Indians alleged that the work had not been done by them; but a tradition among them attributed it to a larger and more powerful race of inhabitants, who had pre-occupied the country.

The following accounts are of canals that were observed with no explanation as to whom the builders were. It is assumed that these works were the efforts of the Allegewi-Hopewell mound builders. The historical accounts would add to the belief that a vigorous economy was in place that would necessitiate such an undertaking as building a canal.

Biographical and Historical Memoirs of the Mississippi, 1891

The great ditch extending from a point below Cape Girardeau, Mo., to the headwaters of the White and St. Francis rivers was excavated in prehistoric days, and was old when Indian legend first refers to it. Whether the object was to use this great canal for purposes of navigation or simply for drainage can never be known. The ancient inhabitants, whether Mobolians or Peruvians, may have known the rich valley of the Nile, of the artificial ponds or lakes and canals used at the time to regulate its high waters and resorted to the same plan here for controlling the mighty Mississippi.

History of Mifflin County Illinois, 1905

On the banks of Green River, in Henry County in Illinois, are traces of an ancient city, which was once the abode of a commercial people, and points to a time when the Rock River was a navigable stream of some commercial importance. A canal connected these two rivers some three miles above the junction. This canal is about a mile and a half long and is perfectly straight for about one-fourth of a mile from the Green River end; it is then relieved by a perfectly easy curve, reaching the Rock River at a bend, and showing that the engineering was done in a masterly manner. The soil is of a very fine texture, mixed with a ferruginous mineral deposit; hence its firmness, and the reason of it withstanding the washings of rains, for this great lapse of time. About twelve miles back and above this canal is another partly natural and partly artificial connecting Rock and Mississippi Rivers. This is so well preserved that about twelve years ago the "Serling" a small Rock River steamer, passed through it into the Mississippi river. These works are as old as the mountains of Egypt, and were in all probability built by a contemporaneous people.

American Antiquarian, Volume V, 1885
Notes and Queries

Rumors of finds have come to us at various times, which we mention with the suggestion that they be followed up and confirmed. Continuous lines of pavements, or rows of burned stone, forty or fifty yards long, on the banks of the Ohio River, twenty inches to five feet, below the top of the bank. The pavements, called macadamized roads, have been noticed on the banks of the Allegheny River. These are probably the fire beds. Has any one evidence to the contrary? A prehistoric canal, connecting two rivers in Illinois; also, several pieces of masonry, exposed in a sand bank after a storm, and covered up again.

History of Greenup County, Kentucky, 1951

At Springfield there is a large enclosure with walls plainly discernible, and it is said to have been an Indian town having an underground opening to the Ohio River.

North American Megaliths

The lack of megalithic remains in North America may be due to the Beaker Peoples arrival in Britain as early as 2700 B.C., leaving a small window of time for megaliths to have been constructed in North America. Standing stones or menhirs have been observed in close proximity to the Red Paint People's graves in Labrador and British Columbia, but are isolated to the extreme northeast and northwestern coasts of North America. One of the Red Paint People's mounds contained a Mycenean doorway lintal with a passage grave at Nulaik Cove, Labrador. Quartz crystals that are common in megalithic graves across Europe, have also been found early North American, earthen and shell mounds. The megalithic builders believed that departed souls went to the "mother" moon. Quartz rocks were believed to be parts of the moon that had found there way to earth.

The most megalithic-like sites, east of the Mississippi are found in the Ohio Valley, in lands occupied by the Allegewi. A stone circle was described by the Bureau of Ethnology in West Virginia, on the Kanawha River that was unique, with nothing like it described elsewhere. The work described in Ross County, Ohio is reminiscent of Og's Circle in Palestine.

Bureau of Ethnology, 12th Annual Report
West Virginia
Below the mouth of the Kanawha the caving in of the bank of the Ohio had exposed a wall of stone, on some of the slabs of which were rude totemic and other marks made be some pecking tool. Careful excavations revealed a circular enclosure about 100 feet in diameter, inside measurement. The wall was composed of angular slabs of various sizes from the hills near by and averaged 25 feet across the base by 3 1/2 in height. Many of the stones bore evidences of fire, the spaces between them (they were laid flat with joints broken) being filled with charcoal, ashes, and earth, separate or mixed. No gateway was found, though no doubt one exist at some point not excavated. The sediment from the overflows has accumulated to the depth of about 5 feet since the wall was built, and its existence was never suspected until exposed as above stated by the falling in of the bank. This may not be aboriginal work.

To the south of this, about two hundred yards, is a stone circle one hundred feet in diameter and five or six feet high. In the center of this is a large stone mound some ten feet high.

The Allegewi mound in Ross County, Ohio was surrounded by a stone wall and falls within the grey area of whether it is megalithic or more similar to mounds constructed by the Corded People. Of coarse, the Corded People have their roots with the megalithic Amorites. Henges in the Ohio Valley, were a continuation of the megalithic cultures in the England, that was adopted by the Beaker People. Burial mounds surrounded by a curb of stones have their origins with the Amorites. Some Hopewell mounds have been discovered which contained vaults containing numerous burials, that are similar to the European long mound, megalithic vaults.

West of the Mississippi, contains more evidence of megalithic builders that may stem from the continued migrations from the Island of Hokkaido, Japan; home of the Joman and Ainu. As in Europe, the Joman transitioned from internments in shell mounds to megalithic burials, with the arrival of the Amorites. A connection between the megalithic remains in the British Isles and Japan was reported in *Archaic England,* Bayley, 1920. "Josephus mentions that the Scythians were called Magogoei by the Greeks: by some authorities the Scythians [Corded People] are equated with the Scotto or Scots. There are still living in Cornwall the presumed descendants of what have been termed the "bedrock" race, and that these people still exhibit in their physiogomies the traces of Oriental or Mongoloid blood. The early passage tombs of Japan are, according to Borlase, literally counterpart in plan and construction to those giant-graves or passage tombs which are prevalent in Cornwall."

Physical evidence that the megalithic builders had spread their culture into Japan was revealed when stone circles were discovered on the island of Hokkaido.

Stars and Stripes, **March, 27, 1957**
A German Missionary Speaks for the Land of the Ainu

OTURU, JAPAN-HOKKAIDO is a strange fascinating island where legends abound. Here, where the modern Orient meets one of the world's oldest cultures, one finds American silos and red barns mixed with hodgepodge with the homes of the ancient Ainu.

Here, in Japan's Yukon, one finds well -planned cities with broad streets and ginger bread houses just a few miles from the mysterious Ranjima stone circles; circles similar to the Druid stones of ancient England.

Since there is no written Ainu language, the history is vague. Father Huber learned much of their religion from an Ainu youth he befriended. He reconstructed much of their history and culture from the religious songs the boy taught him.

One story concerns the stone circles which, he believes, may be connected with the Ainu fire god Oina-Kamui, supposedly the Ainu creator. This god subdued the Ainu evil spirits, imprisoning them in six stone boxes, according to the story. The Ranjima stones form six concentric circles, indicating a possible correlation.

But, too, he said, they may have been a timing device. Because of the stone's relative geographic position, they form something of a sundial for the seasons

Stone circles have been reported on the islands off of British Columbia in Canada, that are a likely extension of the Megalithic Builders, that had spread into in Japan. The large stones used in these works, the receptacles for the dead, the inclusions of quartz in graves are comparable to the Megalithic Builders.

Bancroft's Native Races, **1882**
British Columbia

In such localities, the general feature of the landscape is very similar to many parts of Devonshire, more especially to that on the eastern escarpment of Dartmoor, and the resemblance is rendered the more striking by the numerous stone circles, which lie scattered around. These stone circles point to a period in ethnological history, which has no longer a place in the memory of man. Scattered in irregular groups of from three or four, to fifty or more, these stone circles are found, crowning the rounded promitories over all the South Eastern end of the Island. Their dimensions vary in diameter from three to eighteen feet; of some, only a simple ring of stones marking the outlines now remains. In other instances the circle is not only complete in outline, but is filled in, built up as it were, to a height of three to four feet, with masses of rock and loose stones, collected from amongst the numerous erratic boulders, which cover the surface of the country, and from the gravel of the boulder drift which fills up many of the hollows. These structures are of considerable antiquity, and whatever they have been intended for, have been long disused, for, through the centre of many, the pine, the oak, and the arbutus have shot up and attained considerable dimensions-a full growth. The Indians when questioned, can give no further account of the matter, than that, "it belonged to the old people" and an examination, by taking some of the largest circles to pieces, and digging beneath, throws no light on the subject.

Other British Columbian antiquities consist of shell mounds, burial mounds, and earth-works, chiefly confined to Vancouver Island, and known to me through the investigations and writing of Mr. James Dean."

Burial mounds on Vancouver Island are of two classes, according as they are constructed chiefly of sand and gravel or of stones. One of the first class opened by Mr. Deans in 1871, will illustrate the construction of all. It was located on the second terrace from the sea, the terraces having nearly perpendicular banks of fifty and sixty feet respectively. By carefully cut drift through the center, it was ascertained to have been made in the following manner. First, a circle sixteen feet in diameter was marked out, and the top soil cleared off within the circle; then a basin-shaped hole, six feet in diameter, smaller at the bottom than at the top, was dug the centre, in which the skull, face down, and the larger unburned bones were placed and covered with six inches of earth. On the layer of earth rested a large flat stone, on which were heaped up loose stones, the heap extending about a foot beyond the circumference of the central hole. Outside of this heap, on the surface, a space two feet wide extending round the whole circumference was sprinkled with ashes, and contained a few bones also. Outside of this space again, large stones two or three feet long were set up in the ground like pillars, five feet apart, round the circumference; and finally the earth dug from the central hole, or receptacle for the bones, was thrown into the outer circle, and gravel and sand added to the whole until the mound was five feet high, having rounded form. Four smaller mounds, six and ten feet in diameter, were opened in the same group, showing the same mode of construction, but somewhat less order."

The second class, or stone mounds, which are much more numerous than those of earth, differ but little from the others in their construction, except that the final additions to the mound were of stones instead of earth, and the stones about the circumference were flat and set up close together. A piece of quartz sometimes accompanies the bones, but no other relics are found. When the skeleton is deposited face down, as is usually the case, the skull is placed toward the south, or when in a sitting position, it faces the south, seeming in some cases to have been burned where it sat. In a few instances the skeleton, when it was but little burned, was lying on the left side. Some stones weighting over a ton are found over the human remains. Traces of cedar bark or boards are found in some of the cairns, in which the bone s were apparently enclosed; and in a few others a small empty chamber was formed over the flat covering stone.

The following article describes a stone circle near Mt. Shasta in California. This is more accurately described as a mound, encircled by a ditch with the bottom of the ditch being curbed with stones. No burials were found within the interior mound and the earthwork would appear to be "ceremonial" in purpose. The raised area being surrounded by a ditch is similar to, if not a henge.

Despite the poor quality of the photo, copied from microfiche, the central mound can be seen surrounded by a ditch. Within the ditch was the berm of stones. Also described were stone roads that connected the mounds in this group. This is similar to the earliest mounds in North America at Watson Break, Louisianna, dating as early as 3000 B.C.

Oakland Tribune, **December 22, 1976**

Legends of 'Ghosts' Mount Shasta Race Revived as U.C. Scientist Puzzle Origin of Mounds

MOUNT SHASTA,-Hoary legends of a supernatural race of people said to inhabit the icy slopes of Mount Shasta-tales long since thought discredited, were revived here today by the superstitious to interpret the mystery of the "Siskiyou Stone Circles," which science thus far had been unable to do.

Anthropologists from the University of California examined the mounds and their perimeters of stone mosaic this weekend and came away shaking their heads.

The Indians of the region deny they know anything of the mounds or their history. White men who were born here and remember the mounds from childhood can give no plausible explanation. A foot by foot search of the mound areas by men trained in tracing the activities of the ancients offered no clue. At least one local geologist insists that nature did not make the mounds.

A arrowhead has been picked up in the mound area, which lies near Bray and Tennant, north of Mount Shasta. It is a small arrow-tip chipped from obsidian, which was quarried from Glass Mountain, many miles to the east.

Each mound was the same, generally 60 feet in diameter with the earth rising in a near-perfect circle to the crest, approximately two feet above the surface of the surrounding terrain. Completely circling each mound was the stone path or mosaic.

The rocks of the mosaic unquestionably were picked up from among thousands of volcanic stones sprinkled over all the area. The curious thing is the manner in which they have been brought together around each mound. They seem to be set in a trench around the mound, with gravel and small rocks at the bottom of the trench graduating up to boulder size at the surface

The surface of the rock circle or path, is almost smooth, the rocks fitting like stepping stones. It is far easier to walk on the paths around the mounds, or the paths connecting one series of mounds with another, than to strike off across the fields where the rocks lay naturally.

Also pictured in this article was this stone covered with cup marks.

A similar structure was described by Bancroft in Colorado. A central mound that was surrounded by stones, that were embedded in the ground. The discovery of mica within the mound shows that the builders of this work, had contact with the east, where mica is found.

Bancroft's Native Races, 1882

About half a mile west of Golden City, Jefferson County, Colorado, Mr. Berthoud reports to the Smithsonian Institution the existence of some ancient remains, at the junction of two ravines. They consist of a central mound of granite and sand not over twelve inches high, with traces of five or six shallow pits about it; all surrounded by traces of a wall consisting of a circle of moss-covered rough stones partially imbeded in the soil. South of the central mound is a saucer shaped pit, measuring twelve feet in width and from fifteen to eighteen inches in depth. At this point buffalo-bones and fragments of antlers are plentiful, and pieces of flint with plates of mica have also been discovered..

Mr. Foster quotes from a Denver newspaper a report of large granite blocks, of the nature of 'dolmans' standing in an upright position, on the summit of the Snowy Range; and Taylor had heard through the newspapers of pyramids and bridges in this territory."

North of Colorado, in Saskatchewan another circle was reported. It is dated as early as 2500 B.C. , when the bolder mosaics and medicine wheels of Alberta, Manitoba, Saskatchewan, Wyoming and Montana were constructed.

Spade and Screen from the Saskatchewan Archaeological Society, October 1948

"In Saskatchewen, we have two places called Stonehenge and Standing Rock, but neither of them lives up to its name. However, in Manitoba, about 100 miles east of Winnepeg, there is a circle of large stones and in a pasture south-east of Swift Current there is an alignment of heaps of stones arranged in rows in two different directions. This land has not been broken and is still full of stones, so it is not likely that these structures were set up by early white settlers."

Medicine wheels followed a general plan of consisting of a central rock mound, with spokes or lines of stones coming out of the mound an extending to an outer circle of stones. The structure looking like a bicycle tire from the air. These spokes have been found to align to both solar, lunar and steller events.

Big Horn Medicine Wheel has been determined to have been constructed as a calender. The number of spokes extending from the central cairn are 28. This is an interesting similarity with Stonehenge that contained 56 Aubrey Holes. Professors Atkins and Hoyle of Cambridge University claimed that the Aubrey holes were used in computing the eclipses of the moon, that have a cycle of 56.

Some of the Medicine Wheels have been attributed to the Ojibwa Indians. There traditional homelands being far to the north would be consistent with the presence of the genetic marker of Haplo X within their population. It is possible that this genetic marker was spread by the Hunters and Fishers of Europe and Asia.

The standing stones along the northeast coasts and the passage grave with a Mycenean doorway lintel is evidence of a limited cultural influence from or by the Megalithic builders on the northeast coast. The megalithic culture is evident in the Japanese island of Hokkaido; home of the Joman and Ainu. With the identification of some of the skeletal remains in this region as Ainu, it suggests that migrations were taking place at least as long as the Megalithic period.

The mounds on Mt. Shasta and in Colorado, with their stone berm and ditch are most similar to the Corded People or Amorites. The use of stones weighing over a ton for the burial chamber has no equivalents, except with the megalithic builders. The discovery of quartz within the burial chamber is yet another similarity with the megalithic burials in Europe, and is persuasive evidence of a cultural exchange between the megalithic people of Europe and Asia and the North American continent.

The Sacred Marriage of Opposites
The Etymology of Og and Awa

Clues to the existence of an ancient people's presence or cultural influence can be revealed in language and place names they left behind. Ancient peoples named themselves, their cities and rivers after the gods and goddesses they worshiped. The most common and most ancient of the gods and goddesses was Hawwah/Avvah as the female Earth Goddess (Eve). The principle Earth Goddess names also appear to be Ge, Ra, Ma and Wa/Va.

The Father is Og/Oc Eloah/Ala as the Sky or Sun, male god. "In Hebrew Og is also understood to mean "he who goes in a circle," which is suggestive of the Sun or Eye of Heaven. The sun was the mighty, all-seeing *ogler* or *goggler* of the universe. Many times Og/Oc and Hawa/ Awa/ Ma are combined in words representing the *Sacred Marriage of Opposites; the Earth and Sky.*

The Key, 1969, " In attempting to forge a chain between the British Isles and the Medeteranean, in particular a strong bond between the Stonehenge/Avebury complex and the golden culture of Mycenae and Crete, there are two potentially vital links to examine. If you take the two separated words or names originally in "Havoc," Haue and Oc, the same names that predominate at Stonehenge and Avebury, and reverse them, you have, Ochaue, pronounced Oc-ha-wa.

The name of the mysterious "Sea People" who invaded the Medeteranean in the thirteenth century B.C. has been preserved in an extant letter and in several inscriptions. The letter was sent by the King of the Hittites, whose empire was later smashed into bits by the "Sea People," to their ruler, whom he addressed on equal terms and affectionately as "brother." The name he wrote as the country inhabited by the "Sea People" as, "Akhaiwa, or Akhhiyawa."

Akhaioi, or Akhaiowi, was the name in Greek for those whom we call the Acheans, the race of people who traditionally once occupied Cyprus, Phoenicia, and Egypt; who created the Mycenaean/Cretan culture."

Og

Evidence that the Amorites and their cultural influence was moving west across the Medeteranean and into the Aegean (Og.) was summarized by *Rushing to Eve, Names of the Goddess,* "The history of Og/Oc is also a fascinating one, and one very often associated with Awa or Hawa. Tradition lists the only other human survivor of the flood beside Noah and his family as Og. In Greek legend, the first king of both Attica and Boeotia, founder of Thrace and of the Achaean League, builder of Thebes, is Ogygios! He is said to be responsible for a series of floods in Boeotia and elsewhere, and is confused by later sources with Noah. Another Greek tradition deals with Aigeus, also said to be a founder of Greece, specifically of Athens, who come from Asia Minor via Cyprus and Crete, and whose son Medus is said to be the father of Medes. He is known as the Goat King (from the Greek word for goat, aig), and is said to have brought to Greece both the goat and the cult of Aphrodite, from the older fertility cult of Astarte or Ashtaroth. Cohane, *The Key,* believes that these two figures, Ogygios and Aigeus, can be traced to a single original source known in the Old Testament and Rabbinical tradition as Og, who as king of Bashan, was a giant who was saved from the flood by climbing on the roof of the ark. As founder of the fertility cult at the city of Ashteroth, he was worshipped throughout the Medeteranean region. More than this, however, Cohane believes his memory is preserved in place names throughout the world!"

The etymological links between Og and the Amorites and the later northern branch of the Amorites, collectively known as the Beaker People, is revealed in the number of Og place names in and around Stonehenge. *The Key,* Cohane "The Ock River, a tributary of the Thames flows across Salisbury Plain near Stonehenge. Over the centuries the spelling of this name has also alternated between Og and Ock.

The Og River runs more or less parallel to the Ock River past Stonehenge and the Ogbourne villages. It is a tributary of the Bourne River, the second element in the *Ogbourne* name. This name has also swung back and forth between Og and Ock.

Less than three miles south of Stonehenge is one of the oldest and largest prehistoric sites in England. It is more than a mile in circuit, covering an area of sixty-two acres. The ramparts are thirty-three feet high. Its name is no longer listed in most standard archaeological works, and it has blended to such a degree into the surrounding countryside it is all but forgotten even by local inhabitants. Its name since the beginning of the historic period has been, and still is, Ogbury Camp.

Before there was a village-*bury* on the site, as in the case of Ava at Avebury, this would have been called the plain Og. Or, alternately, on the basis of the evidence from the Ogbourne villages, the Ock River, and the Og River, just plain Oc. As one studies the region covered by these six names, the three *Og/Och*bourne villages, the two *Og/Ock* rivers, and *Og*bury Camp, it would appear logical that in prehistoric times this entire area around Stonehenge was dedicated to some personage or deity named Og/Oc."

Additional attributes of Og can be found in the English language. We know Og is associated with Caves with the word *Troglodyte*. Cave people are been are typified as being *Ogre*s, that are seen as *Ugly*. Additional Og titles, attributes and place names in the British Isles and northern Europe are listed.

The God -Kings and Titans, Bailey, 1973, "In Irish mythology, Og becomes the supreme ruler of the universe and through *Ogma*, with whom his character is intertwined, the especial guardian of the Tuatha De Danaan, one of the mining and seafaring groups of Ireland. It is Og who gives to the Irish the only prehistoric script the British Isles have used, *Ogham*. Bailey quotes the *Key,* Cohan, 1969, with additional place names in Ireland of, " and the *Dun Aongus* on the Aran Islands, one is called *Oghill Fort* and the other *Dun Aongus,* Colhane says is an accepted epithet for Og. The parish of

Aughaval, Og-awa, contains the holiest site of pagan Ireland."

Og place names in the British Isles

Ogilvy	Scotland
Ogmer	South wales
Ogo	Wales
England	
Ogbourne	St. Andrews
Ogbourne	St. George
Ockbrook	Derby
Ogwell	Devon
Ogmore	Glamorganshire
Gog Magog	Rocks at Lands End
Ogwen	Carnarvonshire
Goginan	Wales lead mine
Ogwen River	Caernarvonshire
Oglo Cave	Salop
Dolanog River	Montgomeryshire
Ogo cave	Shropshire
City of Cogg	Oxforsshire

Og survives the flood

Og	Biblical and Rabbinical giant who survives the flood
Oceanus	Greek God of the Sea, who was Titan
Ogyges	He built the first city of Actae, later to be known as Athens
Oegyptus	The namesake of Egypt
Aegir	Swedish God of the Ocean

Og the giant

Og	Giant King of Bashan,
Osogo	Greek, another name for Zeus
Hagan	Norwegian and German legendary giant
Ogmios	Giant simular to Hercules in Gaul
Goginians	According to Homer were the giants who invaded Ireland
Gog Magog	Ancient giant protectors of London
Gog Magog	Biblical Northern tribes, sons of Japhet
Ogier	A giant Danish Warrior
Ogman	Irish champion of the goddess Dana
Tokaki	A Japanese mythological giant

Og the Sun God

Akaku	Egyptian god of light
Igigi	Babylonian spirit of heaven
Agag	Amelekites also used this word as "flame"

Og, God of Writing
Ogham   Irish God of poetry
Ogmios   France, God of Eloquence

Og in North America

The name Allegewi is very similar to the person or tribe, Tawagalawas, who were Amorite vassals of a Hittite king who escaped by sea. Allegewi is *Al* (consort) *og* (sun god) *awa* (earth mother). The same elements are found in Tawagalawas ag (sun god) al (consort) awa (earth mother). Various historian have seen parallels with the Sea People, Amorites and the Hykos who called themselves Akhaiwa ak (og) aiwa (earth mother). On an Egyptian glyph listing kings with the Sea People, it identifies one of them as a king of the Amorites.

The most interesting assimilated uses Og in North America, is its meaning of *"chief." Og* means "chief" or "most high" in Hebrew, *Aga* means "chief" in Turkey, *Aga* was Akkadian for "chief," *Agag* was the usual title of the Amelekite king, The Amelekites, one of the accursed nations conquered by Joshua Numbers *24:7 "*He shall pour the water out of his buckets, and his seed shall be in many waters, and his king shall be higher than Agag, and his kingdom shall be exalted."

The name Og, also means "chief" in many of the Native American languages. *A Concise Dictionary of Minnesota Ojibwa*, by J. Nichols and Earl Nyholm, "Ogmia " means "chief", "boss" or "leader." *The Viking and the Red Man, The Old Norse Orgin of the Alonquin Language,* 1894, "Ogima" means "chief" or "chieftan." "Ogimakewe" means "chiefs wife." *Handbook of Tribal Names of Pennsylania Together with Signification of Indian Words (Iroquois)* 1908, "O-gee-chee" means "head chief", "Oge-ma" is "a chief." *A Dictionary of the Cree Language, as spoken by the Indians in the Provinces of Quebec, Ontario, Manitoba, Saskatchewan and Alberta,* 1986 "Okimaw" means, "honor," "aristocracy," "chieftanship" *American Antiquarian* May, 1885, "Ock" or "Og" was The Chief God of the Alonkins, "the common father" *Archaeologia Americana* 1836, Eskimaux (Tshuktchi, Asia, E. Shore) "Aghatt" is "God"" Ogima" is Chippewa for "chief,", "Okemah" is Ottawa for "chief,"

"Okima" is Sauk for "chief,"and "Okamow" is the Menomenies name for "chief."

Within the Lakota Sioux pantheaon of Gods, the Sun God was known as Wi and the Earth Mother was Maka-aken. *Maka,* the Earth Mother was grandmother of all things. She was created by *Inyan,* who gave spirit to Maka-akan (Earth Goddess).

Within the language of the Algonquin and Iroquois tribes the use of the word "awa" to denote the Earth Mother is replaced with "Ma" The word "Ma" has Sumerian roots, and was used used most extensively in the British Isles.

Hawwah/ Ma, The Earth Mother

Hawwah/Avvah is present at Avebury, revealing it as a ceremonial center dedicated to the Earth Mother. More common in England is the use of "Ma" title for the Earth Mother. Ma-a is the Sumerian source of Earth Mother worship and their word *Madur* is the source for the English word, "Mother." *Midwestern Epigraphic Society, Earth Mother Sacred Language: A Key to Ancient Names Worldwide,* Dr. John J. White, III. *EMSL,* "The word "Ma" for mother, woman, female, and the care giving role is one of the most universal words of mankind."

Archaic England, "One of the legends of the arrival of Magog to Britain was chronicled in, Archaic England, Harold Bayley 1920, "Gog, Magog, Termagol, and the rest of the terrible tribe sprang, according to Scottish myth, from thirty-three daughters of Diocletian, a King of Syria, or Tyria. These thirty-three primeval women drifted in a ship to Britain, then uninhabited, where they lived in solitude, until an order of demons becoming enamoured of them, took them to wife and begot a race of giants. Anthropology and tradition thus alike refer to Magogoei of Syria, or Phoenicia, and there would seem to be numerous indications that between these people and the ethereal, romantic, and artistic Cretans there existed a racial, integral, antipathy."

The British Chronicles relate that when Brute an his companions reached these shores the island was then inhabited, save only for a few giants, called Goemagog. The untoward Gogmagog was one of an elementary big-boned tribe whose divinities were Gog and Magog. According to De Jubainville, "the varios races that have successively inhabited Ireland trace themselves back to common ancestors descended from Magog or Gomer, son of Japhet, so that the Irish genealogy traditions are in perfect harmony with those of the Bible.

A similar story is found in Ireland, that includes stories of the Fomorians (Muru or Amorites) that were legened to have been gigantic with a double row of teeth. *A History of the County of Down, Ireland,* 1875 "The Partholonian was followed by the Nedian and Fomhoraic settlements, the former consisting of the Clanna Neimhidh or followers of Neimhidh, a mythological personage, signifying poetry and alleged like their predecessors to have deduced their origin from Gog and Magog. The Nemedians are ther Scythians of the ancient bards, generally held to have been an eastern people, they are reported to have arrived on the Euxine sea in the time of Jacob, and to have landed in the vicinity of Dalraida, in the County of Down, where they took up residence until they were expelled by their Fomhoraic enemies who succeeded them in the possession. The Fomorians were syled sea champions, as they bore a character of a piratical tribe."

Archaic England, "The figures of Gog and Magog used recently to be cut into the slope of Plymouth Hoe: in Cambridgeshire, are the Gogmagog hills; at the extremity of Land's End are two rocks known respectively as Gog and Magog, and there is an unfavourable allusion to the same twain in Revelation. Gog and Magog are the "protectors" of London, and at civic festivals their images used with pomp and circumstance to be paraded through the City.

In some part of Europe the civic giants were represented as being eight in number, and the Christian Clergy inherited with their office the incongruous duty of keeping them in good order. One of these ceremonials is described by an eye-witness writing in 1809, who tells us that in Valencia no procession of however little importance took place, without being preceded by eight statues of giants of prodigious height.

Four pairs of elemental gods were similarly worshipped in Egypt, each pair male and female, and these eight prime evil beings were known as the Ogdoad or Octet. In Scotland, the Earth Goddess who is said to have existed "from the long eternity of the world," is sometimes described as being the chief of eight "big old women," at other times as "a great big old wife," and with this untoward *Hag* we may equate the English *"Awd Goggie"* who was supposed to guard orchards.

The London figures of *Gog and Magog*--constructed of wicker work- had movable eyes which, to the great joy of the populace, were caused to roll or goggle as the images were preambulated. Skeat thinks the word *gog* is "of imitative origin," but it is more likely that *goggle* was originally *Gog oeuil* or *Gog Eye.* The Irish and Gaelic for Goggle-eyed is *gogshuileach,*which the authorities refer to *gog,* "to move slightly" and suil "an eye".

At Gigglewick or Giggles-fort in Yorkshire there is a celebrated well of which the famed peculiarity is its eightfold flow, and it was of this Giggle Well that Drayton wrote in Polyolbion:- *"At Giggleswick where a fountain can you show."*

The following article expands the argument of cultural diffusion being found within Native American language. The results of this study enhances the belief that the Sioux and Iroquois Hopewell Mound Builders were intermarrying with the Allegewi. Adopting not only the practice of constructing burial mounds and earthworks, but had also incorporated parts of their language, with its etymological roots, stemming from the Sumerians and Amorites of ancient Babylon and the Levant. It is significant that these root words can be traced across the Medeteranean and into the British Isles. These, being the same paths where mining and the introduction of metals are found along with giant skeletons. The number of Sioux tribal and place names and deities bering Earth Mother words is evidence of migrations to North America. To what extent the Og and Earth Mother root words were introduced by the Allegewi or were already being used by the Hunters and Fishers of northern Europe before their migrations to North America is problematical, but would not to have been likely to manifest independently. The list of Earth Mother names and their uses by Native American is a combination of both the *Midwestern Epigraphic Society,* by Dr. White and *The Key,* by Cohane.

Midwestern Epigraphic Society, Earth Mother Sacred Language: A Key to Ancient Names Worldwide, Dr. John J. White, III. " EMSL is an abbreviation of the name Earth Mother Sacred Language. It is a scientific model for an ancient Monosyllabic naming language, meaning that each syllable of a compound name is a word or idea in itself. He name EMSL was selected because the Gods referred to are logically related to our current perception of the Earth Mother Religion and Culture. We find that ancient names plausibly based on this naming system occur worldwide.

In general terms, the Earth Mother Religion can be called a Nature Religion, meaning that the Deity is an all knowing presence existing in the Earth, its plants, animals, and waters and the nearby sky. The female or Mother Principle was emphasized initially due to the Earth's capability to generate living organisms and this food and materials for mankind."

Ge "The Gea/Gaea word is well known, and it is the most common EMSL word if all of the phonetic variations are allowed. We hear the "ge" sound daily in words, such as geodesy, geography, geology, and genealogy, and the Ge-sound is likely the root for words God?Goddess."

Re/Ra "We are not accustomed to the Re/Ta word, due largely to a highly different Sun-meaning in Egyptian Culture, but this is clearly the Greek Earth Goddess Rhea. We start with the names Terra/Thera/Tyre from the Greek world and the names Ere (Ireland)/Earth/Erde (Earth – German) from the English world. The above findings that "E" is a god-word and that "Te/THe/Ty/ De" is a definite article leaves no choice but that R# means "Earth Goddess."
Significant RE/Ra words are Iran, (Lake) Erie, Siberia, Syria, Tara, Tyre

La "Sun God, (Late comer as a consort?)
EMSL The Word "La" for the male sun-god
"It took some time to find the candidate Sun-god word La/L, due to the cultural influence of Egypt, which uses the well known sound Ra. There is also the false notion that L to R to L shifts abound in the ancient words. In any event, there exist many ancient L-containing words like, Italy, Illyria, Illinois, Laconia, Lakota, and Libya with no apparent translation.
The initial clue comes from the Sun-names "Sol/Solar," which we take to be much older than "Sun/Sonne." If "So-" from above is a definite article, then "La" is our candidate. "So-ne" in EMSL would mean "the people" and is far off the mark. We then looked at the Moon name "luna," which our culture takes to be feminine. If "na" means consort of," the "Luna" means "Consort of Lu." Our limited knowledge of ancient religion often specifies that the Moon is the wife of the Sun, and further

we read that "Lugh" is the name of a Celtic Sun-God.

We are thus in a position to consider L#, where # is any vowel, to be a male Sun-god name. We can check this conclusion against the name of the Greek Sun God "Helios." Ignoring the initial 'H" and the Greek suffix "-os" the word strips down to eli. We can interpret eli as eli/a-la, that is, *god-sun*. As with "Luna," we encounter with "eli" a reverse word order from our logical preference, but this appears characteristic of many ancient names.

I direct your attention to the god-words "El," Baa;," and "Beli." We may view the old Semitic god name "El (i)" to be the same word as "helios," that is, "Eli." The EMSL translation of simple L-words, such as "Apollo," "El," "Baal," "Beli," have nearly identical meanings on the surface, that is, they recognize the Sun God"

Bases on these insights, we can translate "Italy" as I-TA-LY, *god-the-sun* or *the-sun-god.* "Hellas" (the correct name for Greece) as E-LA, *god-sun* or *sun god;*
"Lydia" as Ly-DI-A, *sun-the-god* or *the-sun-god;* "Libya" as LI-BY-A, *sun-father-god* or *father-sun-god.* "Illini" as I-LI-NI, *god-sun-people* or *sun-god people.*
"EMSL" connections:
Alabama, Albania, Ocala (FL), Okaloacoochee Swamp (FL), Oklawha City and River, (FL), Alachua (FL), Avalick River (AK) Oglvik (AK), Allagash River (ME)"

Og-La names
"Ogallala, (NE), Ogallah, (KS) Ogahalla, (Ont.) Allegan (MI)
Oklawha River and town in Florida
Okinawa, Japan"

Wa Water/Wind/Spirit (Mother Goddess)
"The interior origin of the Wa-goddess is the apparent from the finding that the vast majority of the Wa body-of water names are rivers and lakes and seldom seas and oceans. Ancient Iberian translation for water was Agewa. A-Ge-Wa, meaning Water (Earth) Goddess. In the "*History of Essex County, Massachusetts,* 1888 "Prehistoric Discoveries, Phoenicians and Norwegians," "This territory, once the abode of the red man, and known to him by the name Agawam, was settled by our ancestors some more than two hundred and fifty years ago. It was, however known to the white race, no doubt, at a very much earlier period."
"EMSL" connections: Each of these place names are located on a lake or a river.
Iowa, Walla Walla, River (Wa) Wawasee (IN) , Wabash River, (Awabasha) (IN), Milwaukee (WI), Oswego (NY), Wabash (IN), Watauga (TN), Waukegan (IL), Washburn (WI) Kanawha (WV). Ottawa, Etowah, (GA)
Awahili, The sacred great Eagle of the Cherokee
Heloha, The great Thunder Bird of the Choctaw tribe, who like the Cherokees, originally lived in the southeast. Heloha, a female deity, bore a name phonetically identical with Eloah."

O One of the most significant findings of the EMSL is a recognition of the meaining of the word O. There are several clues that the meaning of O is blood kinship. Thus it is possible to encounter a tribe with a name like Otey, Oddi, Oge, Ogea or perhaps Ogeni, who wil call themselves The Blood (s) as we translate it. The correct name of such a tribe should be "Blood of the Goddess," which in turn should be interpreted to say "Kin (Children) of the Goddess" and mean "Followers of (Believers in) the Goddess."

We started long ago with the word "Oklahoma." Mr Victor Kachur suggested that "okla" might represent an L to R shift from "okra" and thus "ochre." We have had the discussion of the possible association of "redman," "red paint people," and red ochre usage in ancient burials with EMC., it thus would be reasonable to anticipate that the word "ochre" could be associated with EMSL words for "red" and "ceremonial Earth Mother blood"

Returning now to the word "Oklahoma," we see that it is the EMSL name OK-LA-(H)O-Ma. The name shows us that the word Ocan modify the Goddess word Ma and possibly others. American Indian like Omaha, Ogala, Oconee, Osage and Ocala are now straightforward to interpret.

Na, EMSL "The Na-word is a specific interpretation of EMSL that could be in error in the s ense that Na itself may be a God-word, perhaps even the name of the Earth-God. The usages taken collectively seem to imply the concepts of snakes, serpent, king, and Earth God. Thus we concluded that the likely interpretation of the word Na is "Consort of (the Goddess) or various (Gods).

NaGa, Serpent (Earth God, Goddess consort)

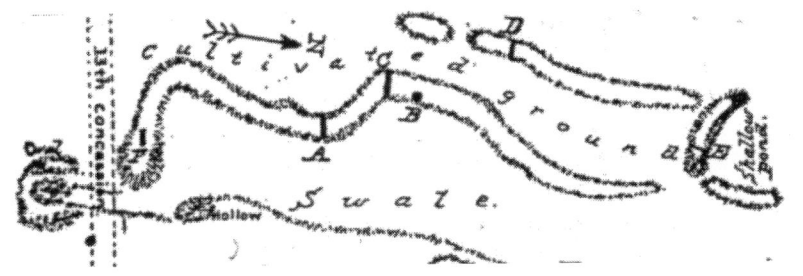

Serpent Mound at Naga Well, in Ontario

The Sacred Marriage of the Earth and Sky; Symbolism in Earthworks and Burial Mounds

Within the ruins of the Allegewi and Hopewell Sioux, are geometric shapes with symbolism that originated from Babylon, Egypt and the Levant. It is written in the Bible that shortly after the Noachian flood, Nimrod and his mother-wife, Semiramis, founded the Pagan religions. From the tower of Babel, sun worship and the worship of gods and goddesses spread across the world.

The most basic shapes used to invoke these gods were the circle and the square. The circle is seen as being everlasting, having neither a beginning or end and is symbolic of the sun and the male principle. The square represents the Earth Mother, the four winds and the cardinal points.

The circle and square are combined in many of the earthworks, that is evidence that these temples were dedicated to the annual mating of the God of the Sky and Goddess of the Earth. The annual rituals dedicated to the mating of the God and Goddess, guaranteed their future success, safety and fertility of the people.

The concept of the *Marriage of the Gods* was a belief that was widespread throughout the ancient world and most evident in the earthworks in Britain and the Ohio Valley. *A Sacred Geometry Prime,* Graham Gardner, "Sacred Geometry deals with our perception and definition of space. It is the Universal framework whereby the spiritual manifests into the material. Spaces constructed using principles of sacred geometry act as a bridge between the worlds, and sacred geometric forms naturally produce dowseable energy fields."

"In ancient times it was believed that numbers are the underlying reality behind all things. All things were linked through number and could be manifested through number."

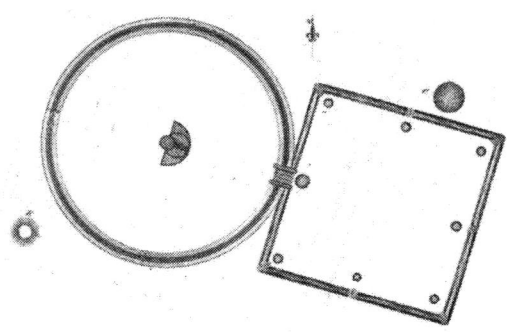

The Circleville, Ohio earthwork, consisting of a circle and a square, was constructed to venerate the Sky God and the Earth Goddess. The alignment of the earthwork was to the winter solstice sunset and the summer solstice sunrise. The square is symbolic of the four winds, the cardinal points and the Earth Mother. The eight mounds within the square are symbolic of creation or reincarnation. The circle is the sun, with a bird effigy that is symbolic of the Earth Mother.

A Sacred Geometry Prime, Graham Gardner, "The combination of the square and circle represents the fusion of heaven and earth, and 'squaring the circle' is regarded as the pinnacle of the sacred geometers art. This means producing a circle overlaying a square such that either the circumference of the circle equals the perimeter of the square or the area of the circle equals that of the square."

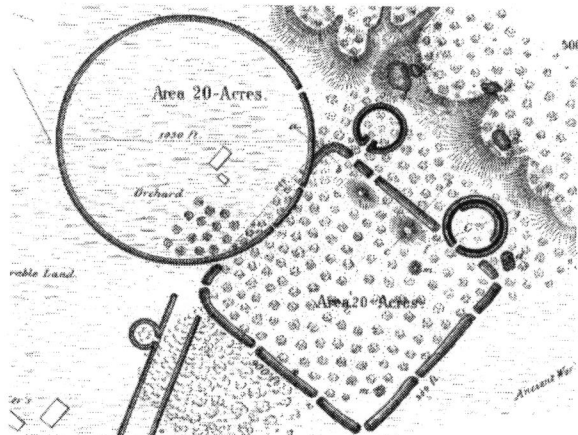

Hopeton Earthworks in Chillicothe, Ohio with a square and circle both encompassing 20 acres representing the fusion of heaven and earth.

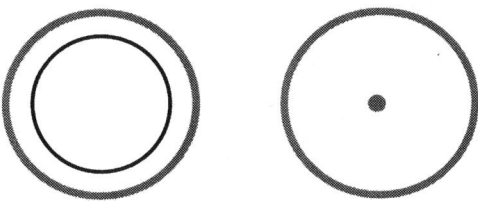

Sun Symbols used by the Egyptians, Sumerians and Hitto-Phoenicians.

The symbolism of a mound surrounded by an earthwork or ditch, represented Earth, Heaven, the God Sun and the Sun God. It also could be symbolic of perfection and totality. A plain circle also represented the number one. The plain circle was symbolic of the one Sun god called by the Sumerians *Ana* or *Un,* which is the Sumerian origin of our English, *One.*

On Hakpen Hill, to the east of Avebury is a henge, with a small burial mound within its center. In the background is the Avebury, stone avenue.

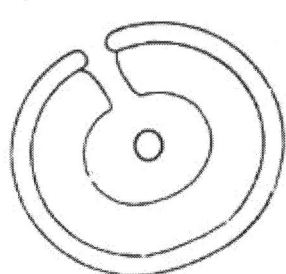 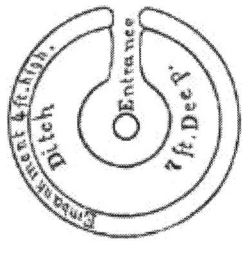

Henges located north of Cambridge City, Indiana. Many of the henges in the Ohio Valley had mounds in their center representing the "Eye of Heaven." The Sun God Og was the all seeing and knowing, *ogler or goggler.*

Phoenician Origins of the Britions and Scots, 1925, " We thus find that the Father-god of the Sumerians (and of the Hitto-Phoenicians), whose earliest-known name, as recorded on the Udug trophy Bowl of the fourth millennium B. C., is *"Zagg" (* or *Za-ga-ga,* which, with a soft *g* gives us the original of *"Zeus,"* the *Dyaus* and *Sakka* of the Vedas and Pali, and the "Father *Sig"* or *Ygg* of the Gothic Eddas) is recorded by a single circle sign as having the equivalent of *Ia* or *Bel,* thus giving the Aryan origin of *"Iah"* or (Jehovah") of the Hebrews and the "Father *Ju (*or Ju-piter" or *Jove* of the Romans.

This title of Ia (or Jove) for the Father-god (Bel), as represented by a single circle, is defined as meaning "God of the House of Waters," which is to disclose the Sumerian source of the conception of Jove as "Jupiter Pluvis" of the Romans. This special aspect and function of the Father-god was obviously conditioned by the popular need of the Early Aryans in settled agricultural life for timely rain and irrigation."

The ditch within the henge at Mounds State Park, retains water after moderate rains. Other henges have also been observed holding water within their ditches. Is this evidence that the Sun also represented the *"God of the House of Waters?"* Note, the serpentine undulations of the outer wall. *The Nephilim Chronicles, A Travel Guide to the Ancient Ruins in the Ohio Valley,* 2010.

Evident in all earthwork complexes are springs or an avenue or sacred via to the water. The sacred waters may have been used to purify, cleanse and restore the soul and body of sins before entering the earthen temples. In many of the burial mounds, conch shells are present that were emblematic of the sacred waters that would restore and cleanse the soul of the dead.

Also found throughout Indiana, Ohio, Kentucky and Tennessee are wells that were walled up with stone. In *Isis Unveiled*, H. P. Blavatsky writes of the connection with the wells and the Sacred Marriage. "A well is "the foundation of salvation" mentioned in *Isaiah* (xii. 3). The water is the *male principle* in its spiritual sense. In the physical relation in the allegory of creation, the water is chaos, and chaos is the female principle vivified by the Spirit of God-the male principle."

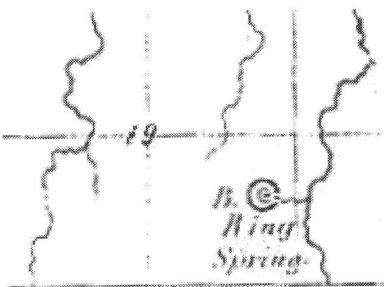

Venerated spring located south of the square earthwork located in Winchester, Indiana.

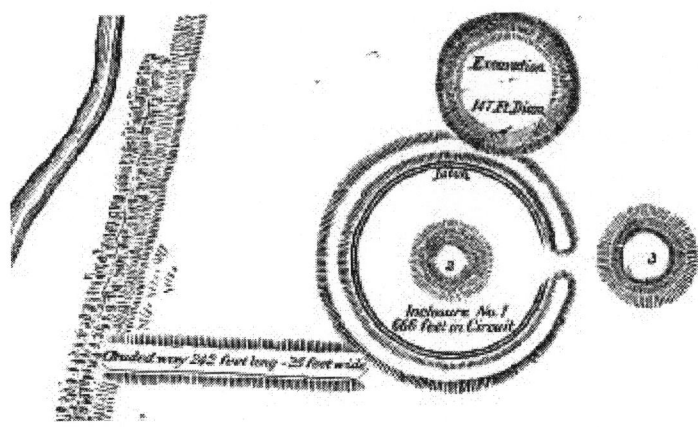

Henge and mounds at Charleston West Virgina with sacred via to the Kanawha River.

3. BELL BARROW.   2. BOWL BARROW.

Mounds in Wiltshire England, from "*Ancient History of Wiltshire,* 1812. The Bowl Barrow is surrounded by an earthwork and ditch. The Bell Barrow is surrounded by only a ditch.

The mound at Marrietta, Ohio is a a Bell Barrow with no external earthwork, outside of the ditch that surrounds the mound. *The Nephilim Chronicles, A Travel Guide to the Ancient Ruins in the Ohio Valley,* **2010.**

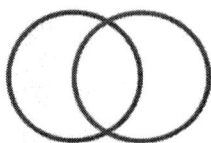

The Vesica Piscis is defined by the Statemaster Encyclopedia as " A symbol made from from two circles of the same radius, intersecting in such a way that the center of each circle lies in the circumference of the other. The name literally means the *bladder of the fish.*" The length to height ratio of the fish is 265:153. "This ratio, equal to 1.73203, was believed to be a holy number called the "measure of the fish." The ratio 265:153 is an approximation of the square root of 3. with the property that can not be obtained with smaller whole numbers. The number 153 appears in the Gospel of John 21:11 " Simon Peter went up, and drew the net to land full of great fishes, and hundred fifty three: and all there were so many, yet was not the net broken." This is believed to be coded message in reference to Pythagorean beliefs.

An earthwork in the shape of a Vesica Piscis was described in Randolph County, Indiana, but has since been destroyed by farming.

History of Randolph County, Indiana, **1885** "There are some circular embankments of the Bales farm (now owned by Mr. Branson), not far from Cedar (Friends) Meeting House, in Stony Creek Township, a little north of Cabin Creek. In one place there are two circular embankments together. The circles cut each other. A mound is in the center of each circle, higher than the embankment. The earth for both the wall and the mound would seem to have been taken from the space between the two. The embankments are now about three feet higher than the level of the ground outside. The central mounds are perhaps ten feet across and four feet high. The ground enclosed in both is about three acres, two acres in the large and one acre in the smaller. There is an opening like a wagon-way on the east Side of each enclosure."

Earthworks in the shape of the vesica are visible at Mounds State Park, only about 30 miles distant.

The alignment with the winter solstice sunrise, is evidence that these shapes were constructed to represent the vulva of the Earth Mother.

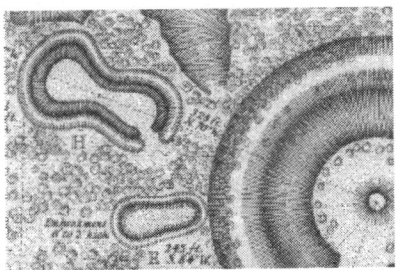

Earthwork at Mounds State Park in Anderson, Indiana has a constricted middle that represents the Vesica Piscis or vulva of the Earth Mother. It's alignment with the winter solstice sunrise was symbolic of a portal for the birth of the sun. There were are eight earthworks in this group, which is a good clue that the works were constructed to celebrate the reincarnation or birth of the sun on the daybreak of December 22. From *Prehistoric Antiquities of Indiana*, Eli Lilly, 1937

Photo shows that the earthwork intersects the interior ditch, at more of an acute angle than Lilly's drawing. *The Nephilim Chronicles, A Travel Guide to the Ancient Ruins in the Ohio Valley, 2010.*

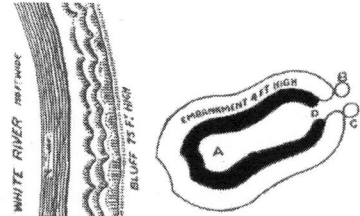

The north henge within the Park is aligned to the May 1st sunrise; to venerate the Earth Mother and her rejuvinating powers that has transformed the landscape from the death of winter to the lushness of spring. She has been celebrated by agricultural peoples throughout history. She was was *Mad-dur* of the ancient Sumerians and the root of our English "*Mother*;" *Maia* of the Greeks, *Mahi* and *Maya* of the Vedas and the goddess *Queen May* of the ancient Britons and *Maka* to the Sioux.

The interpretation of the Swastika seems to vary in different countries. It is held to be the symbol of the Sun-god; Agni, The Fire-god; the Rain-god, Indra, the Sky-god, and the god of light and forked lightnings; the generative principle, the fire generator and the birth of fire. The most common form found within Allegewi and Hopewell Sioux burial mounds is called the hook cross is called the *Ogee*, in the Siouan language.

The American Antiquarian Vol., 20 1897, " The Swastika in America" "It is difficult to decide as to the significance in America, though judging from its shape and its association with other symbols, especially the sun circle and the cross, we conclude it is designed to represent the revolution of the sky and is in reality a revolving cross. In favor of this supposition is the fact, that the Swastika and the Triskelis are frequently seen on the inside of a circle; sometimes in the center of disks, and are arranged in such a way to convey the idea of motion, the symbols of the sun and moon and the serpent all conveying the same idea of revolution."

Madame Blavatsky, *Isis Unveiled, 1877,* "The swastika is an equal-armed cross- a symbol much older than Christianity. The upright line represents the masuline influence, the horizontal line the feminine one. From the union of two opposites, masuline and feminine, positive and negative, comes all manifestation. The lines are added to the cross, signifying motion, the *Wheel of Life.* The swastika with its four arms symbolizes birth, life, death and immortality. There are four winds, the four seasons and the four elements- all of which are summed up in the Swastika."

Two swastikas were found in the Hopewell Group mounds in Ohio made from sheet copper. From *The Mound Builders,* Shetrone, 1941.

The swastika or ogee continued to be used by the Sioux Indians until historic times. Photo is of a Sioux basketball team, August of 1909.

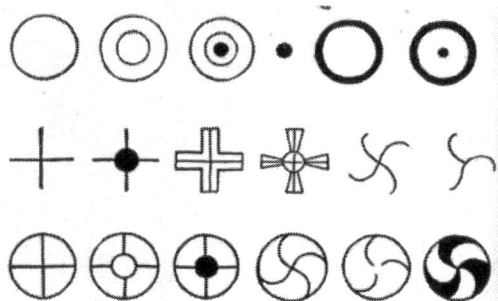

Additional sun symbols of the Allegewi and Hopewell Sioux that were used as late as the Mississippian Era, 1400 A.D. From *The Mound Builders,* Shetrone, 1941.

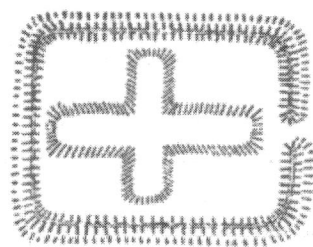 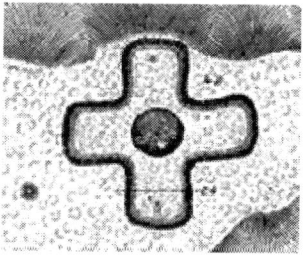

Earthwork on the left is located in Banwell, Somerset county, England. On the right is the Tarleton Cross located in Fairfield County, Ohio. The depression in the center of the Tarleton Cross is where fires were lit. Fire and sun-worship, were combined to venerate the sun as the source of light and heat. When the cross was within a circle it stood for the sun and the tree of life and is comparable to the yin-yang symbol of the Eastern world.

Photo shows two of the arms of the Tarleton Cross, that is still visible in a public park in Ohio. *The Nephilim Chronicles, A Travel Guide to the Ancient Ruins in the Ohio Valley,* 2010.

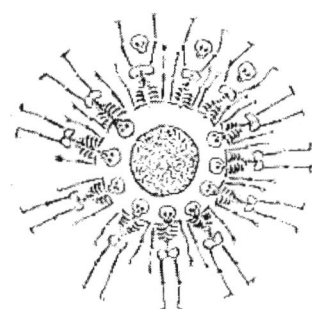

Wheel of life, A symbol for the *Eight folded Way* **is symbolic of reincarnation. Spoked burials are found in the Late Archaic and Woodland Periods within Adena and Hopewell burial mounds. This type of burial could also be interpreted as the spokes representing the Father Sun god incorporated within the nave of the Mother the Earth Mother, representing the** *Sacred Marriage.*

The following list is of burials that were placed in a spoked postion within a mound. In some cases, all the skeletons were of large size that are listed within the chapter on the Allegewi Giants. Many of the skeletal remains listed are described with "archaic" type features. Some of the burials have conch shells within the center of the mound in which all the skeletons radiated from. The conch shell or stones that were retrieved from an adjoining river and incorporated into the mound construction were symbolic of purification and absolution. In some mounds the central point contained evidence of fires that were built, prior to the mound being capped with dirt. Fire was the agent that would transport the souls in to the afterlife.

History of Miami County, Ohio, 1898

About a mile south of Piqua, on the point of the hill, is a mound described by Mr. Wilthe as being 240 feet in circumference, six feet in height, and surrounded by a ditch with pebbles. One mile southeast of the main fort is another, 160 feet in circumference, with ditch on the inside and entrance on the east and west. One mile south on Section 7 is another, 300 feet in circumference, with a southeast entrance, gravel embankment and ditch inside. Three hundred yards to the northeast is another, 250 feet in circumference and nine feet in height. Excavations showed this also to contain a sacrificial alter, made of clay burnt red, and covered with ashes; charcoal and burnt bone three inches thick. […] Southeast of this, between the river and canal, was an ancient burial ground. Ten skeletons were exhumed by Mr. J. Reyt; they were buried in a circle, with their feet toward the center, which was occupied by a beautifully ornamented piece of pottery.

Horseshoe shaped henge near Piqua, Ohio that has several interpretations; in Assyrian and Egyptian sculptures, it signified the mystical door of life, or was symbolic of the vulva of the Earth Mother. The embankment extends on the right a short distance down the creek bank implying a connection between the earthwork and the waters. *The Nephilim Chronicles, A Travel Guide to the Ancient Ruins in the Ohio Valley.*

History of Coshocton County, Ohio, 1881

Among the mounds plowed down years ago was one in Oxford township thirty feet wide. A circle enclosing three acres north of West Lafayette and several mounds of Lafayette township were obliterated by the plow, one on the Shaw Estate, one cut away by the railroad on the Ferguson farm, and another leveled on the Higbee place. Seventy years ago the river road in Franklin township leveled a mound containing half a dozen skeletons arranged like the radi of a circle with heads toward the center.

North American Burial Customs, Dr. H.C. Yarrow, 1879
Burial near the Scioto River, Ohio

To the southwest of this tumulus, about 40 rods from it, is another, more than 90 feet in height. It stands on a large hill, which appears to be artificial. This must have been the common cemetery, as it contains an immense number of human skeletons of all sizes and ages. The skeletons are laid horizontally, with their heads generally towards the center and the feet towards the outside of the tumulus. A considerable part of this work still stands uninjured, except by time.

History of Darke County Ohio, 1880

One of the most interesting burial spots was discovered on the farm of Jesse Woods in German Township. In digging the cellar under the house where he lives, Mr. Woods discovered a skeleton in a sitting posture. It was covered with plates of Mica and was the central figure in a group of other skeletons arranged in a circle around it. The skeletons in the circle were lying at full length.

History of Pickaway Ohio, **1885**

   To the southwest of this tumulus, about forty rods from it, is another, more than sixty feet in height... it stands on a large hill, which appears to be artificial. This must have been the common cemetery, as it contains an immense number of human skeletons, of all sizes and ages. The skeletons are laid horizontally, with their heads generally towards the center and feet towards the outside of the tumulus. A considerable part of their work still stands uninjured.

Indiana Geological Survey **1884**
Hamilton County, Indiana

   But Strawtown has an antiquity evidently higher than the days of the Delaware Indians. The mound builders have left their foot-prints in this vicinity by the numerous relics of the Stone age that have been picked up by the present inhabitants. A little west of the present village ther is a burial mound about six feet high; it has been plowed over fir a number of years, so that not only its height has been reduced, but its base rendered so indistinct that its diameter can not be accurately measured; it is, however, between seventy and eighty feet. It was opened in 1882 by Judge Overman, of Tipton, and four skeletons were found lying on the original surface of the ground, with their heads together and their feet directed to the cardinal points of the compass.

On the same plat of ground as the burial mound is this large henge that was plowed for many years, but the outlines of the inner ditch and outer wall are still visible. The spoked burial pointing to the cardinal points is symbolic of the Earth Mother. Archaeologist determined this work was constructed by the Oto Sioux. The Oto Sioux ranged the extent of the Mississippi River to Louisianna. *The Nephilim Chronicles, A Travel Guide to the Ancient Ruins in the Ohio Valley, 2010.*

Ophiolatrea, **1896**

   At Cappile Bluffs, on the Mississippi River, was found a conical, truncated mound, surrounded by nine radiating effigies of men, the heads pointing inward

Louisiana Historical Society Vol., II,, 1902

The work was begun June 22, on a mound located on the property of Mr. C.H. Snyder of Como, Franklin parish. The mound is on the edge of a small cypress brake and a short distance from the left bank of Brown's Bayou. It was of medium size, having but eighty-five feet base diameter and a height of eight feet, with a surface area of fifty feet.

After penetrating the superficial layer of clay, I found human bones at a depth of about fifteen or eighteen inches; they were badly decayed and scarcely permitted touching. The shaft being a little west of the centre, and about five feet square, encountered only the skulls, some of which were lying very close together. I removed the earth further around, and found the bodies extended toward the circumference of the mound. All bodies in this layer were buried in a circular manner.

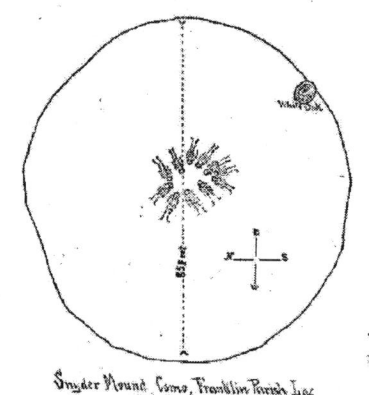

Snyder Mound, Como, Franklin Parish, La.

History of Parke and Vermillion Counties, Indiana 1913

Another on the Head farm, near Newport, had copper rods, or spearheads and smaller stone implements. A burial mound near the northeast corner, contained a chief in a sitting position, in the center. Radiating from his body, like the spokes of a wheel, were five persons, slaves or wives, to wait upon him in the other world. His useful implements for the other world were a great number of copper beads from a half inch to an inch and a quarter in diameter, seven copper axes, on contained unmelted virgin silver.

Indiana Geological Survey, 1875
Owen County

On the Mcbride farm, adjoining the last, is a mound 12 feet high and 150 feet in diameter, with two smaller ones attached at the southwest. On top of the large mound was a circle of stone vaults 6 feet long, 2 wide and deep, covering the whole top, each grave containing two or more skeletons, with heads toward the center ans bodies radiating out like the spokes of a wheel. The bones indicated persons from four ten inches to five and a half high.

History of Rush County, Indiana, 1888

Two nearly perfect skeletons and parts of a third were found in another (section 27, township 12, range 9) buried with the heads turned toward a common center; also copper and bone beads.

Smithsonian Institutes Bureau of Ethnology, **1890- 1891**
Lawrence County, Indiana
Another mound on the Lawrenceville road, about 3 miles southeast of Russellville, had also been opened and several skeletons found about 2 feet below the surface, with heads outward and feet toward the center. No articles of any kind were with them.

Fall Creek Townshio, Early history of Madison County, Indiana, **1949**
Section 1, Township 17 Range 7 East. No. 9 is a sort of sugar loaf mound. In early settler days this was surrounded by swamps and low marshy ground and it is improbable that any trail or road passed near this location. Several years ago Cash Keller excavated part of this mound for gravel and in near the center he uncovered three human skeletons. Two were of matured persons and the other was of a child. The manner of their burial indicated that they may have been white people as they were laid out in fan shape with their feet close together. All known Indian burials in this community are in a sitting position, but whether white or Indian time has erased all trace except that found in their graves.

History of Preston County, West Virginia **1912**
  On the Croff bottom, seven miles above Rowlesburg, the pioneers found three burial mounds. On a ridge three miles east of Fellowsville is a mound which originally was twenty-five feet in diameter and fifteen feet high. One more lies near Pringles Run, five miles south of Kingwood, and there seems to be still another on Roaring Creek, two miles below Albright. On opening these tumuli, the bones of men, women and children were found at the base. The corpses were arranged in a circle, and were placed in a sitting posture with the feet pointing toward the center

Smithsonian Institutes Bureau of Ethnology **1890- 1891**
Fayette County, West Virginia
  Many of the stones used in these heaps have evidently been obtained by rude quarrying in the stratified cliffs, often half a mile distant. Some of them measure from 4 to 6 feet in length, half as wide, and of a thickness which renders them so heavy as to require from two to four stout men to handle them. Beneath the somewhat upturned edges of many of these stones in the different layers are frequently found the decayed (and often charred) remains of human skeletons, usually horizontal, with the head or feet (generally the latter) toward the central "wellhole."

American Antiquarian, March **1886**
Mound Excavation in Tennessee
The most interesting skeleton to me was the one numbered 49 (This is situated on the south bank of Little Tennessee River, not very far from the site of Old Fort London)---this was nearly in the center of the mound, lying with the head southwest, and eight other skeletons lying close around it, all with their heads turned toward it. With one of the surrounding skeletons I found a celt and a discoidial stone, two bone implements, a soapstone pipe, an engraved shell and a bear's tooth, all lying about the head...

The Prehistoric Men of Kentucky, **Colonel Bennett H. Young 1910**

   What is known as the Lindsay Mound, on Buffal Creek, four miles from Raleigh, explored by Mr. Lyon, revealed many interesting features in regard to burial customs. The bodies were placed in a circle upon their backs, with head directed toward the center and faces turned upon the left side, the arrangement being similar to the spokes of a wheel.

Smithsonian Institutes Bureau of Ethnology **1890- 1891**
Chautauqua County, New York

   In front of Mr. Gould's residence and 80 rods to the east a bone pit was opened several years ago. Remains of skeletons of each sex and of all ages were found; the number could not be determined. a grave had been previously opened near the above and five skeletons found in a circle with feet outward. The position and size of the bones corresponded closely with those on Mr. Cowans place near Rutledge.

History of Erie County Pennsylvania **1925**

   A large mound near the New York Central R. R. tracks in North East township, about three miles east of the borough of North East, was opened many years ago by Dr. Heard, a prominent physician and surgeon of that place, and several skeletons uncovered, all with the feet pointed towards the center as in the spokes of a wheel

History of Cass County Michigan, **1902**

   A large number of human skeletons were found (over a hundred, it is said), buried in a circle with their heads toward a common center. Many of the skulls bore the marks of weapons, which indicated that death had ensued from violence. Those who saw the inferred that the skeletons were those of men who had died in battle. All had evidently been buried at the same time.

Smithsonian Institutes Bureau of Ethnology **1890- 1891**
Dunleith Illinois

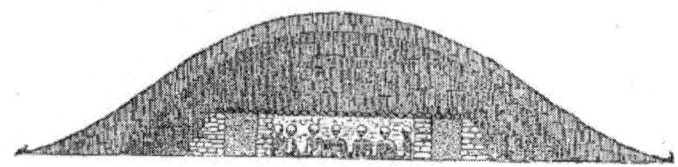

   The most interesting feature of the group was found in No., 16 a symmetrical mound 65 feet in diameter and 10 feet high. The first 6 feet from the top consisted of hard gray earth, seemingly a mortar-like composition, which required the use of the pic. This covered vault built in part of stone and in part of round logs. When fully uncovered this was found to be a rectangular crypt, inside measurement showing it to be 13 feet long and 7 feet wide. The four straight, surrounding walls were built of small unhewn stones to the height of 3 feet and a foot or more in thickness. Three feet from each end was a cross wall or partition of like character, thus leaving a central chamber 7 feet square, and a narrow cell at each end about 2 feet wide and 7 feet long. This had been entirely covered with a single layer of round logs, varying in diameter from 6 to 12 inches, laid close together side by side across the width of the vault, the ends resting upon and extending to uneven lengths beyond the side walls.

In the cenral chamber were 11 skeletons, 6 adults, 4 children of different sizes, and 1 infant, the last evidently buried in the arms of one of the adults, presumably its mother. They had all apparently been interred at one time as they were found arranged in a circle in a sitting posture, with backs against the walls. In the center of the space around which they were grouped was a fine large shell, Busyon perversum, which had been converted into a drinking cup by removing the columella.

Fig. 57, shows the plan of the vault, the positions of the skeletons, and the projecting ends of the logs on the outside.

American Antiquarian, Oct., 1879
Peoria County, Illinois

In the northwest part of Peoria County, in the state of Illinois, is quite an interesting group of mounds. They are scattered along the north bank of Spoon River, for a distance of five miles. There is one peculiarity that have never noticed in other mounds, and that is that the "finds" are all deposited in "warpaint," or red ochre

These mounds are 45 in number, mostly round burial mounds, about 40 feet in diameter, but one mound is a cross, shaped thus t, being 45 feet long, the cross being 33 feet long... the whole mound is about 2 feet high... One mound was highly interesting, being beyond a doubt a "cremation mound"--the only one I ever saw. Five bodies were burnt laid across one another in the form of a star, the skulls being to the S., S-W., N-W., N., and N-E. At a point between the ear and point of the shoulders was a small jasper pebble, except the skull to the south. The heat of the five had burnt the clay to brickred to the depth of 18 to 20 inches. The bones were all burnt from the middle of the pile, the head and shoulders alone remaining with a few fragments of charred bone.

American Antiquarian, April 1878
Geneva Illinois

One written by James Maitland, makes mention of the exploration of several mounds near Geneva, Illinois. There were found in the mounds a number of skeletons and skulls and other relics. Some of the skeletons are described as "lying side by side but having their bones in a position to describe the arc of a circle." Some bone needles and pieces of pottery were found with them, but what is noticeable especially as associated with the circular position of the bones, were the evident tokens of fire in the mounds. "Three feet below the surface was a considerable amount of charcoal extending through the mound at about the same depth. Below the charcoal, the earth was caked as if by a long continued fire, and a considerable deposit of ashes."

Gematria and the Babylonian and Biblical Codex

Gematria is the act of assigning a numerical value to letters that exposes hidden meanings of words or phrases. Numbers in of themselves held sacred meanings and attributes that could be attached to phrases or to the measurements of a structure or object. This practice has its origins around 1800 B.C on the Medeteranean coasts with the development of the alphabet by the Amorites.

Like the later Romans, each letter also had a numerical equivalent. A concept that most are familiar with when presented with the letters XVI, it can readily be deciphered as meaning 16. Also, about 1800 B.C. The Babylonians developed a base six method of counting, called a duodecimal system. Our current clocks and measure of time have their roots on these same divisions of 12 and 6. The Babylonians also introduced advanced mathematical formulas such as multiplication, division, pi and square roots.

The cornerstone numbers of Gematria are 660 or 666 and 1080. 660 and 666 represent the sun, the light, and the male principle. 1080 represents the moon, the Earth Mother, darkness and the female principle. 1746 is the number of fusion that is achieved by adding 666 + 1080 = 1746. 1746 represents the balance of the powers of light and dark, the *Yin and Yang,* also known as *The Treasury of Spirit* and the *Pearl of Wisdom* and the balance of powers between Heaven and Earth. 555.5 times pi is 1746. 555 is a numerical anagram for the *Sacred Marriage* of the sun and earth. The square root of 1080 is 33, that is the number for the *gateway* or *path.* Christ was crucified at age 33; he is the gateway or path to heaven. The number 8 transgresses many religions with the same meaning of a *new beginning* or *resurrection.* and also that of the *Supreme Being.*

These numbers can also be found in their use as cycles of time that was formulated by the Babylonians and Greeks. These periods of time used to forecast the dates when great cataclysms would occur upon the earth. According to H. P. Balvatsky in *Isis Unveiled, A Master-Key to the Mysteries of Ancient and Modern Science and Theology,* originally published in 1877. "In order to demonstrate that the notions which the ancients entertained about dividing human history into cycles were not utterly devoid of a philosophical basis, we will close this chapter by introducing to the reader one of the oldest traditions of antiquity as to the evolution of our planet.

At the close of each "great year," called by Aristotle-according to Censorinus-the *greatest*, and which consists of six sars. Our planet is subjected to a thorough physical revolution.

Berosses, himself a Chaldean astrologer, at the temple of Belus, at Babylon, gives the duration of a sar, or sarsus, 3,600 years; a neros 600; and a sossus 60.

The polar and equatorial climates gradually exchange places; the former moving slowly toward the line, and the tropical zone, with its exuberant vegetation and swarming animal life, replacing the forbidding wastes of the icy poles. This change in climate is necessarily attended by cataclysms, earthquakes, and other cosmical throes. As the beds of the ocean are displaced, at he end of every decimillenium and about a neros, a semi-universal deluge like the legendary Noachian flood is brought about.

So uncertain were the commentators about the length of this year, that none except Herodotus and Linus, who assigned to it 10,800 years, [3600 sars X 3 =1080]. According to the claims of the Babylonian priests, corroborated by Eupolemus, the city of Babylon, owes its foundation to those who were saved from the catastrophe of the deluge; *they were the giants* and they built the tower which is noticed in history. (This is in flat contradiction of the Bible narrative, which tells us that the deluge was sent for the special destruction of these *giants. (*The Babylonian priest had no object to invent lies.) These giants who were great astrologers and had received moreover from their fathers, "sons of God," every instruction pertaining to secret matters."

The number 1080 is the number that is symbolic of the S*pirit of the Earth,* and can be associated with the Earth Mother. The earliest use of the number 1080 was found in a list compiled by Berosses, A Babylonian priest who died about 260 B.C., who compiled a list of Babylonian kings who lived before the flood.

Kings	Length of reign
Alorus	36,000
Alaparos	10,800
Amelon	46,800
Ammenon	43,200
Megalaros	64,800
Daos	36,000
Euedorachos	64,800
Amempsinos	36,000
Otiartes	28,800
Xisouthros	64,800
Total	**432,000 years**

.   Numbers of interest within the Babylonian Kings lists before the flood are the 10,800, 43,200, 64,800 because they are all divisible by 1080. If the planet is subjected to a thorough physical revolution, every 6 sars or 216,000 years; the Babylonian kings would constitute that the earth had gone through a prior catastrophe before the Noachian flood.   The number 432,000 years as the total amount of years prior to the flood is divisible by 3600 and also 1080. $432,000 \div 3600 = 120$, $432,000 \div 1080 = 400$. $432,000 \div 2 = 216,000 \div 2 = 10,800$.   216 is the sum total of $6 \times 6 \times 6 = 216$. It is also found by doubling 1080 to 2160.

This number 216 is again evident in the *Bible*. Within the *Book of Revelations*, following the identification of the number of the beast, in 13:18 ,"and his number is Six hundred, three score [60] and six." This is followed by the quote in the next chapter Revelations 14:1 "And I looked, and, lo, a Lamb stood on the mount Sion, and with him a hundred forty and four thousand, having his Father's name written in their foreheads." When 144,000 is divided by 666 the resulting number is 216.216216!

6	32	3	34	35	1
7	11	27	28	8	30
24	14	16	15	23	19
13	20	22	21	17	18
25	29	10	9	26	12
36	5	33	4	2	31

The number 666 was derived from what is called the *Magic Square Of The Sun*. It was formulated by the Babylonians. Adding the numbers across or diagonal will result in the sum of 111. Resulting in each side totalling 666.

The number 666 is recognized from "Revelations," in the Bible. It represents the number of the "Beast." But, if we look at the political climate in the days of Paul and Peter and the other of the followers of Christ, it is one dominated by Rome. The newly formed Christians were being persecuted throughout the Roman provinces. All of the evil of the world to the early Christian writers could be encapsulated into one word, 'Rome.' One of the most misunderstood quotes in the Bible is *Revelation 13:18* "Let him that hath understanding *count* the number of the beast: for it is the number of a man; and his number is six hundred threescore and six." The key word is *count*. The number 666 was an anagram for the Roman numerical system. If you add the Roman numerical sequence of D=500 + C=100 + L=50 + X=10 + V= 5 + I = 1, it equals the total of 666. The number 666 is used several times in the Old Testament with no satanic or evil connotations, but with its accurate ancient meaning of the "Sun."

In Chronicles II, 9:13 they bring Solomon "in one year, six hundred, three score and six talents of gold." Evidence that the Bible is using 666 as a reference to the Sun, can be found in the name of Solomon itself. *"Sol"* is Latin for Sun, *"Om"* is Hindu for Sun, and *"On"* is Egyptian for Sun. There is also a Gematria codex within the name Solomon in that it contains the letter " O," three times. " O" is the 15th letter in both the current and ancient Greek alphabet. Written, 1+5, 1+5, 1+5, the total is

666.

In Ezra 2:13 in describing the tribes who were returning from captivity in Babylon by the hand of Nebuchadnezzar. "The children of Adonikam, six hundred and sixty six." "Adoni" is a reference to the Phoenician Sun God, Adonis.

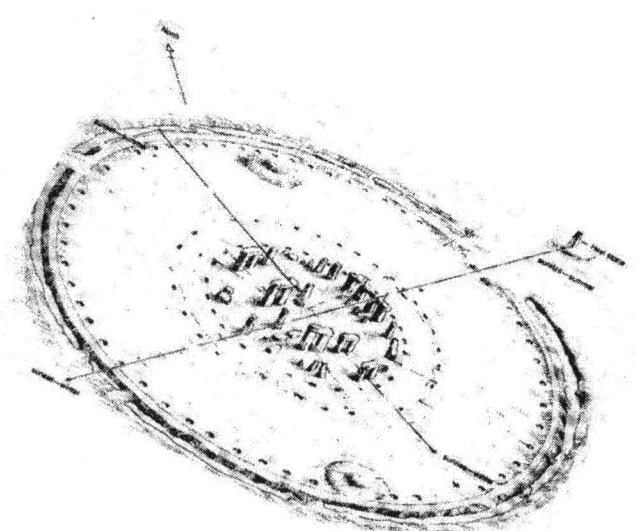

In *Alchemy, The Great Work. An Investigation into the relationship between Astrology, Alchemy and the Mystical, 1998,* by Allan Tidmarsh, gives the area of the Sarsen circles=1080 megalithic yards and the area of the Aubrey hole circle as 6660 square yards. Tidmarsh writes, "The holy sanctuary or temple was laid out, with measurements built into it to symbolize the marriage of the solar (male) and the lunar (female), as well as the union of heaven and earth."

In another work, *StonehengeDance of the Giants,* by Ross Nichols, from *Greater Sites of the Britannic Islands in the Book of Druidry.* A series of vesicas are imposed on Stonehenge with the following results. "Stonehenge is the archetype of the Henge's in the Ohio Valley. This open air Pagan temple is associated with giants on both continents. Stonehenge, was built using the ancient principles of Gematria. "There are three vesicas within Stonehenge: the first encloses the sarsen circle, the third the circle fitting within the bluestone horseshoe. This third vesica = 61.2 ft x 35.24 ft and the diamond = 1,080 ft sq. area, the number of Hermes. The circumference of 2 large circies is that of the square of the sun. The middle circie coincides with the circle of 30 'y' holes now invisible. At Midsummer's

Solstice the sun penetrates the first and third vesicas-the sacred number 666 penetrating the earth-spirit mercury's number 1,080. Added, they make the number of perfections 1,746, by gematria the *Pearl of Wisdom.*

The number 33 occurs several times in the Bible. There were 33 descendants from Adam to David. Christ lived 33 years. Christ is the gateway. It is through Christ (33), that you must pass to reach the spiritual or Holy place.

The number 33 also occurs, associated with giants, Dr. Beedo is quoted in ("Archaic England" Harold Baily 1920, pg. 192) "Our British Giants, Gog, Magog, Termagol, and the rest of the terrible tribe, sprang, according to Scottish myth, from the thirty three daughters of Diocletian, a King of Syria, or Tyria. These thirty-three primeval women drifted in a ship to Britain, then uninhabited, where they lived in solitude, until an order of demons becoming enamoured of them, took them to wife and begot a race of giants."

There is also an ancient Icelandic myth of Magus, who at the age of 330 years magically cast his skin and transformed himself into a youthful man. Maga is sometimes associated with Hercules or Og. The "casting his skin" may refer to December 20, when the sun of old is destroyed to be reborn 3 days later. The Gnostics believed in 30 divine rulers that represented each day of the month. They called these Aeons. Aeons was a British term for the sun.

The original assumption of 33 meaning path or gateway has to be expanded to the image of the sun on its yearly path in the sky.

Long Meg is said to be one of the last Titans that inhabited the British Isles. The diameter of Long Meg is 330 feet that is reminiscent of Magus. However, her stone being the largest is 18 feet high that is believed to be an anagram for 1080. Long Meg has 72 stones that make up the stone circle, which Bayley wrote in *Archaic England,* had a realtion to the seventy-two dodecans into which the Chaldean and Egyptian zodiac was divided. Christ, is said to of had 72 disciples. 72 X 15 = 1080.

The image that is repeated in these earthworks and is identified with 'beginning' is the picture of the world enveloped in the primeval waters from which arises the primordial mound. In the theology of Egypt, this mound gave birth to 8 deities, known as the Ogdoad. These deities placed an egg on the mound that would be the birth of the Sun-God. There is some dispute on whether this was Thoth or Amum, but the important concept is that they were both represented as birds.

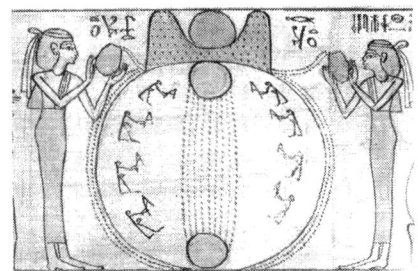

Egyptian Ogdoad

The story in the Bible of Noah is similar in that 8 people survive the flood in a world inundated with water and they send out a bird that brings back evidence of a mound springing from the waters. Noah is identified with the number 8 in Peter 2:5 "And he spared not the old world, but saved Noah the eighth person" In Hebraic writings it said that Noah kept the giant Og alive and fed him from a hole that he made in the roof. Since Og is an ancient Sun God, this story is even closer to the original Egyptian creation story.

This image of God , seated in the north sky occurs several times in the Bible. This originates from the north-star that never moves in the night sky, yet all of the other stars spin around it. These stars are more specifically the Little Dipper, or Ursa Minor. The seven stars of the dipper are terminated at the north-star. This 8th star was seen as the sanctuary of God, that all others revolved around. This 8 stages of heaven is still referred to today when someone says "they are in 7th heaven" they are actually making a reference to long lost concept of God residing in the night sky at the north-star. The name *El* for God, which meant "Strong One," and its subsequent usage in words like Jeruselum, Bethel, Ezekiel has its etymological roots in the Sumerian word *"Ool"* which meant "to spin." Ool was in reference to the stars spinning around the North-star. It was El, the "Strong One" who held up and turned the fabric of the heavens with his great strength. The Egyptians held sacred the rolling beatle for this same reason. *Ool*, is the root of our present word "roll" and the "eel."

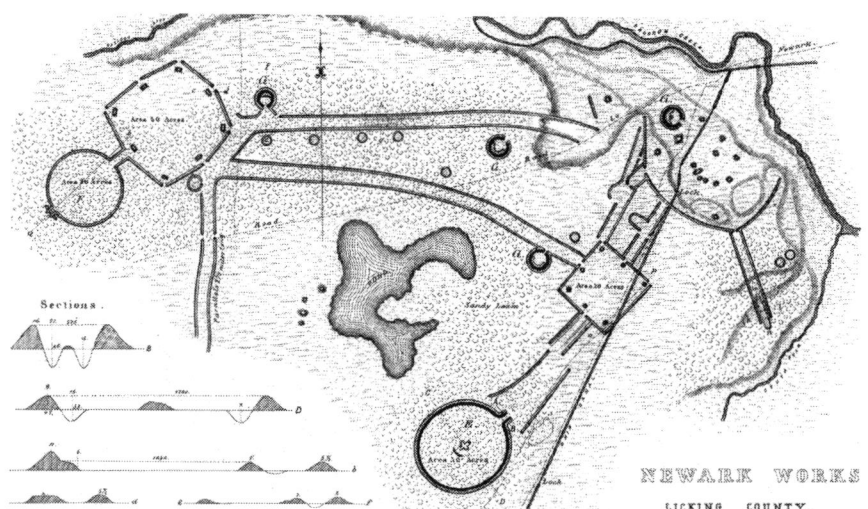

The number 8 repeats itself many times within the earthworks in the Ohio Valley. The octagon at Newark is aligned to the minimum and maximum moon sets and rises. The octagon and the number 8 represent, "birth," or "resurrection," and tied to the prehistoric belief that the moon was the "keeper of souls." The May 1st alignment of the henge with the bird effigy in its center is symbolic of the Earth Mother. The square earthwork is also an earth Mother symbol. The henge is 1250 feet in diameter, representing the largest henge constructed by the Allegewi Hopewell. The largest henge in Britain was at Avebury that also measured 1250 in diameter.

666, 1080, 555 ,33 and 8 are prominent in the ancient world of the Amorites, Babylonians and the Greeks. It is also found within the measurements of the earthworks of the mound builders of the Ohio Valley. If we are to believe that the Amorite giants migrated to North America and were the builders of the earthworks in Ohio. Then, is it not probable that complex mathematics and gematria that originated in Babylon and was used by the Amorites, would also be be present in the Ohio Valley?

Allegewi-Hopewell Numerology and Gematria

The evidence of a religious cult based on numerology is theorized on the occurrence of certain lengths or numbers within the Allegewi Hopewell earthworks. These various lengths appear with such regularity to conclude that the earthworks were methodically measured using both simple and complex mathematics in conjunction with a numerical canon that specified the lengths and diameters of various circles and square earthworks, prior to their construction. Certain numbers or lengths must have exposed hidden words, phrases or images. In other words, numbers in of themselves held sacred meanings and attributes that were encompassed within the construction of an earthwork. The earthworks were then, not only held sacred within their shape and function, but also the numbers or lengths used in their construction

The foot is a standard measurement throughout history each version being within about an inch or so from one another. There is however, what was called the Olympian foot from Greece that is identical to our standard foot. Since the Greeks adopted the use of Gematria and the alphabet from the Amorites, this may have been where this measurement was diffused to the British Isles and North America. The question of measurements of the earthworks in the Ohio Valley being 660, 666, 1080, 555, 33 is not mathematically possible without the mound builders having the identical form of measurement in the use today, of the foot, the rod of 16.5 feet and the furlong of 660 feet.

. William Romain, author of the, *Mysteries of the Hopewell, Astronomers, Geometers, and Magicians of the Eastern Woodlands*, 2000, examined the mathematical relationships between the circles and square earthworks and octagons and shows the complex arithmetic that was utilized in their construction. Romain explains the spatial relationships of these earthworks in mathematical formulas and shows that the Allegewi and Hopewell knew how to "square a circle," meaning the square and circle had equal perimeters. Also, they understood nested squares, where the diagonal of a square earthwork conjoined with a circular earthwork had a corresponding diameter. The Allegewi Hopewell earthen octagons are also explained geometrically as formulated from truncated squares and the mathematics involved in the making of these imposing earthworks.

This investigation explores the lengths, diameters and circumferences of earthworks and the contexts in which certain numbers reoccur. It is within these repeating numbers that a knowledge of multiplication, division, pi and square roots becomes apparent. The numbers or lengths that occur with the most regularity within Allegewi and Hopewell works are 3, 30, 33,120, 210, 212, 215, 240, 250, 420, 660, 666, 555, 1050,1080, 800, 1720 and 8. The interrelation of these lengths within the earthworks reveals the meanings of the symbolism contained within the earthworks shapes and their function.

Theoretically, the numerical codex used by the Allegewi Hopewell was that the lengths of 660 and 666 feet, were symbolic of the male or solar deity.

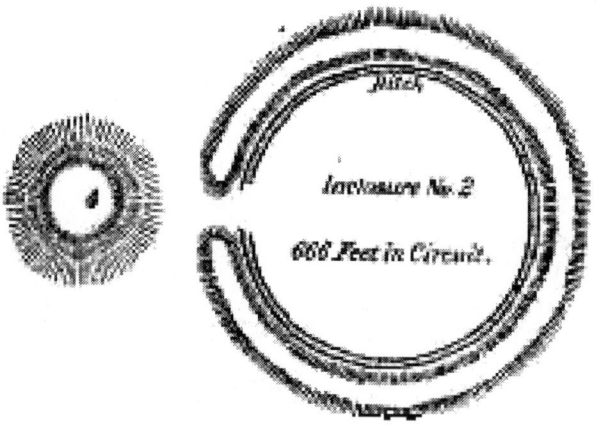

Two henges at Charleston, West Virginia measured 212 feet in diameter or 666 feet in circumference. 666 being symbolic of the Sun within the Gematria codex.

1080 feet or length was symbolic of the female and the Earth Mother. The two cornerstone numbers or lengths, representing the male and female aspects of 1080 and 660 can be further reduced or simplified, as evidenced by the many works that contain the numbers 240 and 420. It is from these two numbers that the lengths, diameters, and circumferences of many earthworks are derived.

The numerical codex was utilized in the construction of many earthworks, and can be viewed mathematically as:

420 +240 =660. 420 + 420 + 240 = 1080. 1080 – 660 = 420. 660 – 420 = 240. The lengths of 420 and 240 occur singularly and combined within earthworks.

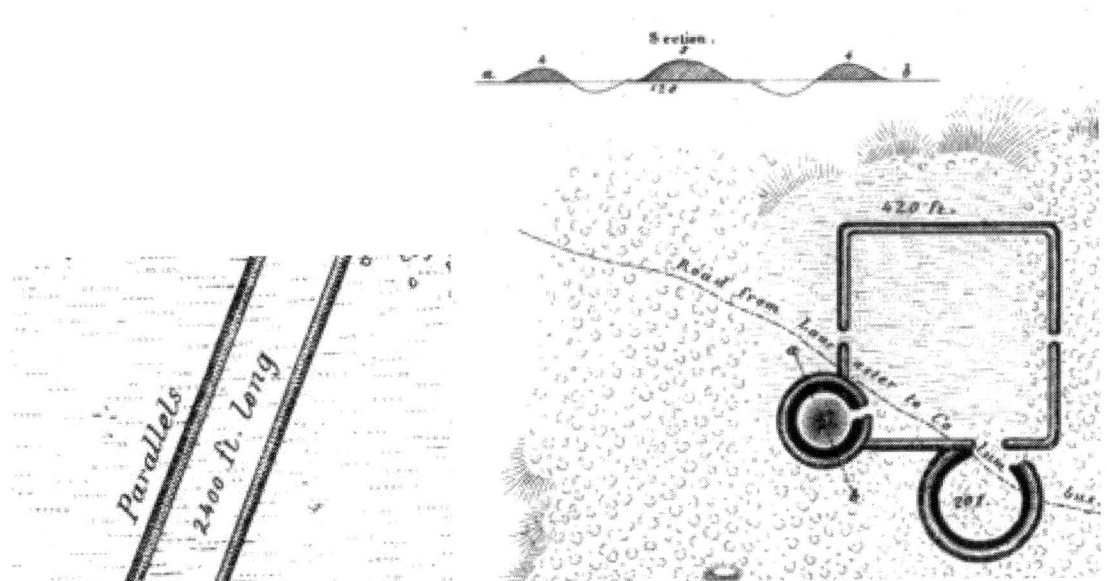

The length measured of the sacred via at Hopeton was 2400 feet. To the right is a square earthwork that measured 420 feet per side. 420 could be expressed as either, 1080-660=420 or 660-240 =420. Adding two sides of the square. 420+420=840, 1080-840=240. The adjoining henge on the southwest corner measured 120 feet in diameter. 120 is half of 240 or 120 X 9 = 1080. If you add the length of one side of the square, with the diameter of the circle, 420 + 120 = 540, 540 X 2 = 1080.

Graded Way at Piketon measured 1080 feet in length with walls separated by 215 feet.

The works at Piketon incorporate more of the numbers within the Allegewi Hopewell canon than any other earthwork site. The length of the parallel walls of the sacred via are 1080 feet, separated in the middle by 215 feet. An earthwork that was located south of the via, in what is now the city cemetery measured 420 in two sections and 240 on another. The end wall measured 212 feet, 212 X pi (3.141) = 666.

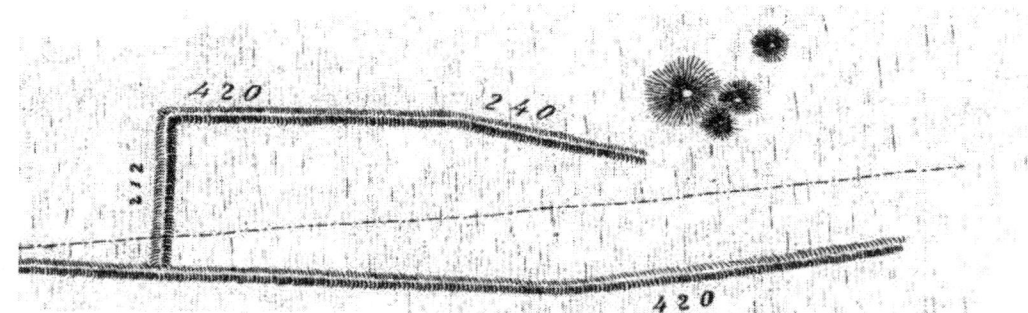

At Piketon the sacred via was 1080 feet. The end wall measures 212 feet, 212 X pi or 3.141 = 666. Adding the lengths of the separate angles, 420 + 240 = 660. 420 + 420 + 240 = 1080.

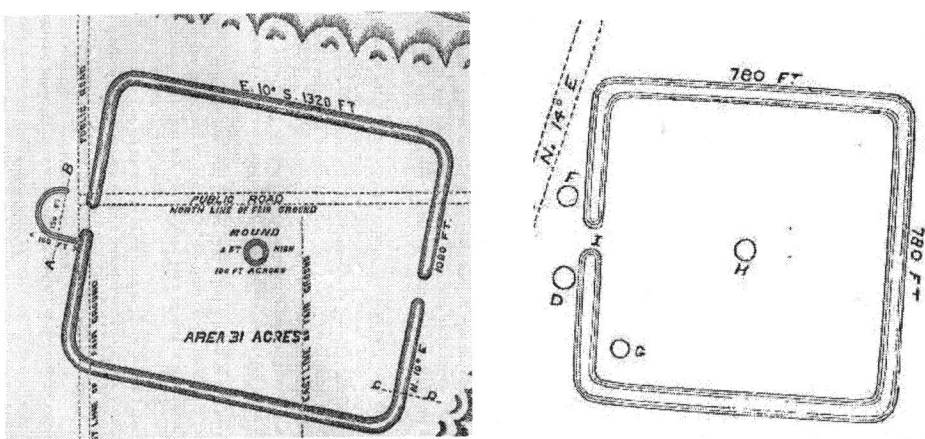

Located within seven miles of each other are one rectangular and one square earthwork in southeast Indiana. The work at Winchester measured 1080 feet on the north-south walls, and 1320 feet on the east-west, or 660 X 2 = 1320. The earthwork seven miles south in Fountain City measured 780 feet per side, which is 540 feet less than the 1320 foot walls at Winchester. 1320 – 780 = 540. 540 x 2 = 1080. Also, it is notable that 780 is 120 more than 660. 120 X 9 = 1080. These two works were constructed to be numerically harmonic.

In geometry an octagon is a polygon that has eight sides. An octagon is defined by *Wikepedia* as, "A regular octagon has all sides the same length and internal angles are all the same size. It has eight lines of reflective symmetry and rotational symmetry of order 8. The internal angle of each vertex of a regular octagon is 135 degrees and the sum of all the internal angles is 1080 degrees."

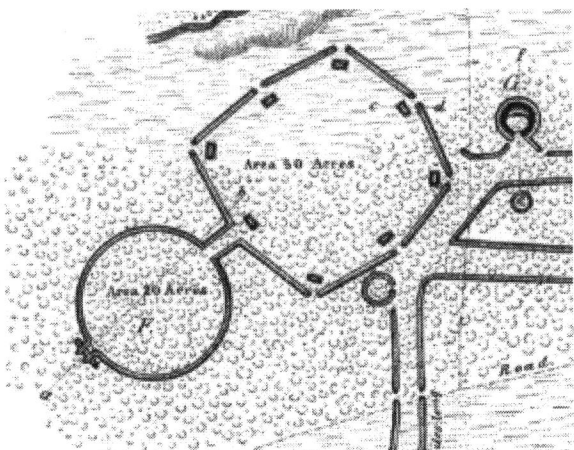

Astronomers, R. Horn and R. Hively, in A *Statistical Study of Lunar Alignments at the Newark Earthworks,* 2006, determined that the axis of the octagon was aligned to the minimum southern and northern moonrise.

Photo is the wall of the octagon with interior mound visible at the gateway. The circle and octagon are preserved by a golf coarse that is within the earthen walls. From, *The Nephilim Chronicles, A Travel Guide to the Ancient Ruins in the Ohio Valley*, 2010.

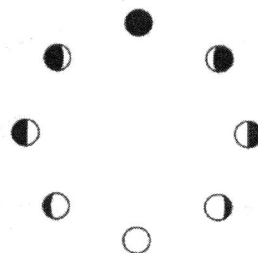

The eight mounds at the gates of the octagon may also be symbolic of the eight phases of the moon. Evidence that the earthworks at Newark were dedicated to the worship of the Earth Mother is furthered by the alignment of the large henge to the May 1 sunrise. The effigy in the center of the henge is a bird that is a traditional symbol of the Earth Mother.

The number 555 is evident within the circumference of several henges located at Charleston West Virginia and north of Lexington, Kentucky. The significance of the works north of Lexington is that they lend further evidence of the Allegewi Hopewell's knowledge of square roots. The henge has a circumference of 555 feet, which is significant because if you stretch 555.5 and pi out far enough and multiply them, the total is 1746 which is the sum total of 666 + 1080; reresenting the *Sacred Marriage.*

232

Another work adjacent to the henge had an earthen wall that measured 1080 feet. The gateway to the henge is 33 feet. 33 is the square root of 1080.

Henge, located north of Lexington, Kentucky with a circumference of 555 feet.

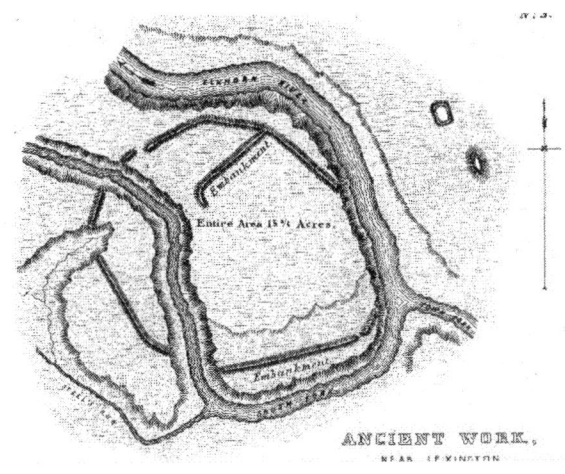

Henge at Lexington Kentucky with a circumference of 555 feet. 555.5 X pi (3.141) =1746 or the sum total of 666 + 1080. The square root of 1080 is 33, the width of the gateway. The numerology suggest that this the henge represented both the male and female aspects and was used to venerate the sun and the earth, representing the *Sacred Marriage of Opposites*. The earthwork above was located on an adjacent hill from the henge with one of the walls measuring 1080 feet in length.

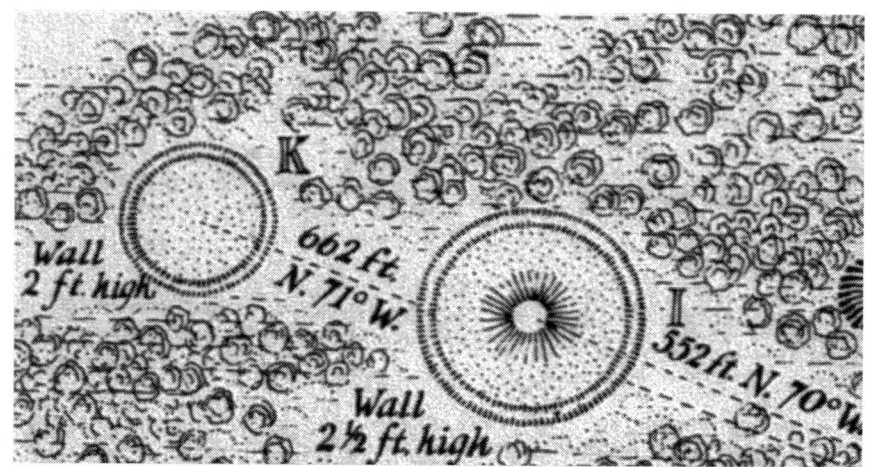

The number 555 and 666 also occur together at Mounds State Park, in Anderson, Indiana as measured by Lilly in *Prehistoric Antiquities of Indiana,* 1937. Two earthworks aligned to the summer solstice sunset and the winter solstice sunrise as positioned and measured from the central mound, formerly located within the henge. The lengths from the central mound to earthwork "K" was 662 (666) feet and to earthwork "I" was 552 (555) feet. The largest henge of the group is 660 feet in circumference. Another henge within this group was 120 feet in diameter. 555 X 120 = 66600

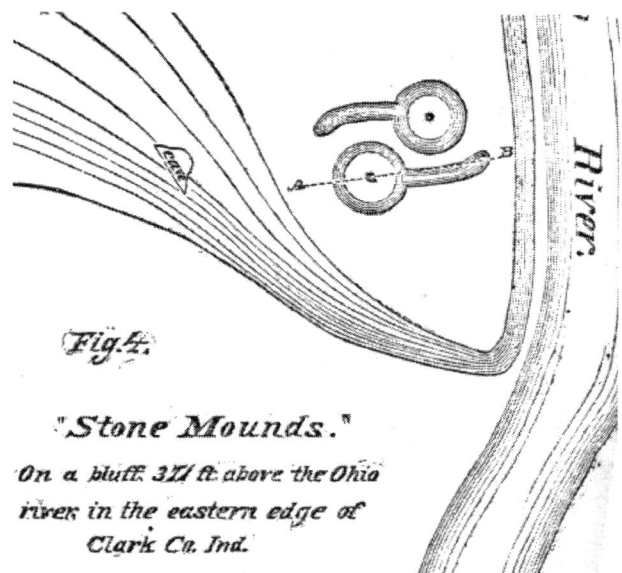

Fig.4.

"Stone Mounds."

On a bluff 371 ft. above the Ohio river in the eastern edge of Clark Co. Ind.

This symbol has historically represented the marriage of the opposites, the yin and yang or the "Holy Union of Opposites." It represents and equilibrium of natural forces. This notion of these shapes representing balance is in harmony with the apparent alignment of the works with the equinox sunrise and sunset, when day and night are equal. Also note the cave at the bottom of the bluff; this is evidence that underworld spirits were also venerated at this site.

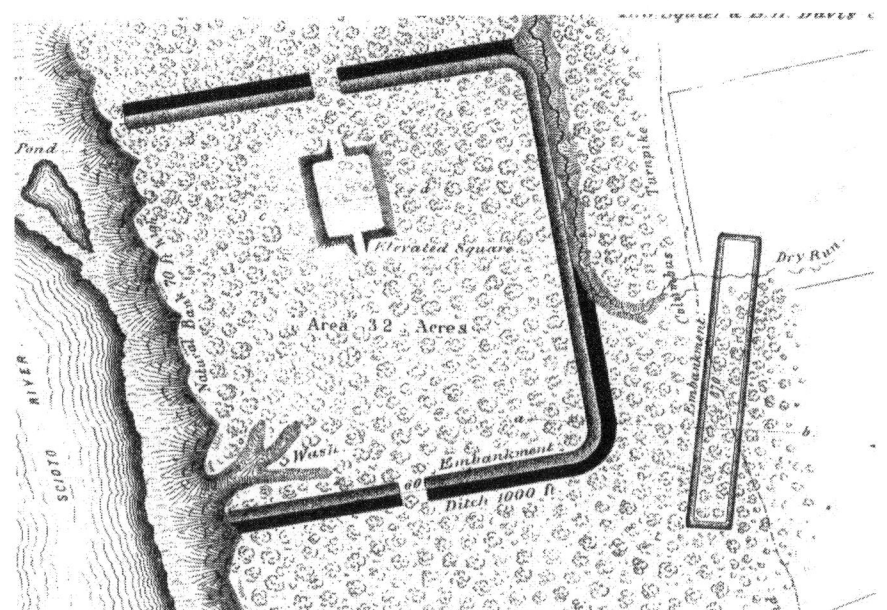

The Cedar Banks Works at Chillicothe featured a platform mound that may have been the seat of power for the Capital City. Of interest is the earthwork on the right that measured 870 feet in length. Both sides of the work measured, 870 + 870 = 1740 as does 660 + 1080 = 1740. The importance of the length of 870 feet is revealed in the following equations. 870-660=210, 1080-870=210.

The most common diameter of an Allegewi henge is 210 or 212 feet in diameter, or a circumference of 660 or 666 feet. Henges that are 660 feet in circumference occur at the Junction Group in Chillicothe and Athens in Ohio, Cambridge City and Mounds State Park in Indiana and at Camden South Carolina. With the apparent solar alignments of these works, the numbers 660 and 666 are easily associated with the Sun.

The derivation of 210 can be achieved two ways, either as ½ of 420 or as the the number times pi to achieve 660. 210 X pi (3.141)= 660, 212 X pi (3.141)=666. With the lengths of 210 and 212 being found in linear works as well as the diameters within henges, it is evident that the Allegewi and Hopewell were using pi to formulate the construction of many of the circular works.

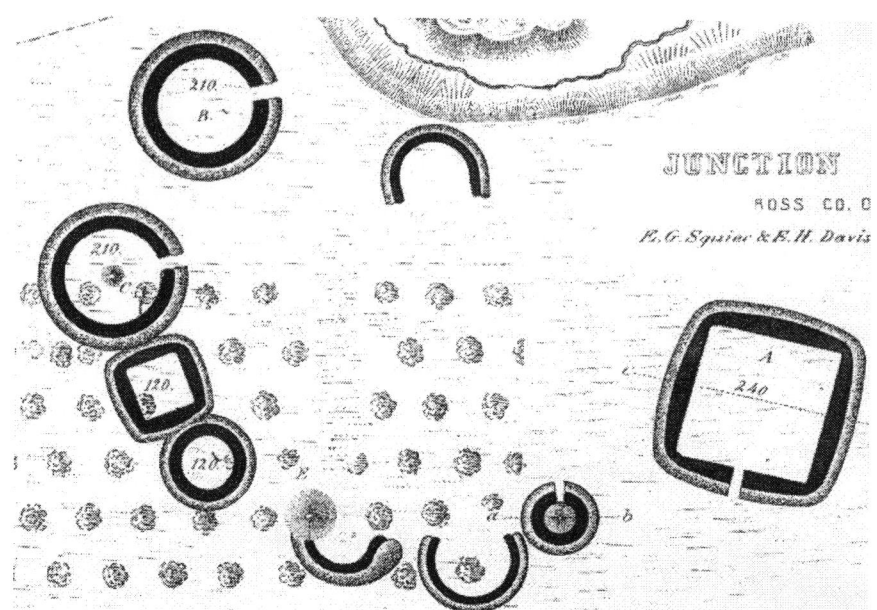

The Junction Group is located at the mouth of the North Fork of Pant Creek and Paint Creek, in Chillicothe, Ohio. Within the is group are two henges with a diameter of 210 feet, this times pi, (3.141) equals a circumference of 660 feet. A joined circle and square are each 120 feet in diameter. The combined lengths of the diameters 210 + 210 +120 +120 = 660. 660 can also be derived by adding the diameter of the two largest henges with the diameter of the square, or 210 + 210 + 240 = 660. There are 8 earthworks in this group, symbolic of "resurrection." The two horseshoe shaped works signify the mystical door of life, or was symbolic of the vulva of the Earth Mother. The burial mound with a serpent emerging from it is also an Earth Mother symbol in this context.

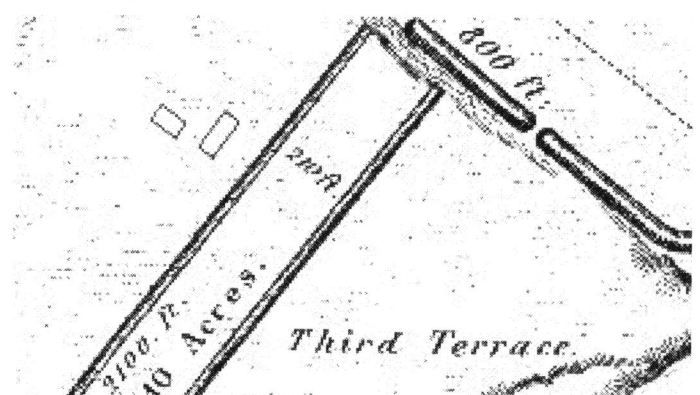

The measure of 210 feet occurs in the diameter of numerous henges, it also can be found within linear earthworks constructed by the Allegewi. A sacred via that extends from the square, opposite Portsmouth, Ohio was 210 feet wide and 2100 feet long. 2100 could also be represented as 1050+1050 =2100.

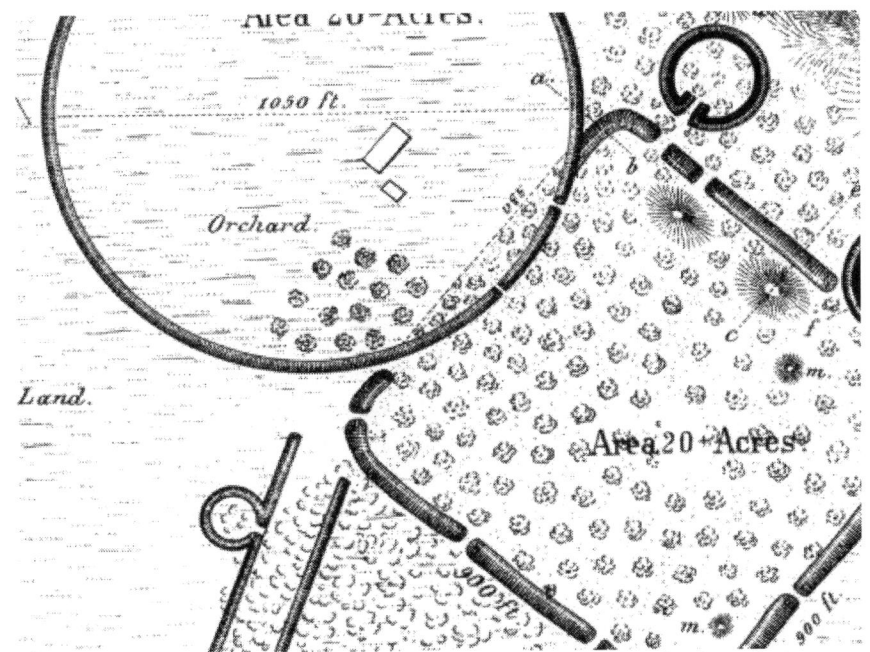

The measurement of the diameters of the circular earthworks attached to the octagons at Newark, High Bank and Hopeton are 1050 feet or 210 X 5 = 1050. The lengths of the sides of the square work are 900 feet or the sum total of 660 + 240 = 900. The length of the sacred via was 2400 feet.

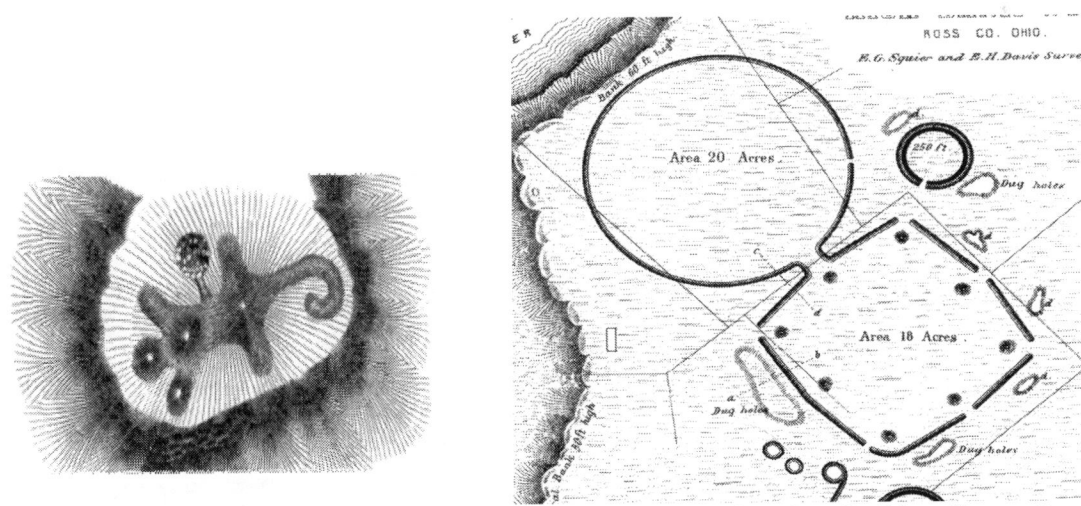

Effigy at Granville from head to tail measured 210 feet and is aligned to the May 1st, sunrise. Highbank's circle is 1050 feet in diameter. 210 x 5 = 1050. Highbank is aligned to the winter solstice sunrise and the summer solstice sunset.

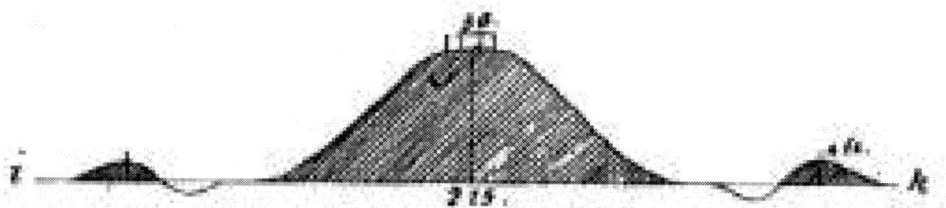

As at Piketon and the earthwork complex at Marrietta, Ohio included the length of 215 feet, shown above as the diameter of the mound and henge. The mound and earthwork were attached to a square work that had sides that measured 1080 feet. Also at this site are several platform mounds, one of which was 210 feet in diameter.

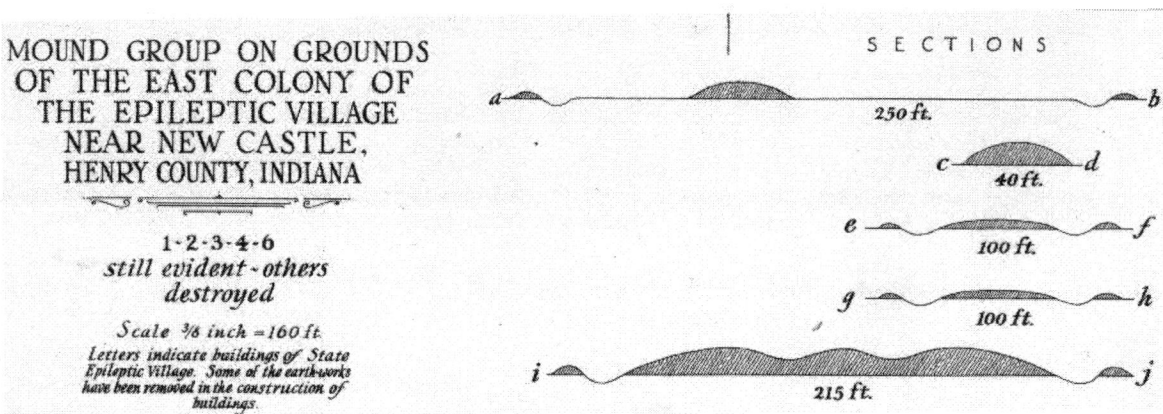

The large mound at New Castle was 215 feet and was surrounded by an earthwork and ditch with a fiddle shape, similar to Earthwork "H" at at Mounds State Park in Anderson, Indiana, that was also 215 feet in length. The length of 215 must have been symbolic of the Earth Mother. Also note the combined lengths of the two smaller henges and mound are 240 feet. From *Prehistoric Antiquities of Indiana*, 1937.

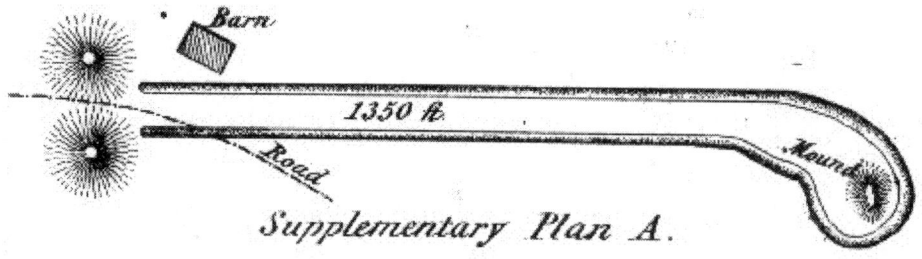

Sacred via at Ft. Ancient was 1350 feet in length. 215 X pi (3.141) = 675 X 2 =1350.

A similar work called, "The Cursus" is located near Stonehenge.

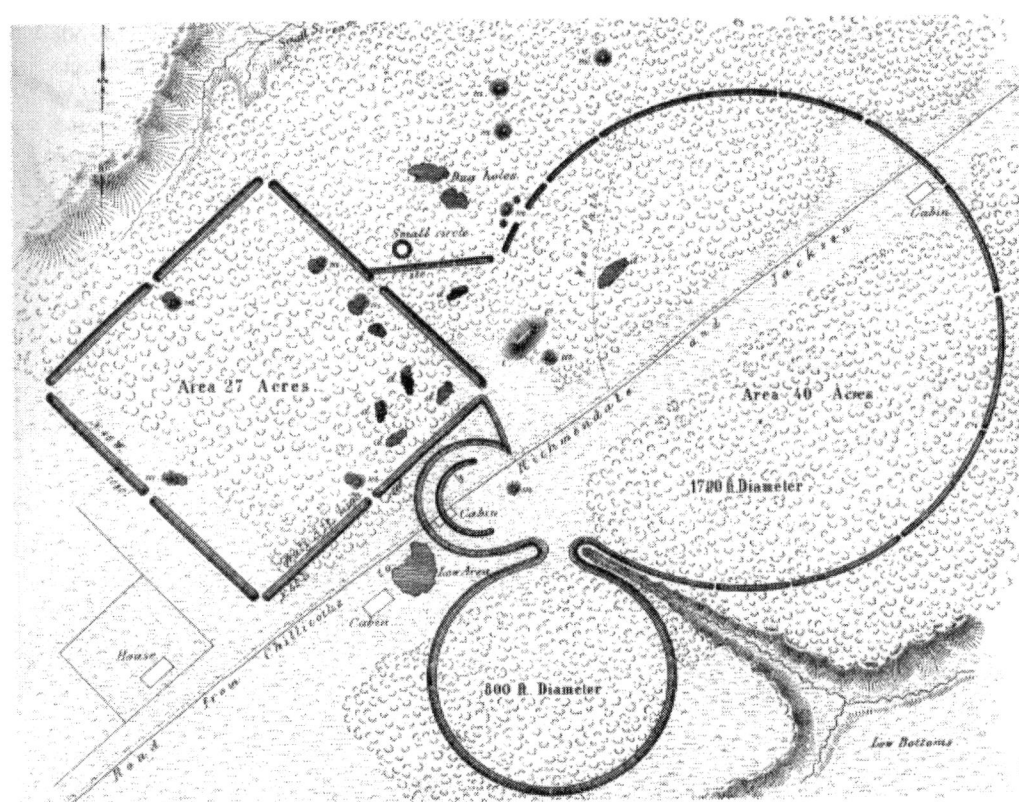

Liberty Works located south of Chillicothe, Ohio, dimensionally the same as four other earthworks around Chillicothe. The large circle measured, 1720 feet or 215 X 8 = 1720. The smaller circle is 800 feet in diameter. The square earthwork has sides that measure 1080 feet.

The most common number that occurs within the Hopewell numerical canon is 250. This number is most evident within the diameter of henges that differ from the Allegewi's in that the circular platform is larger because the surrounding ditch is less wide and outer earthwork is usually only several feet. The lowest repeating multiple of 250 is 50 that had been measured around some Hopewellian earthworks that were thought to be the base of a structure. Other significant multiples of 50 occur at the Highbanks works near Chillicothe that had several henges that were 300 feet in diameter. The largest henge built was at Newark, measuring 1250 feet in diameter.

Many of the henges that measured 250 feet in diameter were aligned to the equinox, but not all. In one work 250 represents the serpent, that was also a common attribute of the undulating Allegewi henges. It could be deduced that the length of 250 feet represented the serpent and the serpent is symbolic of the Sun.

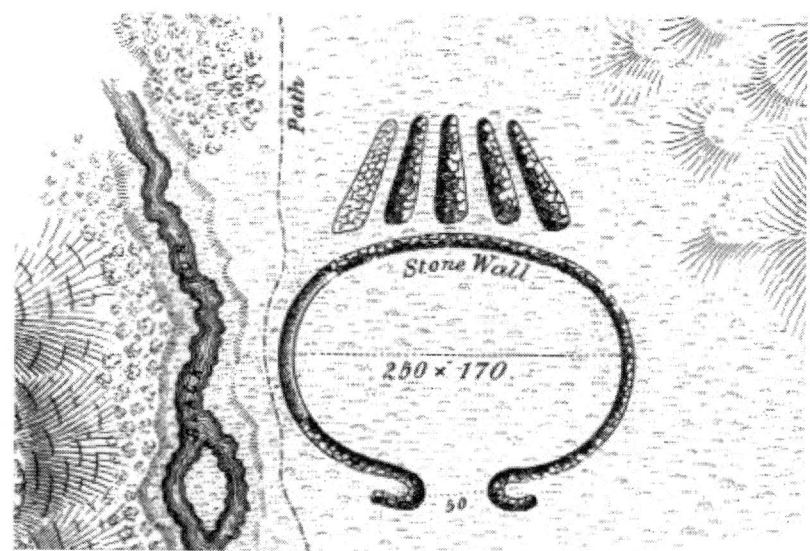

This stone work is evidence that the length of 250 was associated with the serpent. The gateway was 50 feet wide. It was located near, and visible from Spruce Hill that was also serpentine in shape and also constructed from stones.

Another serpentine work was at Fort Hill that had 33 gateways within its undulating stone walls. The central platform adjacent to the gateway depicting two serpent's heads was 250 feet in length as diagrammed in *Ancient Monuments.*

One of the 33 gateways at Fort Hill, whose stone walls undulate like a massive serpent enclosing this large hilltop. A short distance from this work is the serpent mound that measured 1330 feet, with 3 bends in the body, 3 coils in the tail and was aligned to the confluence of 3 creeks. *The Nephilim Chronicles, A Travel Guide to the Ancient Ruins in the Ohio Valley.*

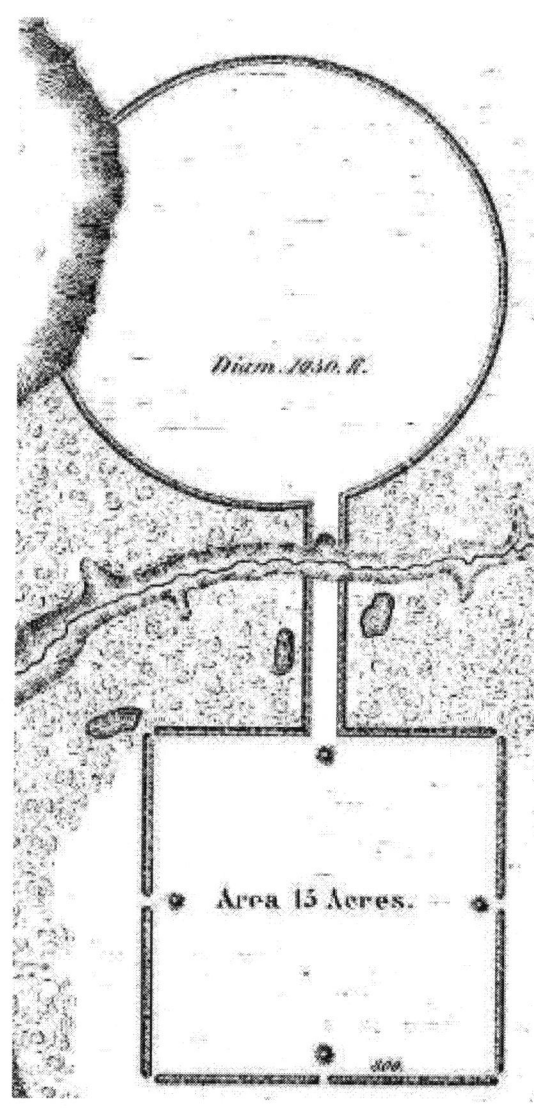

Diam. 1050. ft.

Area 15 Acres.

The occurrence of the number 250 feet reveals the significance of 800 feet that is present within square and circular works. The earthwork to the left is called the Seal Township Works, in Pike County, Ohio. The circle is 1050 feet in diameter and the square has sides of 800 feet. 1050-800=250.

It is evident that the Allegewi Hopewell utilized a numerical canon expressed in lengths that were incorporated in the construction of the earthworks found primarily in the Ohio Valley. With the numerical repetition of two numbers, 660 and 1080, 660 being the sun or male principle and 1080 representing the Earth and the female principle. Many of the earthworks were dedicated to both Sun and the Earth. The earthworks were constructed, to achieve the *Sacred Marriage,* between these two powers that were the most revered.

Serpents

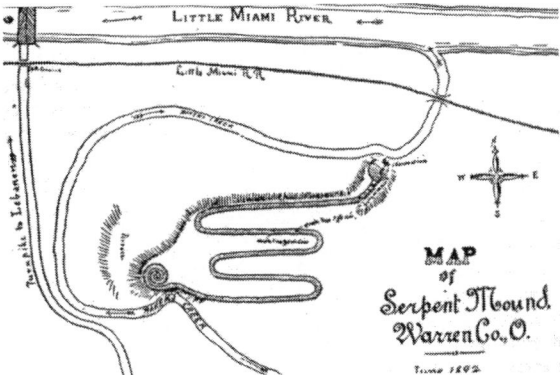

LITTLE MIAMI RIVER

MAP
of
Serpent Mound
Warren Co., O.

Serpent Mound in Warren County, Ohio, that was destroyed by a gravel company. A similar serpent effigy is still visible north of Holton, Indiana.

The image of the serpent can be seen within many of the earthworks in the Ohio Valley. Henges with undulating outer walls were symbolic of the serpent. *Ancient Monuments of the Mississippi Valley,* 1848. " The serpent, separate or in combination with the circle, egg, or globe has been a predominate symbol among many primitive nations. It prevailed in Egypt, Greece, and Assyria, and entered widely into the superstitions of the Celts, the Hindu, and the Chinese. It even penetrated into America; and was conspicuous in the mythology of the ancient Mexicans, among whom its significance does not seem to have differed materially from that which it possessed in the old world."

There are unique similarities in the serpent effigies in Oban Scotland, (*Ob* is Egyptian for serpent*)* and the serpent in Adams County, Ohio. Both serpents are depicted with mouths open swallowing an egg. Both of the "eggs" showed traces of sacrificial fires within their interior. The serpent at Oban is facing three peaks in the distance. The Ohio serpent is facing the confluence of three creeks in addition to the body having three bends and the tail coiling three times. The Ohio serpent is aligned to the summer solstice sunset, with the bends in the body oriented towards the 18.6 year minimum and maximum moon sets.

Serpent Mound at Oban Scotland

History of Powys Fadog Vol., I, **1885**

The Sun-God, as the giver of life, was represented under the type of the serpent. This animal readily forms a circle, and a circle was the emblem of eternity. The serpent was also celebrated for its wisdom (Gen. 111, 1; Matt x, 16) It has been said with considerable truth that, "in the mythology of the primitive world, the serpent is universally the symbol of the sun, and the generative power of the solar beams is always typified by pendent Uraei. The Uraeus is the basilik or cobra di capello. The Basilisk-Arau of Kam is stlyed in an ancient papyrus, 'Soul of the body of Ra', the sun."

"When the Egyptians wish to express extended period (aion) they depict a serpent whose tail is concealed by the rest of his body, which they call Ouraios or Uraeus. The serpent is exeedingly long-lived, and not only retains its youth by putting off old age, but also it is wont to receive a greater increase of strength. The serpent, again, forms a circle, and was so represented with its tail in its mouth by the Phoenicians, and thus appears on numerous Gnostic gems; a circle-formed serpent often of eternity and of God.

"About three miles on the other side of Oban is Glen Feochan. Here lies the huge serpent-shaped mound, the very existence of which, strange to say, was utterly unknown to the scientific world till discovered by Mr. Phene, and by him revealed to the Antiquarian Society in the summer of 1871. Being in Oban soon afterwards,..

"Finding ourselves thus unconsciously in the very presence of the Great Dragon, we hastened to improve our acquaintance, and in a couple of minutes has scrambled on the ridge which forms his backbone, and thence perceived that we were standing on an artificial mound three hundred feet in length, forming a double curve, a huge letter S, and wonderfully perfect in anatomical outline. This we perceived the more perfectly on reaching the head, which lies at the western end, whence diverge small ridges, which may have represented the paws of the reptile. On the head rest a circle of stones, supposed to be emblematic of the solar disk, and exactly corresponding with the solar circle as represented on the head of the mystic serpents of Egypt and Phoenicia, and in the Great American Serpent Mound. At the time of Mr. Phene's first visit to this spot there still remained in the center of this circle some traces of an alter, which, thanks to the depredations of cattle and herd-boys have wholly disappeared.

"The circle was excavated on the 12th October, 1871, and within it were found three large stones, forming a chamber which contained burnt human bones, charcoal, and charred hazel nuts. Surely the spirits of our pagan ancestors must rejoice to see how faithfully we, their descendants, continue to burn our hazel nuts on Hallow E'en, their Autumnal Fire Festival, though our modern divination is paracticed only with reference to such a trivial matter as the faith of sweethearts...

One of the serpent-mounds discovered in North America, described by the Messrs. Squire and Davis, represents a serpent 700 feet long as he lies with his tail curled up into a spiral form, and his mouth gaping to swallow an egg 160 feet long by 60 feet across. At the Edinburgh meeting of the British Association, in 1871, Mr. Phene gave an account of his discovery in Argyleshire of a similar mound several hundred feet long, and about 15 feet high by 30 feet broad, tapering gradually to the tail, the head being surrounded by a circular cairn, which he supposes to answer to the solar disc above the head of the Egyptian Uraeus or Araius, the position of which, with head crest, answers the form of the Oban serpent-mound. All the great myths are manifold in meaning, and replete with complex significations.

Mr. Phene, likewise found several other serpent mounds surrounded with so-called Druidical remains, among the Eildon and Arran hills. All these are more or less akin to the reptile mounds discovered by Messrs. Squire and Lapham, always in connection with sacrificial or sepulchral remains. The position of the alter in the circle or oval at the head of the serpent is identical with that of the Argyleshire mound, the head in each case lying towards the west. The American mound is, however, on a larger scale than its Scotch cousin, being altogether a thousand feet long. It points towards three rivers, thus indicating the reverence of the triple symbol-another instance of which occurs on the hill known as Lapham's Peak, on which lofty summit three artificial mounds were found constructed of earth and stone.

Drawing of the Serpent Mound in Adams County, Ohio.

Photo of the Serpent Mound in Adams County, Ohio. *The Nephilim Chronicles, A Travel Guide to the Ancient Ruins of the Ohio Valley.*

All of the sun-gods had serpents for symbols, the Canaanite's Baal, Phoenician's Adonis and the Greek's Bacchus. The serpent was the symbol for Divine Wisdom of the Egyptians. When depicted with it's tail in its tail in its mouth it denoted, Eternity. The serpent is also the symbol of time, from which wisdom springs.

Isis Unveiled: A Master Key to the Mysteries of Ancients and Modern Science and Theology, H.P. Blavatdky, 1877. "This tradition of the Dragon and the Sun-occasionally replaced by the Moon-has awakened echoes in the remotest parts of the world. It may be accounted for with perfect readiness by the once universal heliolatrous religion. There was a time when Asia, Europe, Africa, and America were covered with temples sacred to the sun and the dragons. The priests assumed the names of their deities, and thus the tradition of these spread like a net-work all over the globe."

. "There is sufficient evidence that the religious customs of the Mexicans, Peruvians, and other American races are nearly identical with those of the ancient Phoenicians, Babylonians, and Egyptians; and if, moreover, we discover that their religious terms have etymologically the same origin; how are we to avoid believing that they are the descendants of those whose forefathers "fled before the brigand, Joshua."

The serpent mounds found in the Britain and the Ohio Valley are reproductions of the ancient symbolism found in Babylon and Egypt. *Here Be Dragons: The Strange Enigma of Serpent Mounds,* Phillip Gardner, " Posidonius also tells of one [serpent mound] in Syria, which was so large that horse riders on either side could not see each other."

The two similar serpent mounds in Oban Scotland and in Adams County, Ohio, are unique in that the serpents mouths is open and are depicted as swallowing an egg. The serpent's head and egg in Adams County, Ohio is aligned to the summer solstice sunset. Gardener writes, "In Wales the people were said to emerge and congregate on Midsummer's Eve to blow into the Serpent Stone-Eggs. [...] These serpent stones were said to be coloured pebbles, which gave 'second sight' and healing. Midsummer's Eve was the night when the serpents would role themselves into hissing balls and create the glain egg, also known as 'snake stone' or 'Druids egg.' What is an egg? Simply an 'entry portal' into this world. A device to give life."

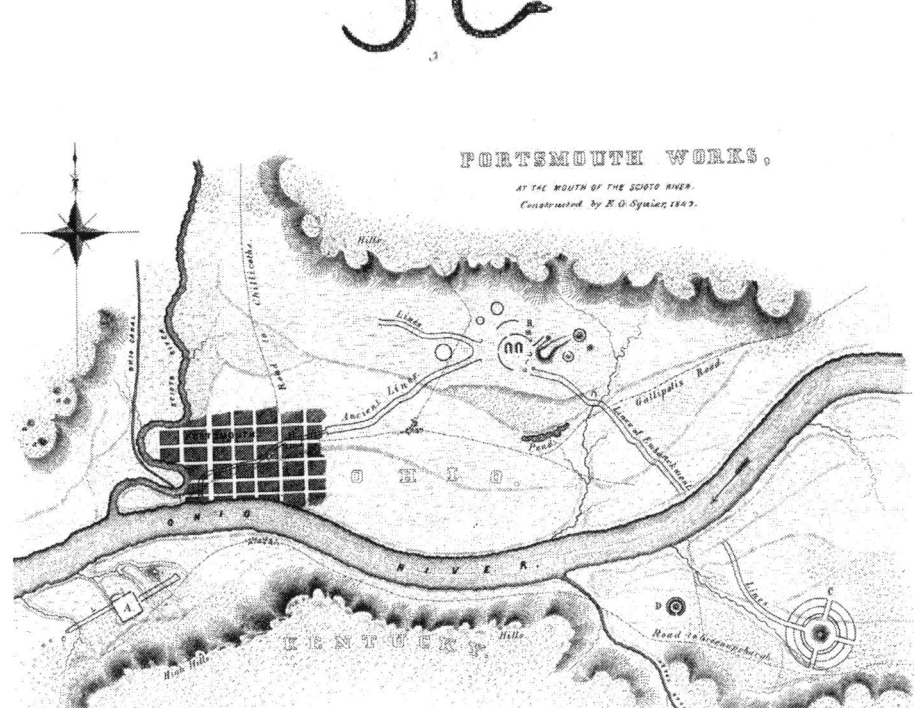

Portsmouth, Ohio Earthworks

To the casual observer, the similarities of the Portsmouth, Ohio Works and those at Avebury are obvious. The sacred vias, culminating with the circular head of the serpent are identical. Avebury has been compared with the Egyptian Ankh, the Egyptian symbol of life. Two stone circles are within the circular henge at Avebury, while at Portsmouth they contain two horse-shoe shaped earthworks. The symbol of the horse-shoe is believed to be a survival of an ancient religious symbol often seen in Assyrian and Egyptian sculptures, signifying the mystical door of life.

Little remains of the Portsmouth Earthworks. One of the horseshoe shaped earthworks is preserved in a city park. Comparing the size of the horseshoe work in the map to the size in this photograph, gives an idea of the immensity of the earthwork complex. From *The Nephilim Chronicles, A Travel Guide to the Ancient Ruins in the Ohio Valley*, 2010.

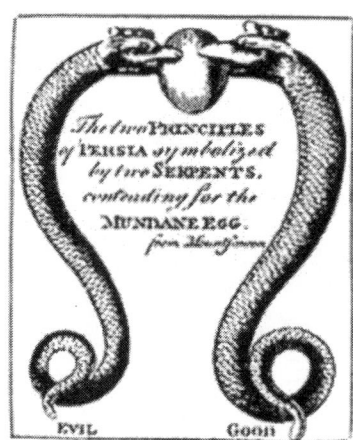

The Sun was the great enlightener of the world. Symbolic of the sun was the serpent who gave to mankind the knowledge of good and evil. This drawing from Persia shows the serpent contending for the egg of the earth.

The Serpentine forms facing a sun disc or egg was created in mica and found in a Allegewi Hopewell mound in Ohio.

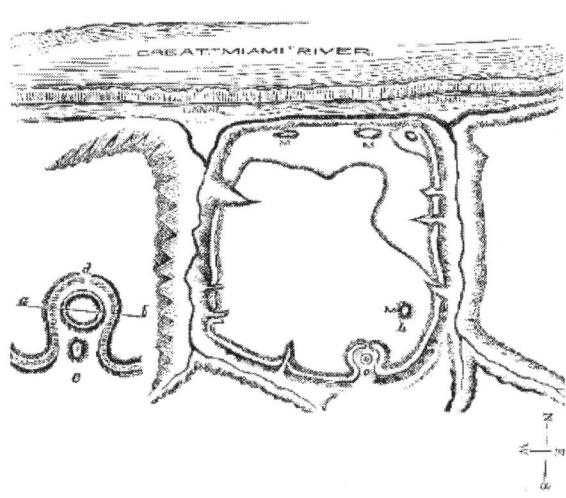

Two serpents facing the Sun Disk represent the double principle of good and evil. "These two principles are immutable, and existed from all eternity, as they will ever continue to exist." *Isis Unveiled*

From the body of the serpent looking towards the center sun disk. This earthwork in Ohio is one of the best preserved and most remarkable in the State. From, *The Nephilim Chronicles, A Travel Guide to the Ancient Ruins in the Ohio Valley, 2010.*

Tablets

Few items have been more scoffed at and called hoaxes by archaeologists than the various tablets that have been found within burial mounds. Strangely, all these tablets have been found with languages that were contemporary when the mounds were erected. I would agree that by themselves they would make a difficult argument for foreign contact with or by the Allegewi Hopewell. However, they are one more piece of evidence that is hard to explain, except within the context that migrations took place, by a people who came to North America and had knowledge of the ancient scripts.

The tablet found at Grave Creek in Moundsville, West Virginia is significant because it reveals that the Allegewi Hopewell were ruled by both Kings and Queens. Women rulers would be consistent with Earth Mother worship that is found within the shapes and lengths of the earthworks. It is also of interest that Grave Creek was the largest mound erected by the Allegewi.

Grave Creek Mound from *The Nephilim Chronicles, A Travel Guide to the Ancient Ruins in the Ohio Valley*, 2010.

Charleston Daily Mail, **October 22, 1922**
"Skeletons in Mound"

   One of the most interesting of the five state parks is Mound Park, at Moundsville from which that city derived its name. Probably no other relic of pre-historic origin has attracted as wide study among archaeologists as the Grave Creeks mound which has given up skeletons of the ancients who constructed it.

   Aside from the mammoth tumulus, itself 69 feet high and 900 feet in circumference, there were originally no fewer than seven mounds situated in the broad plain at the point. None was nearly equal to the one now standing, and the locations of most of the smaller ones are now lost to all excepting a few.

   Archaeologists investigating the mound some years ago dug out a skeleton said to be that of a female because of the formation of the bones. The skeleton was seven feet four inches tall and the jawbone would easily fit over the face of a man weighing 160 pounds.

   Seventeen hundred ivory beads, 500 seashells of an involute species and five copper bracelets were found in the vault. The beads and shells were about the neck and breast of the skeleton while the bracelets were upon the arms.

   There was also taken from the mound the skeleton of a man eight feet tall. There were no ornaments beside it. These skeletons were sent to the Smithsonian Institution in Washington.

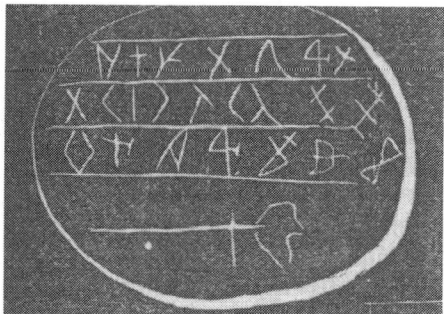

Grave Creek Tablet from,*The Mound Builders,* **MacLean 1879**

   In *Ancient America, 1976,* Berry Fell translated the inscription, stating that the script was Iberian and the language was Punic. Punic is derived from the Phoenician Semitic language. It was brought to North Africa from the city of Tyre in the period of the Roman Empire. It's earliest form dates to the first century A.D.

   Fells trasslation:

The mound raised-on-high for Tasach
This tile
(His) queen caused-to-be-made.

Stone tablet discovered within a burial mound in Muskingum County, Ohio. The mound was unique in that the largest skeletons uncovered in Ohio were found within the mound. The description of the gigantic skeletons found within this mound is listed in the Allegewi Giants.

History of Muskingum County, Ohio, 1882

"The State of Ohio, Muskingum county, ss:

Willima T Lewis, being first by me duly sworn, deposeth and saith: I began work on the Smith Gallery on September 2d, 1879, and continued to work there until June 14, 1880; and that between December 20, 1879, and January 10, 1880, I photographed for Dr. J. F. Everhart an engraved stone, said to have been exhumed from a mound in Brush Creek Township, and that I have this day identified the negative that I then took, in the Gallery No. 101, Main street, Zanesville, Ohio; that when I was about to print the picture for Dr. Everhart I assured him I could, by retouching the negative, make the characters on the stone appear plainer, and that Dr. Everhart objected, saying he wanted nothing more or less than an exact copy of the stone, with out any alterations whatever, and that I am prepared to identify the stone from which the negative referred to was taken, and that there was no sign of any recent engraving or marking on the engraved side of the stone.

W.T. Lewis

Sworn to before me and subscribed in my presence this 16th day of March, A. D. 1881.

Wm. H. Cunningham, Jr.,

Notary Public in and for said County and State.

The reader will observe in the Report the absence of scientific precautions, and perhaps the scientist who expects to find things in a scientific way may censure us for this, but when it is remembered that the object in this, as in every effort in exploring hidden things is to read the facts discovered, without the shackles of theory, it will be conceded that this could not have been accomplished better than leaving the exploration to those who had no theoretic knowledge on the subject.

And that whatever the inscription might mean remained for development by research, as no one could decipher characters as old as these have been found to be, and the inscription had not been viewed by an archaeologist, or one acquainted with the character.

The position of the stone indicated that it had once been erected with parallel lines perpendicular. Observing the right angle marks, however, and remembering that "angle stones" were found upon the Great Pyramid, and they were placed with the vertex of the angle uppermost, the writer postured the stone accordingly, and recognizing certain of the characters as Greek, and that, according to many writers, characters of ideation have been postured differently in different ages, evidenced especially in Webster's Dictionary of the English Language, 1879, P. 1762: Chart of "Ancient Alphabets," it was deemed legitimate to adopt the same course.

The first left hand character between the upper parallel lines in Alpha, the second is Omega, the third a spot, a numeral, the next a scepter with a numeral above, the next numerals of order, the next a serpent-symbol of life-spirit, the next the sign of addition, the next Delta, the next the ligatured Greek sign of the infinitive; the cavity between the upper and lower rows of characters to be grouped with those below the lower row, and represents sun, moon and stars, or heavenly bodies; the first left hand character in the lower row represents a seal or stamp in use the third century B.C. The next is another form of serpent, associated with a numeral, the next the ligatured character repeated, the next numerals of order, the last the angle marks, corresponding with the "angle stones."

The discovery that "Alpha and Omega" are the first two characters of the inscription was as startling as it is true. And the connection with the Great Pyramid as indicated by the corresponding signs, "the angle stones," found only on the Pyramids, and upon this grave stone, as far is known, began to loom up, and Mr. Smyth's three keys for the opening of the Great Pyramid seemed to have a bearing upon this inscription; so that they are here quoted for the benefit of the reader. "Key first: The key of pure mathematics." "Key the second: The of applied mathematics-of astronomical and physical science." "Key the third: The key of positive human history,-past, present and future, as supplied in some of its leading points and chief religious connections by Divine Revelation to certain chosen and inspired men...

From the foregoing we reach the following translation:

I am the Alpha and the Omega, saith the Lord God, which is and which was, and which is to come, the Almighty; giving first, power on earth; secondly, the spirit, added from heaven without ending.

Bat Creek Stone

The Bat Creek Stone was excavated in 1889 from an undisturbed burial at Bat Creek, Tennessee by the Smithsonian Institute. The stone was found underneath the head of one of the skeletons. Wooden earspools and bronze bracelets. A renowned Hebrew scholar, Cyrus Gordon, confirmed that the writing was Paleo-Hebrew, dating from the first to second century A.D.

The longest phrase, (LYHWD) was translated as *"for Juda."* This was the Jewish Aramaic spelling of *Judea.*

Carbon dating was performed on the earspools that confirmed their date was from the first century

A.D. Test were also done on the bracelets, that confirmed that were made of brass. Brass is an alloy of copper and zinc. The composition of the alloy was similar to that used by the Romans.

Round stone discs called *spindle whorls* are also found in the Medeteranean and the British Isles. They have been considered by some archaeologist as evidence of weaving, since similar pieces are found within ancient looms. However, many of these whorls have characters inscribed on them, and they probably were used as good luck fetishes, and worn as a necklace or attached to clothing.

Their shape of a circle with a hole in the center is the same as the Allegewi Hopewell icon for the sun and may have been used in sun worshipping ceremonies. Adding to this supposition are the characters that are inscribed on the stones that consists of the sun's rays.

Spindle whorls found in ancient Troy were described in *Archaic England,* Bayley, 1920. "Egypt was known as "The Land of the Eye" the amulet of the All-seeing Eye was perhaps even more popular in Egypt than in Eturia, and the mysterious and unaccountable objects called "spindle whorls," which occur so profoundly in British tombs, and which also have been found in countless numbers underneath Troy, were probably Eye amulets, rudely representative of the human iris. The Trojan examples here illustrated are conspicuously decorated with the British Broad Arrow, which is said to have been the symbol of the Awen or Holy Spirit. In their accounts of the traditional symbols, speech, letters, and signs of Britain, according to their preservation by means of memory, voice and usages, the Welsh Bards asserted that the three strokes of the Broad Arrow or bardic hieroglyph for God originated from three diverging rays of light seen descending towards the earth.

Out of these strokes were constituted all the letters of the bardic alphabet, the three strokes / | \ reading in these characters respectively 0 1 0, and thus spelling the mystic OHIO."

Spindle Whorls from Troy from *Archaic England,* 1920, Bayley. The three lines / | \, believed to be symbolic of 0 1 0 and the origin of the mystic word, OHIO.

Spindle Whorls from southern Ohio from Allegewi Hopewell burial mounds. The whorl on the left bering the ancient mark of / | \. The spoked pattern is recognized as being symbolic of the sun or the Egyptian hieroglyph for "the earth" or "underworld."

The Allegewi Giants

Physical similarities exists between Shell Mound People, the Glacial Kame, and the Allegewi mound builders in the Ohio Valley. All were tall, with archaic physical features that are closely tied to the Cro-Magnon of the Upper Paleolithic. Cultural similarities of these groups may stem, in part from the Amorite Beaker People spreading north into Finland and the shores of the Baltic sea, the isles of Denmark, and the Scandinavian countries. Gimbutus pointed out that the Beaker People had developed maritime propensities and had amalgamated with culture of the Hunters and Fishers.

The cultural exchanges between the Beaker People and the Hunters and Fishers may have occurred on northern European soil prior to coming to North America. It is also possible that cultural exchanges and assimilation in Europe, solidified a more cohesive cultural element in North America. Migrations from northern Europe and Northern Japan would eventually cease and leave the inhabitants of North America isolated.

The isolation was the result of the Celts migrating into the British Isles, from 1500-1200 B.C and ending the domination by the Beaker Peoples in England. At the same time, the Amorites were also being displaced from the eastern Medeteranean by the Hittites, Egyptians and Hebrews. These dates, also corresponding to the end of the manufacturing of copper weapons and mining operations on Lake Superior.

Archaic type skulls of the Amorites and the Hunters and Fishers of northern Europe is evident in North America. Since both populations were of the Upper Paleolithic variety, there are few physical differences between some of the skulls found in conical mounds in England and those found in the glacial kames around the Great Lakes.

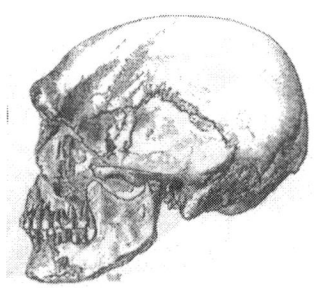 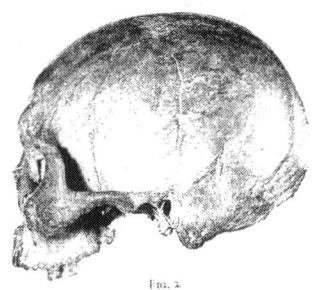

On the left is a skull from a conical mound in southwestern England, attributed to the Beaker People. On the right is a Glacial Kame skull from Michigan. The shape, brow ridge and sloping forehead are similar in both skulls.

Trading routes extended from the Atlantic to the Pacific and brought obsidian, galena, mica, shells, copper and other trade goods to the Ohio Valley be crafted into various objects. A burial mound on the gulf coast of Alabama is evidence of the Dinaric/Allegewi's extensive range across North America. The mound's proximity to a shell mound is evidence that the Allegewi were geographically and culturally connected with earlier populations that are similar to the Hunters and Fishers of Eurasia. Hog Island mound in Alabama had distinctive Allegewi elements of cupstones that were included with the burials.

Smithsonian Institutions' Bureau of Ethnology 44th Annual Report
The Hog Island Mound

On the bank of the river, a mile below the shell heap just described is an earth mound about 50 by 60 feet, longest north and south. As the ground has long been cultivated, it is probably the shape has been somewhat changed; it was no doubt practically circular when built.

On the east side, mostly in the trench but extending a short distance under the outside wall, was a grave 8 1/2 feet long, 4 feet wide, and dug 2 feet deep into the natural soil. In this were four skeletons, two lying sise by side on the bottom, the other two directly on these. The bodies were extended, heads to the northeast. One of those on the bottom was about 6 feet 4 inches long, the bones very large, the tibia had pronounced anterior curvature while the process for the attachment of muscles on the femurs were large and rugged. With this skeleton, near the neck or breast, were several copper beads, on the right side of the pelvis was a double crescent sheet of copper. The skeleton immediately under it had a similar object of the same pattern, similarly placed.

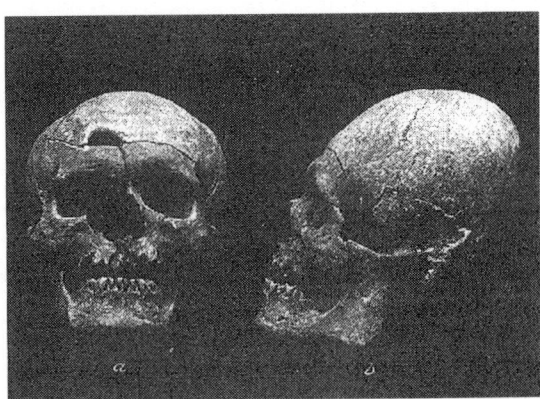

Dinaric/Alligewi type skull with flattened back of the head. From the burial mound on Hog Island, Alabama.

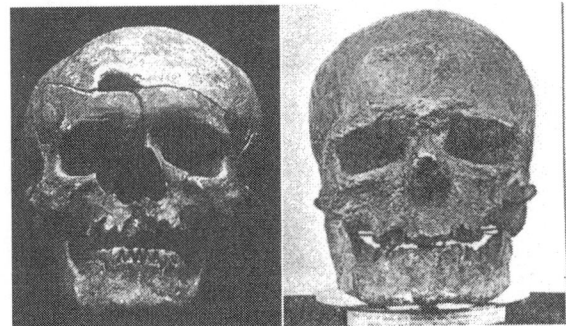

This is the frontal view of the skull from Hog Island compared to the similar Cro-Magnon skull found at Dordogne, France that was also 6 feet 4 inches tall.

The Dinaric skulls with their their heads flattened in the back are easily recognized. This along with their archaic features of protruding brow ridge, thick skulls, massive jaws and large height make them unique to any other skeletal remains. The Dinaric type is found most extensively in the Ohio Valley.

Stephens Coons, wrote of the skeletal remains in *The Bronze Age of Brition*, " The Beaker skulls as a whole are large, long and high vaulted, what ever their shape. They form one of the rare groups in the world with a cranial length of 184 mm. And and index over 80. This peculiarity they share with the few known brachycephalic crania of the Upper Palaeolithic" The only other known people with this "rare" head type were the Allegewi.

Comparing skull types of the Beaker People of the British Isles that were a combination of Corded People, Dinaric and Borreby Cro-Magnon with those of the Allegewi, shows some striking similarities. Both of which show more affinities towards Upper Paleolithic Cro-Magnon than to modern skulls with protruding brow ridges, thick skull wall and large mandibles, in addition to the large size of the skeletal remains.

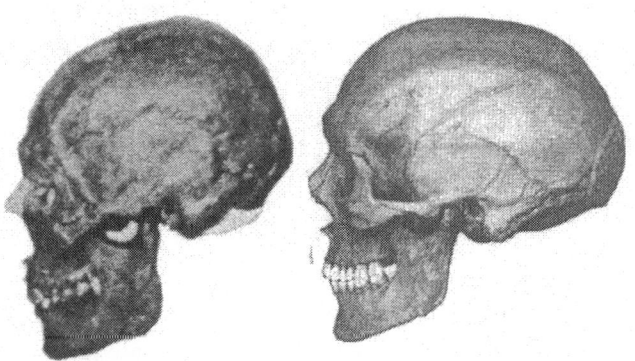

Side view of an Allegewi skull from *The Adena*, Webb and Snow, 1974, on the left and a Dinaric skull on the right. The furrowed brow, nasal notch, facial prognathism or protruding upper jaw and defined chin are identical, with both showing close affinities to Upper Paleolithic Cro-Magnon skulls.

The earliest burial mounds in the Ohio Valley were more distinctive Allegewi, but this isolation was short lived. As the Shell Mound People began to move into the interior they adopted many of the Allegewi burial and religious traits. In the Great Lakes region, the Glacial Kame and Point Peninsula Iroquois, also became very Allegewi-like. By about 100 A.D., the combination of Allegewi, Sioux, Cherokee and Iroquois manifests into the Allegewi Hopewell Culture.

Also found within the heart of the Allegewi Hopewell mound culture in Ohio are skulls that exhibit a double row of teeth. The occurrence of this trait along the extent of the Mississippi River may indicate the general movement of the coastal populations moving into the interior.

The burial mound at Fair Play, Louisianna contained skulls that exhibited all of the traits of archaic type skulls including having an occipital bun and a double row of teeth.

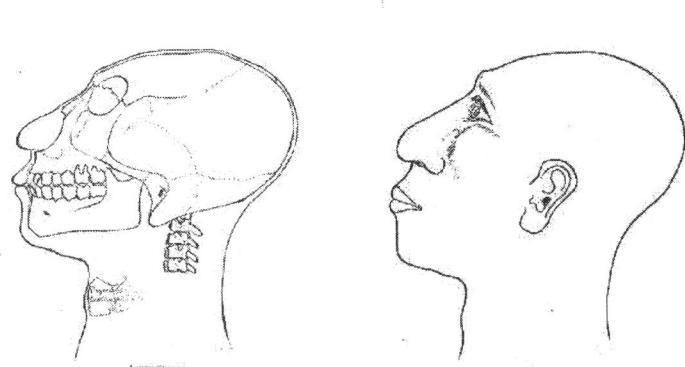

Archaic type skull from the Latro Mound with facial reconstruction by the Louisianna Historical Society.

Publications Louisiana Historical Society Vol., I, **1901**
The Mounds of Lousiana **by Prof. George Beyer**
The Larto Group

   On the morning of my arrival at Fair Play Landing, I found quite a number of gentlemen ready to accompany me to the mounds. These were nearly ten miles distant, and situated on the banks of a horseshoe-shaped Lake Larto. The road led for some distance along Black river, then out across country to the head of the lake, but ended shortly after passing the mounds.

   In the preceding pages I have somewhat transgressed, and I must now take up my report in regard to the location and formation of the group of mounds, which has given us, not only so much material for reflection and speculation, but has brought us face to face with the facts of the aborigines of Louisiana, of which we knew but little heretofore. I found the group consisting of four mounds, situated immediately on the banks of the lake. Three of them are of about the same size, while the fourth is smaller by one-half-in every respect, with the exception of the height-which about equals that of the others, at least at the present time. The diameter of the larger mound at the base ranges from 150 and 160 feet. The height of all of them is now only between seven and eight feet. Their original height has been probably twice that much, but continued use by men and cattle has tended not only to compress the earth, but has also caused the wearing away of considerable of its material. The mounds are about 100 feet apart, and are connected by ridges, which to some extent still remain. These ridges were, I suppose, from ten to fifteen feet wide at their base, but only about three or feet high. Mounds and connecting ridges were well covered with vegetation; on nearly all fairly large trees were growing. On one of the mounds a planter had built his residence and on another his corn crib; and during overflows, that gentleman informed me, both remained high and dry. While the first larger one, coming from the direction of Black river.

   The bones were lying close together, and the bodies had been buried side by side-head toward the south, feet to the north. It was practically impossible to obtain an entire skull or skeleton, and every single bone had to be cut out of the hard clay with the knife. On the left side of some of the skulls I found the fragments of vases or bowls, also a few arrowheads, and with one of the bodies a medium sized axe had been buried.

   After the removal of the remains I continued the excavation, and suddenly I came upon a bed of ashes. This bed covered a small area of about four feet square, and was about an inch and a half in thickness. Examination of the ashes revealed charred catfish ribs and garfish scales, but to say whether these were the remains of sacrificial rites or of a repast would be assuming to much...While yet removing some traces of the ash bed I came upon another skeleton, and proceeding with the greatest caution I discovered two others lying close to it

In regard to their physical standard it is well conceded that the mound builders were a fairly large race generally, although by no means of such gigantic proportions as some writers would lead us to believe. From the remains of a large number of bodies which I examined at the Larto mounds, I would judge that full grown men might have averaged nearly six feet. Quite a sensation was created by the fact that part of the skull (frontal bone) was found which measured actually 7/8 of an inch in thickness. Upon further search I found other portions of the same skull of corresponding thickness. This, however, proved to be only an exceptional case, for the majority of the other skulls present no such abnormal development; but upon comparing them in this respect with so-called mound builders' skulls from other sections, an excessive thickness is noticeable.

In the same line of abnormality was the finding of one of one skull in which the dentition reached the unusual number of forty teeth, the increase consisting of *eight additional incisors.* The remarkable preservation of the teeth is noteworthy.

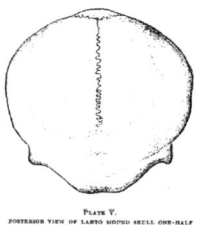

PLATE V.
POSTERIOR VIEW OF LARTO MOUND SKULL ONE-HALF NATURAL SIZE.

Latro skull from above showing the occipital bun at the rear of the skull.

The formation of the skulls found in the Lartro mounds, as compared with those of the other localities, is highly anomalous. I have given an illustration of one which, with the exception of the facial bones is nearly perfect. The approach to the Neanderthal skull is in this instance even closer than the one Dr. Foster outlines for comparison, which, according to that author, was exhumed by Dr. Campbell from a mound opposite Dunleith, Illinois. Our Indians of to-day possess a formation of skull which can in no way compare with those under consideration.

Ironton Register, **May 5, 1892,**
Where Proctorville now stands was one-day part of a well-paved city, but I think the greater part of it is now in the Ohio River. [Sic] Only a few mounds, there; one of which was near the C. Wilgus mansion and contained a skeleton of a very large person, *with all double teeth,* and sound, in a jaw bone that would go over the jaw with the flesh on, of a large man; the common burying ground was well filled with skeletons at a depth of about 6 feet. Part of the pavement was of boulder stone and part of well-preserved brick.

History of Morrow County, Ohio, **Vol., I, 1911**
In 1829, when the hotel was built in Chesterville, a mound nearby was made to furnish the material for the brick. In digging it away, a large human skeleton was found, but no measurements were made. It is related that the jaw bone was found to fit easily over that of a citizen of the village, who was remarkable for his large jaw. The local physicians examined the cranium and found it proportionately large, *with more teeth than the white race of today.* The skeleton was taken to Mansfield, and has been lost sight of entirely.

Historical Collections of Ohio, Vol., II

Huge skeletons--In Seneca township (Noble County, Ohio) was opened, in 1872, one of numerous Indian mounds that abound in the neighborhood. This particular one was locally known as the "Bates" mound. Upon being dug into it was found to contain a few broken pieces of earthenware, a lot of flint-heads and one or two stone implements and the remains of three skeletons, whose size would indicate they measured in life at least eight feet in height. The remarkable feature of these remains was they had *double teeth in front as well as in back of the mouth and in both upper and lower jaws.* Upon exposure to the atmosphere the skeletons soon crumbled back to mother earth.

Wisconsin Decatur Republican (Decatur Iowa) January 13, 1870
Wonderful Discovery
Skeletons of a giant race found near Potosi
From Dubuque Times, Jan. 5.

The evidence appears to be pretty well settled that this whole western country was once inhabited by a race of beings of gigantic stature, who were not only hard working, industrious fellows, but well up in the fine arts. What their laws, institutions and code of morals consisted of, we shall never know, as printing presses and interviewing reporters were scarce in those days, but from the numerous mounds scattered over the country, which learned ...illegible...tell us were the works of their hands, it is quite easy to assume that they were heavy on the dig and took much delight in wielding the spade and shovel. They would be useful fellows to have in these days of railroads and canals and its to be regretted that the race died out before the present system of internal improvements commenced.

From time to time the skeletons of an unknown race have been discovered in the different mounds mentioned, up and down the Mississippi River, the last discovery of the kind being made near Potosi, Wisconsin, a little over a week ago. A young man by the name of Patterson, brother-in-law to S.M. Langworthy, Esq., of this city was engaged with a number of men digging out the foundation of a saw mill, near the bank of the river. In digging out this, it became necessary to remove one of these mounds or tumuli. The workman had descended to the depth of about seven feet, when they unearthed two human skeletons, the bones of which were almost entire, and in a good state of preservation upon taking them out, an accurate measurement was made of the skeletons, which one of them was found to be seven and a half feet, and the other eight feet in length. The jaws of each were filled with *double rows of teeth,* while the cheek bones were very high and prominent. Under the bones a large collection of arrowheads and strange toys were found, which had evidently been buried with them.

Strange to state, the workman, instead of preserving these bones, carted them off into the road, and it feared that the great majority are now wasted. It is highly probable that other skeletons exist in the vicinity.

Mr. Langworthy, we understand, will soon visit that locality for the purpose of preserving the skeletons already found and pushing further and more persistent search for fresh discoveries.

The following list of giant skeletal remains are from the areas where the mounds associated with ceremonial earthworks are located. The majority of the historical accounts are from the Ohio Valley, with others extending into areas that are noted as being within the Hopewell cultural sphere.

Arkansas

Idaho Daily Statesman, **June 12, 1899**
Food For The Credulous

Remains of a Race of Giants Found in Arkansas-Human Skeletons Unearthed Eight and Ten Feet in Height-Strange and Unknown Pottery-Relics of a Former Age.
(From the Memphis Appeal)

The statements which we make below, and the facts detailed are so strange and almost incredible and so like the many …illegible…and canards that have appeared from time to time appeared in the press of Europe and America, that we premise them with the declaration that they are strictly true, and that we have not exaggerated what we have seen one iota. With this much as a preface we will proceed to our story: -

Chickasawba, two miles west of Battlefield Point, in Arkansas, on the east bank of the lovely stream called Pemiscott Bayou, a tributary of the St. Francis river, stands an Indian mound, some twenty-five feet high and about an acre in area at the top. This mound is called Chickasawba, and from it the high and beautiful country surrounding it, some twelve square miles in area, derives its name Chickasawba. The mound derives its name from Chickasawba, a chief of the Shawnee tribe, who lived, died and was buried there. The chief was one of the last race of hunters who lived in that beautiful region, and who once peopled it quite thickly-for Indians we mean. From 1820 to 1831 he and his hunters assembled annually at …illegible…Point, then, as now, the principle shipping place of the surrounding country, and bartered off their furs, peltries, buffalo robes and honey to the white settlers and the trading boats on the river, receiving in return powder, shot, lead blankets…illegible. Aunt Kitty Williams, who now resides there, relates that Chickasawba would frequently bring in for sale at one time as much as twenty gallons of pure honey in deerskin bags slung to his back. He was always a firm friend of the whites, a man of gigantic stature and Herculean strength. In his nineteenth year he took a young wife, and by her had two children. In 1831 she died, and the old chief did not long survive her, dying in the same year, age ninety-three or four years. Mr. W. Fitzgerald, who moved to that country in 1822, says that up to the time of his death Chickasawba supplied him with game. He was buried at the foot of the mound on which he had lived, by his tribe, most of whom departed for the Nation immediately after performing his funeral rites. A few, however lingered there up to a late date, the last of them, we believe, being John East, who in 1860, at the breaking out of the war, joined Captain Chaily Bowen's company of the late "so-called," and fought the war through, as gallant a 'reb,' as any of them, coming back home in 1865 to return to the arts of peace. Chickasawba was perfectly honest, and best informed chief of his tribe. His contemporary chiefs were Long Knife, Sunshine, Corn Meal, Moonshine (Mike Brennan), &c. Mike Brennan and Quill buried him. He had a son, named John Pennscott. A number of years ago in making an excavation into or near Chickawba's mound a portion of a Gigantic Human Skeleton was found.

The men who digging, becoming interested, unearthed the entire skeleton, and from measurements given us by reliable parties the frame of the man to who it belongs could not have been less than eight or nine feet in height. Under the skull, which easily slipped over the head of our informant (who, we will here state is one of our best citizens), resembling nothing in the way of Indian pottery which had before been seen by them. It was exactly the shape of the round-bodied, long necked carafes, or water decanters, a specimen of which may be seen on Gaston's dining table. The material of which this vase was made was a peculiar kind of clay, and the workmanship was very fine. The belly or body of it was ornamented with figures or hieroglyphics, consisting of a correct delineation of human hands, parallel to each other, open, palms outward, and running up and down the vase, the wrist to the base, and the fingers towards the neck. On either side of these hands were tibiae or thigh bones, also correctly delineated, running around the vase. There were other things found with the skeleton, but this is all our informant remembers. Since that time wherever an excavation has been made in Chickasaba Country in the neighborhood of the mound.

Similar skeletons have been found, under the skull of every one were found similar funeral vases, almost exactly like the described. There are now in this city several of the vases and portions of the huge skeletons. One of the editors of the Appeal yesterday measured a thigh bone, which is fully three feet long. The thigh and shin bones, together with the bones of foot, stood up in proper position, in a physician's office in this city, measured five feet in height, and show the body to which the leg belonged to have been from nine to ten feet in height. At Beaufort's landing, near Barfield, in digging a deep ditch, a skeleton was dug up, the leg of which measured between five and six feet in length, and other bones in proportion. In a very few days we hope to be able to lay before our readers accurate measurement and descriptions of the portions of the skeletons now in the city and of the articles found in the graves.

Georgia

The Middlebury Kentucky News. **December 30, 1930**
SKELETONS OF GIANTS
Remarkable Relics of an Extinct Race Excavated in Georgia

Mr. J. B. Toomer received a letter from Mr. Hazelton, who is on a visit to Gainsville. The letter contained several beads made of bone, and gave an interesting account of the opening of a large Indian mound near that town by a committee of scientist from the Smithsonian Institute. After removing the dirt for some distance, a layer of large flag-stones was found, which had evidently been dressed by hand and showed that the men who quarried this rock understood their business. These stones were removed, when in a kind of vault beneath them, the skeleton of a giant, seven feet two inches was found. His hair was coarse and jet black and hung to the waist, the brow being ornamented with a copper crown. The skeleton was remarkably well preserved and taken from the vault intact. Near this skeleton were found the bodies of several children of various sizes. The remains of the latter were covered with beads, made of bone of some kind. Upon removing these, the bodies were found to be inclosed in a net work made of straw and reeds, and beneath this was a covering of the skin of some animal. In fact, the bodies had been prepared somewhat after the manner of mummies, and will doubtless throw new light on the history of a people who reared these mounds. Upon the stones that covered the vault were carved inscriptions, and if deciphered will probably lift the veil that has enshrouded the history of the race of giants that undoubtedly at one time inhabited this continent. All the relics were carefully packed and forwarded to the Smithsonian Institute, and are said to be the most interesting collection ever found in America.

Illinois

History of Daviess County, Illinois 1879

The mounds on the bluff have nearly all been opened within the last two or three years by Louis A. Rowly, Esq., Mr. W. M. Snyder and Mr. John Dowling, assisted by Sidney Hunkins and Dr. W. S. Crawford. These gentlemen have taken much interest in these prehistoric structures, and have very carefully investigated them. In all that have been opened the excavators have found in the centre a pit that was evidently dug about two and a half feet below the original surface of the ground, about six feet long and four feet wide, in the form of a parallelogram. The bottom and sides of this pit are of hard clay. The bones in this pit indicate a race of gigantic stature, buried in a sitting posture around the sides of the pit, with legs extending towards the centre.

History of Mifflin County, Ohio, 1880

South of this, on the banks of Peoria Lake, near the city of Peoria, Illinois, there were excavated a few years ago by the Scientific Association of Peoria the contents of a very large, oval mound, and in it were found three human skeletons, a man, a woman and a boy, all lying straight beside each other, the boy asleep on the woman's arm. The skeleton of the boy was about three feet long, but the man and the woman had a stature of seven feet. The bones were decomposed rapidly on being exposed to the air, except the skulls, which being of a harder texture had better withstood the tooth of time. Though these figures were of immense stature, their immense skulls were fully in proportion to their frames, and possessed of a frontal development of reasoning powers of immense size.

History of Logan County, Illinois, 1886

It is sometimes difficult to distinguish the place of sepulcher raised by the Mound Builders from the modern graves of the Indians. The tombs of the former were in general larger than those of the latter, and were used as receptacles for a great number of bodies, and contained relics of art, evincing a higher decree of civilization than that attained by the Indians.

The ancient earthworks of the Mound Builders have occasionally been appropriated as burial places by the Indians, but the skeletons of the latter may be distinguished from the osteological remains of the former by their greater stature.

12th Annual Report of the Bureau of Ethnology to the Secretary of the Smithsonian Institution, 1890-1891
Dunleith Illinois

No. 5, the largest of the group was carefully examined. Two feet below the surface, near the apex, was a skeleton, doubtless an intrusive Indian burial...Near the original surface, 10 or 12 feet from the center, on the lower side, lying at full length on its back, was one of the largest skeletons discovered by the Bureau agents, the length as proved by actual measurement being between 7 and 8 feet. It was clearly traceable, but crumbled to pieces immediately after removal from the hard earth in which it was encased.

12th Annual Report of the Bureau of Ethnology to the Secretary of the Smithsonian Institution
1890-1891
Pike County, Illinois

   The other, situated on the point of a commanding bluff, was also conical in form, 50 feet in diameter and about 8 feet high. The outer layer consisted of sandy soil, 2 feet thick, filled with slightly decayed skeletons, probably Indians of intrusive burials. The earth of the main portion of this mound was a very fine yellowish sand which shoveled like ashes and was everywhere,to a depth of 2 to 4 feet, as full of human skeletons as could be stowed away in it, even to two three tiers. Among these were a number of bones not together as skeletons, but mingled in confusion and probably from scaffolds or other localities. Except one, which was rather more than 7 feet long, these skeletons appeared to be of medium size and many of them much decayed.

Iowa State Reporter, **September 10, 1891**
SKELETONS OF GIANTS
Hundreds of Skulls Found-Interesting Discoveries in the Burial Mounds at Carthage, Ill.

   CARTHGAGE, ILL., Sept. 4-Assisted by students, Profs. Dysanger, Hail, Segler and O'Hara, of Carthage college, on Wednesday opened the Sweeney burial mound on the farm of Cyrus Felt, northeast of this city. At first a covering of stone was encountered, all of the red limestone variety . Most of these stones on being removed crumbled away into sand. Under them were found immense quantities of bones, many skulls and several pieces of flint. Some pieces of pottery so badly decayed and broken as to be unrecognizable were found also. Measurements were taken of several skulls. One measured 7 ¾ inches across the parietal bone, another 6 inches. Three femur bones were found measuring 9 inches in length, some that measured 17 ½ inches in length and others measuring from 12 to 14 inches. One measured 3 inches across the lower end of the femur bone. Dr. Veatch says the bones indicate that the men must have been from 8 to 7 feet tall at least. One jawbone was secured that contained a perfect row of teeth which evidently belonged to a middle-aged person.

   Upon digging a few feet farther down another layer of rocks was discovered, and upon removing these several skeletons, perfect in form, lay all huddled together as if they had fallen in battle. Some of the skeletons were preserved almost entire, although most of the bones would crumble away upon being exposed to the air. It is believed that fully 500 corpses were buried here, as basketful after basketful of bones were taken out, while the mound is literally full of them. The skulls were all filled with a peculiarly soft and very black loam, different from the surrounding earth.

Daily Review, **March 15, 1901**
Bones Of A Giant

   Allton Ills., March 15-Workman who were digging on the farm of Z. B. Job at East Alton, yesterday unearthed the skeleton of a man of gigantic stature. The bones had been in the ground many years, and when touched by the workmen crumbled away, but the skull remained intact, and was brought to Alton to be exhibited. The skull is very large and the jawbone is of unusual size. The ground where the skeleton was found had not been disturbed in many years, and there was a mound at the place, which was being leveled off.

Historical Encyclopedia of Illinois and History of Lake County, 1902

Of the early history of the region which now embraces Lake County but little can be written. The Mound Builders had occupied it and passed away, leaving no written language and but little even as tradition...These mounds were quite   numerous...Excavations...have revealed the crumbling bones of a mighty race. Samual Miller, who has resided in the county since 1835, is authority for the statement that one skeleton which he assisted in unearthing was a trifle more than eight feet in length, the skull being correspondingly large, while many other skeletons measured at least seven feet.

The American Antiquarian, Volume 13

We now refer to the discovery which we made in connection with the great serpent effigy near Quincy, Illinois. The serpent is a massive effigy, which conforms to the bluff throughout its entire length. It folds are brought out very forcibly by four conical burial mounds located near the center of the ridge, midway between the head and tail of the serpent. The mounds contained many bodies, none of them remarkable except the one which was cremated at the base of the mound. This was a large body. It was lying on its back, and partially burned. The bones, however, were preserved, and what was the most singular about the case, on the very center of the body, near the secret parts, a skeleton of a serpent was coiled up, as if there was an intention to make it significant. The hands were folded over the body just below this skeleton. The body had it feet to the east, and its face was turned upward, as if to look toward the sun.

American Antiquarian, 1905
Sacrificial mounds- Excavations at Chillicothe, Illinois

A previous exploration had resulted in the discovery of numerous remains, but at three feet below these a well preserved skeleton was found lying on its back, with the head pointing southwest. The form was large, the jaws massive, and the teeth perfect. At the feet lay the bones of an infant, and the skeleton, when living, was probably a female and a mother.

The Washington Post, October 25, 1906
OLD TIME GIANTS
Skeletons Eight Feet Long Unearthed in Illinois
Used Bamboo Implements
Burial Ground of the Ancients is Opened Near Quincy
Relics Brought to Light Which May Solve the Mystery of the Mound Builders-Carving,
Apparently of Ivory Which Rivals from an Artist View-point the Work of the Japanese.

Quincy, Ill., Oct 24-On what is known as one of the Illinois River hills, about midway between Cooperstown and the river, and eight miles from Mount Sterling in Brown County, has just been made one of the richest and possibly the most wonderful of pre-historic finds. A curious resident of the locality.[...] mound in almost the center of the. Illegible... farm of Mrs. Crabtree, a widow, and already the results of the exhumation would make the eyes of the archaeologist dance with delight.
Skeleton Eight Feet Long

With the first day's work the mound began to give up traces of handiwork of past ages and the bones of those who had wrought it, and others immediately joined in the search, still going on. Thus for several skeletons, by actual measurement are eight feet long, and several pieces of remarkable pottery, beads, and curious implements have been taken out. The bones crumble badly almost as soon as they are taken in the open air. They are so numerous that it is believed a prehistoric burying ground has been found, greater in extent and more perfectly preserved than any yet discovered.

Under the bones of each of the ancient dead, were found pieces of pottery, beneath the fragments of the skulls of some of them great vases, the largest of which would easily hold two gallons. Underneath one skeleton was a curious bowl. In the center of whose basin was the well fashioned figure of a king seated upon a log, and it is thought that these bones may be those of a great leader of the race that once ruled this portion of the continent.

Implements of Bamboo

Strangest of all articles found with the bones were implements that are apparently made of bamboo, some of them evidently shaped for purposes of weaving. Countless beads were found in the mound of a strange material, almost white, and possibly made from the best of potter's clay.

Forty of the greater pieces found in the mound were taken to Cooperstown, where they are now temporarily in the possession of H. C. Ren, postmaster. One of the pieces is shell-shaped dish with a wolf's head, the work on which leaves no doubt that it was carved, even the teeth of the wolf gleaming from it, as exquisitely done as some of the ivory carving of the Japanese, and some of those who have seen the piece believed it to ivory. This and what is thought to be implements of bamboo at once caused interesting speculations.

Hundreds of people from Mount Sterling, Cooperstown, and the country around for miles have been attracted to the scene of the excavations by the news of the wonderful find. The soil of the mound has never before been disturbed, and to this fact is attributed the marvelous preservations of things taken from it. The mound is one of a series along the Illinois River hills, which a noted archaeologist pronounced would one day solve, when fully explored, the mystery of the mound builders.

Indiana

The Indiana Gazetteer, 1849
Decatur County, Indiana

On the bottom of Big Flat Rock, in north-west corner of Decatur county, is a mound about eighty feet in diameter, and eight feet high, originally covered with trees, like the forests around. An excavation was made into it a few years since. First there was a mixture of earth, sand and gravel for one foot; then dark earth, charcoal, lime and burnt pebbles were cemented together so as to be penetrated with difficulty; then a bed of loose sand and gravel mixed with charcoal; then were found the bones of a human being, in a reclining position, with a flat stone over the breast and another under the skull. Most of the bones were nearly decomposed, but some of them, and part of the teeth, were quite sound. From the size of such of the bones of the skeleton as remain, it must have once been of gigantic size. A short distance from this mound is a smaller one, which contains a great number of skeletons.

Indiana Geological Survey, 1862
Henry County, Indiana

About seven or eight miles west of New Castle, a number of Indian skeletons were disinterred in the constructing a turnpike, and about the same distance south of town some remarkable humans bones and skeletons of giant size were dug out, with other relics, during the making of the road.

Fifth Annual Report of the Geological Survey of Indiana, 1873
Geology of Lawrence County, Indiana

At the site of the former county seat, Palestine, there was a vaulted tomb containing skeletons of persons of not less than 6 ½ feet. Hammered copper earrings and a globular "war whistle," were also found.

Indianapolis News, **November 11, 1875**
Jennings County Indiana
"Remains Of Vanished Giants Found In State"

One of the strangest contributions ever to come to hand tells of the existence in what is now Indiana, long before statehood and even before the Indians came here, of a mysterious giant mound-builders race whose men were more than nine feet tall.

What's more the contributor of this odd information, Helen W. Ochs of Columbus, Ind., wrote that evidence of their one-time existence here still remains near Brewersville, Jennings County.

She quoted from the geological report many years ago on Jennings County by W.W. Borden that the remains of the largest work of those moundbuilders in that country were to be seen on the bluffs 75 to 100 feet above Sand Creek in Sand Creek Township. The report added:

"It is a stone mound 71 feet in diameter, showing at this time a height of three to five feet above the surrounding surface. The exterior walls appear to be made of stones placed on edge but the central portion did not show any regular arrangement of the stones"

Mrs. Ochs said the first discovery of human skeletal remains in that mound was made in 1865 when a farmer, getting stone for a spring house, dug into "a sort of tomb" in which he found the skeleton of a small child.

She quoted George M. Robison, his son, as saying the top of the mound was not less than 30 feet above the level of the surrounding ground. He added:

"I well remember that several large forest trees were growing on the top. One was a white oak not less than three feet in diameter at the base"

Discovery of the child's skeleton aroused much curiosity, causing several people to dig into the top of the mound and resulting in the finding of several other skeletons. Mrs. Ochs added:

"Some of them were bound with perfectly-preserved bands of cedar wrapped around their chest while others were charred, perhaps in observance of a religious rite. Weapons found with the skeletons were unlike those used by Indians"

She quoted Robison further as saying that no intelligent investigative work was conducted there until 1879, 14 years after the discovery of the mound. He continued:

"The state geologist brought a couple of men here, one from Cincinnati and one from New York, and with Dr. Charles Green of North Vernon, they made quite an extensive examination. Among other things found was the skeleton of a man, it was intact, or rather, I might say, the bones were not scattered. It measured nine feet, eight inches.

"There was sort of necklace of mica lying around the neck and down across the breast. At the feet stood a sort of 'image' made of burned clay with pieces of flint rock imbedded in it"

Robison kept that image and some of the bones. Mrs. Ochs said that as late as 1937 bones of that giant were in a basket in the office of the Kellar Mill along Sand Creek about a mile below the mound's site:

"Kenneth Kellar, grandson of Robison, remembers that basket of bones. He said the bones were lost when the 1937 flood washed out the office"

Robinson who told of seeing the huge skeleton exhumed added that, according to the men of science who were there, they were the remains of a white race that had inhabited that part of the country before the advent of the red man. He said there were no signs of anything like pottery, no signs of metal working of any kind, just simply the bones of a "dead and gone race of human beings that we today know practically nothing about. We know not whence they came or where they went"

Mrs. Ochs said that the giant-like race had worked hard to entomb its dead. The rocks in the mound had been placed end to end with no attempt to plaster or seal them together. She continued:

"Evidence that this mound was dug into has washed out until the one-time graves now are smooth indentations in the leaf-covered ground"

She said Edith Hale, a retired schoolteacher; Beulah Kellar Lowe, granddaughter, and Kenneth

Kellar, the grandson of Robison, remembered the bones and image described by Robison.
"This I feel substantiates the findings under discussion," Mrs. Ochs concluded.

Indiana Geological Survey, 1881
Delaware County, Indiana

The Indians used many of the hills as burial places; bones have been discovered which from their size would indicate that they belonged to a race of giants.

History of Randolph County, Indiana, by E. Tucker, 1882

There are many antiquities in Randolph County, mounds embankments, ect. some of which are described below:

One of the best known is to be seen (partly) in the fair grounds northwest of Winchester. It is an enclosure of forty-three acres in the form of an exact square. The embankment was from seven to ten feet wide, as also having a mound in the center of the area fifteen feet high. The whole enclosure and the embankment also, when found by the settlers, was covered with large forest trees exactly like the adjacent regions. The eastern opening was unprotected, the western one was surrounded outwardly by an embankment shaped like a horse shoe open toward the gate, joined on the north side to the embankment, but left open on the south side of the gate for a passage to the outer grounds.

The embankment has been considerably lowered throughout the greater portion of its extent by cultivation, by the passage of highways, ect., but it is still several feet high, and is very plainly traceable along its entire extent.

Some of the bank on the south side toward the southeast corner still remains, as it existed at the settlement of the county. That part is now some six feet high, and perhaps twenty-five feet wide. A large portion of the eastern bank has lately been dug away for the purpose of brick making, and it is said that charcoal is found scattered throughout the mass of clay composing the embankment.

On the side of the creek not very far distant were gravel banks containing great quantities of human bones, which are said to have been hauled away by wagonloads. These skeletons were many of them large, but bones were much decayed and crumbled readily when disturbed and brought out to the air.

Histories of Pike and Dubois Counties, Indiana, 1885
Pike County, Indiana

John Stucky, Mr. Osborn and a few others, whose names are forgotten, were digging a grave on top of a mound near Siple's, and reaching the depth of about three feet came upon the remains of three persons. The first was a huge being, the lower maxillary being large enough to pass over that of a living person, flesh and all. Mr. Stucky further says that the femur bone was several inches longer than that of an ordinary man. Unfortunately these remains have been neglected and lost. Of the remains of the other two, one seemed to have been a women, the other a child. The skeleton of the women was reclining between the legs of the huge man, and the child between those of the women.

Lima Daily News, (Lima Ohio) July 27, 1892
May Buy The Mounds
Congress To Purchase Prehistoric Works
(Anderson, Ind., Letter)

The questions of converting the Indiana prehistoric mounds into a national park will be revived again this session of congress and more favorable action may be taken. As archaeologists continue the study of the mound builders they find that the Indiana mounds are most remarkable of all in the nation. Recent discoveries have added a great deal of interest to the Indiana mounds and they have again demanded the attention of the Smithsonian Institution, which was one of the prime movers some years ago in the attempt to have the grounds converted into a national park.

A camera cannot do the Indiana mounds justice. They are not great heaps of earth which show well in a photograph, as is the case with those in Ohio and along the Mississippi, and are not even as attractive as those in Illinois and the northwest, which follow the contour of snakes and wild beast, but they posses outlines well defined and precise Scientist are convinced that their builders possessed many of the talents of the ancients of Egypt and Asia. Like the other mounds, they are covered with forest, which show that ages have passed since the builders occupied them.

The precision of the modern surveyor and the methods of the nineteenth century builder have been combined in the Indian mounds and the result is a work of art rather than a crude heap. If it was known that the builders had surveyed Saturn through telescopic lens and beheld the circles around the inner globe, it might be claimed that they had used the planet and its girdle as their pattern for the construction of earthworks. The five great mounds lie just east of the city. The outer circle of the greatest of the five is but ten feet in height, but broad enough to allow teams to pass over its crest. It is 180 feet in diameter, and measured from any point it is identically the same distance from the center of the mound. The precision of these outer ridges is so nice they at once attract attention. With a graceful curve the ridge slopes on an angle of about 120 degrees to a great ditch fifteen feet wide and about fifteen feet deep. Like the ridge, it is perfect circle. From the ditch rises the inner, the great mound. The rise is rounded and evened off as prettily as though it had just been completed. In the very center of this mound, which is fully 100 feet across, is a prominence and this is five feet above the outer circle ridge and twenty feet higher than the inner ditch. From this a path wide enough for teams to pass runs to the outer ridge, where there is an opening. It bridges the ditch. All mounds large and small are built identically this pattern, all of the openings being to the north and on a direct line from the center mound to the North Star. These openings have been much studied, but significance of their direction has not been determined. The recent discoveries, given later, all tend to the belief that all of these mounds are buried deep under the present surface and were built on the strata of shale probably before the alluvial deposits were made.

The great mounds of the Indian group all belong to the Bronnenburg family, which is among the wealthiest and best known in the county. The Bronnenbergs, while enterprising farmers have little idea of the assistance they might give to science by allowing excavation in the mounds. They have persistently refused to allow any excavations made in any of the mounds, but recently a midnight party was organized which dug in the center of the center mound. Although the men went down twenty feet they found nothing but loose alluvial soil that had evidently not been used in the construction of the mounds, but had accumulated later. This strengthens the theory that the real works of primitive art lie far below the present surface of the ground, and are built upon the underlying strata of slate.

Dora Biddle of Anderson a collector of antiques has a skull, and another is on exhibition here, which has been severed just above the ears, in such a manner as to remove the crown of the head and lay the brain bare. These skulls were found with others under conditions, which would indicate that they were those of the mound builders. They are very large, show marked intellectually, and unlike skulls of the present day, or of the Indians, have a fifth skull bone in the back of the head. There can be no doubt that the purpose of removing the tops of these skulls was to remove the brain tissue. The skulls have been severed with some fine instrument, which did the work as precisely as the surgeon's saw of today would do it.

Recently, while making an excavation near the mounds, workmen who did not appreciate the find suddenly came upon a composition, which resembled a baked cement or clay. It was round and secure. They broke into it and found they had opened a hermetically sealed cave, which resembles greatly our cisterns of the present day. It was dry as a powder-house, and the air, which came from its recesses, was sickening and tainted with great age. Here in this small receptacle, scarcely large enough to hold more, were found six skeletons in a sitting position. All six skeletons in a sitting position. All were propped up evidently when first put in. When the fresh air came rolling in they crumbled to pieces and but for a few bones which remain no trace is left of this remarkable find. The bones that are saved, however, indicate a people who were very large-decidedly larger than those of the present day. Parts of the skulls showed that the heads were very large also-the foreheads were very large.

There can be little doubt that this find is closely connected with the mounds and that the skeletons were those of mound builders. It is claimed a similar discovery was made some years ago near the mounds, and that this proves convincingly that mound builders were the occupants of the cells. This mode of burial could not have been that of the modern Indians who occupied this part of the country at the time of the landing of Columbus.

Francis Walker of this city, who has long advocated the converting of the Indian mounds into a national park, says that the mound builders of this section were far advanced in the arts and sciences. If the mounds were as supposed, built upon the shales which underlie the alluvial deposits, a reference to geological data would place the existence of these aborigines back as far as the time of the Pharaohs.

To the east of the mounds is a cave of artificial formation that leads in toward the great mound 150 feet distant, and is fully fifty feet below the present surface of the mounds. There is little doubt that here lies the solving of the great mystery. It is probable that following this would bring a person in the inner chamber of a work of primitive building that would solve the doubts now existing regarding the history of this remarkable people.

Should the movement to convert these lands into a national park be successful the Smithsonian Institute and other Institutions of learning which have been greatly interested in this group will make excavations that are now impossible. They have long regarded the builders of these mounds as those from which they would get most knowledge, owning to the superiority and advancement these people evidently held over other tribes of builders. Many minor discoveries have been made in the past few months that throw additional light upon the mounds and the builders, but they do not differ greatly from the few set out above and simply serve to further the theories, which have recently taken the place of the older ones

History of Blackford County, Indiana, 1895

One Moundbuilders skull, from a subject only twelve years old-as is evident from the stage of development, which the teeth exhibit, is as large as the skull of a full-grown man of modern type. This was found in a mound, accompanied with cup shaped vessels. With such mounds and relics, the Salomonie River abounds

History of Clay County, Indiana, 1909

Sandy knoll, about a mile west of the Eel River, east of a line from Coffee to Howesville, has attracted more attention, as such, than any other or, perhaps, all other points, in the county...All the skeletons discovered were of gigantic proportions, a stature of seven feet, or thereabout, all in the sitting posture, with fractured femurs, or thigh bones, a phenomenon unexplained. John B. Poe, one of the early pioneers, himself six feet in height and proportionately developed, who made excavations and tests, found the tibia (bone of the lower leg) in all cases from one to two inches longer than his own, and could place the maxillary (lower jaw bone) over his own flesh and all.

History of Lawrence, Orange and Monroe Counties, Indiana 1914

A mound similar to the last at the site of the former county, Palestine, or 'old Palestine', as it is called, was explored in 1870, by Messrs. Newland, Dodd and Houston. On the surface of the hill a confused mass of stones, such as a man could conveniently carry, were noticed indicating a circular wall twenty feet in diameter. It was found to be a vaulted tomb. The first or upper vault contained the bones of many women and children; a layer of flat stones divided this from the second which contains the bones of men; another layer of flags, and at the bottom, six feet below the surface, two skeletons were found with their heads placed to the east and faces to the north. The last were persons of great size, being not less than six and a half feet high.

Artisans and Artifacts of Vanished Races, **Theophilus Dickerson, 1915**

PECULIAR GRAVEL MOUND IN HENRY COUNTY, INDIANA
This Isolated Monument of Nature at an Early Period Surrounded by Water-Two Roadways.
HUMAN SKELETON EIGHT FEET IN HEIGHT UNEARTHED TWELVE FEET BENEATH SURFACE-EIGHTY FOUR IVORY BEADS FOUND IN IVORY SAUCER ON THE BREAST OF GIANT.

A few miles north of Kennerd, in Henry county, Indiana, is a remarkable mound that covers an area of five acres.

Unlike other mounds found in Indiana and other states, it is composed primarily of sand and gravel and covered by a forest of native trees of a century's growth.

There is not another deposit of sand or gravel in six or eight miles. The surrounding country is plain.

This p[ile of sand and gravel, as stated in above, covers an area of five acres and is of cone shape.

When first known by white men it had a well defined ditch around it, and two made roadways, wide enough for a wagon, one from the north and the other from the south.

Farmers and road builders that needed gravel and sand found these glacial screenings to come handy in the building of public highways and for a small price per cubic yard paid to the owner of land found it more convenient than going to Springport or Mount Summit, a distance of eight miles.

After opening this deposit to a depth of 12 feet from the top of mound they unearthed a human skeleton whose framework measured nearly eight feet in height.

His skull would fit over the head of a large man; his jaws being massive and teeth in a perfect state of preservation.

On the breast of this big chief was a saucer-shaped vessel of ivory, about six inches in diameter, containing 84 ivory beads, that must have been made from the tusk of a mastodon.

We tried the persuasion of money on the old farmer in order to secure the ivory specimens, but he was invincible. We had no desire to become the possessor of human bones.

Centennial History of Rush County Indiana, **1921**

Forty years ago there was such a mound explored on the old Gary farm, also in Posey Township, and in that were disclosed numerous bits of pottery, a considerable quantity of beads of a variated sort and the skeleton of a gigantic man.

Indiana Progress, **November 9, 1921**
Huge Skeleton Unearthed
Indiana Produces Bones of Man Believed to Have Been Mound Builders

Indianapolis, Ind.-The complete skeleton of one of Indiana's oldest inhabitants, said by Dr. W. N. Logan, state geologist, to be that of a mound builder, has placed in the state museum.

The skeleton, more than six feet in height, was found by T. C. Heistant of Bloomington and Dick Guernsey of Bedford, in excavating a prehistoric mound, near the east fork of White river, in Lawrence county. Doctor Logan says the time when the mound builders inhabited Indiana cannot be determined as to years, but that it was a long period before the first record of the Indians is certain.

The mound from which the body was taken was in the form of a square with a vault system constructed of slabs of limestone. The skeleton was lying as buried, with all parts intact, and in position, with the exception of some of the more fragile parts.

Nevada State Journal **(Reno Nevada) December 30, 1923**
Cave Found in Indiana Hints Age Old Race
Giant Skeletons and Metal Strange to America Seen in Ancient Sepulcher
Blind Snakes are found
Bottomless Pit Temporarily Halts Exploration to be Resumed Later
By Loyd Bollett, International News Service Staff Correspondent

Indianapolis, Dec. 29

Recent discoveries in widely separated localities of the hills of Southern Indiana may be expected to add something to the general store of knowledge of natural history. The topography of Indians is interesting from the sand dunes bordering Lake Michigan to the rolling hills, which occupy the southern part.

The opening of an ancient sepulcher built by a race of men antedating the American Indian and probably not related to the mound builders has aroused much curiosity. This occurred in Jennings County, 10 miles from North Vernon in a bend of a small creek where some excavating was done on a mound built by human hands and reaching 100 feet in height and about the same in diameter

Bodies Guarded

Protected by great stone slabs skeletons of three men the longest of which measured much beyond that of modern men were found. They showed that infinite care had been taken that the remains be preserved against the ravages of the elements. Metals not common in North Americas also were found.

Although the locality has been settled more than 100 years, residents near Folsomville in the extreme southwestern part of the state, did not know until recently that a cave of mammoth proportions existed close by. The discovery was made by a hunter who accidently stumbled upon the entranced. The cave revealed some things which easily outdo the author of "The Arabian Nights" and have the added advantage of more veracity perhaps.

The discovery was Earl J. Nester, a mail carrier of Boonville. The most interesting "find" was a species of blind snakes which hiss so loudly that the gloom of the earth's innermost recesses becomes all the more hideous. Nester and some friends succeeded in capturing two specimens alive, and they were sent to the Smithsonian Institute in Washington.

After Nester and his party had proceeded a distance of 1500 feet they came upon a pit which apparently had no bottom. It was sounded to a depth of 2500 feet. One compartment of the cave was warm and comfortable and another extremely frigid, Nester found.

Nester said he believed a race of prehistoric men knew of the cave's presence and made it their home and added that its size would accommodate thousands as a dwelling place. He found many things to indicate human habitater, including arrow heads and stone implements.

The pit Nester and his party were unable to bridge, and further exploration was halted. At a future date the cave will be thoroughly explored.

Unmistakable evidence abound that Indiana was inhabited long before the dawn of earliest known history. The early men appears to have preferred the hills and caves in their fight against environment and to keep body and soul together, which is not true nowadays, as the most densely populated section is the level country of the north-central part.

Many caves exist along the Ohio River in southern Indiana. This map prepared by the Indiana Geological Society in 1888 shows the Wyandotte caves and the "Giants Ruins."

History of Delaware County, Indiana, 1924

About one mile and a half south of Muncie, in Center Township is another class of these earthworks-- a mound of considerable proportions which is said to have been dug into by some parties in search of relics. The excavation, however developed the fact that it contained, instead of relics, human bones. "One of these skeletons was of gigantic proportions. The jaw and thighbones were in a good state of preservation, and nearly complete. The jaw-bone was so large that it could be easily slipped over the jaw of the largest man of the party--a tall, big boned six footer, and the the thigh bone of the skeleton was three inches longer than his" The discovery of these numerous bones fixed the class under which this specimen should be arranged--the sepulcher--and would also warrant the presumption that there were specimens of some of the other classes not far distant, though investigation has not developed the fact.

Lima News, (Lima, Ohio) December 28, 1933
4 Skeletons Found
Noblesville, Ind., Dec. 28-(AP)-The skeletons of four men, believed to have been pre-historic mound builders, one measuring six feet six inches, were unearthed by road builders.

Iowa

Ft. Wayne Gazette, August 26, 1873
FOUND IN MOUND
Relics of an Ancient Race Discovered in Northeast Iowa

A very interesting archaeological discovery has just been made near Floyd, north of Waterloo Ia., on the Cedar river. For some time past it has been known that several ancient mounds were scattered along the banks of the river, but, though it had been excavated, nothing had been found except near Charles City. A few day ago work was begun excavating the mounds at Floyd. The largest was a circular in shape, thirty feet in diameter and about two feet high. It was situated in a field and had been plowed over for years. After digging down about four feet the bones of five persons were discovered. They were in a sitting posture and faced toward the north. A complet investigation showed that the original excavation had been floored with a layer of gravel, upon which the remains had been placed. The earth was then packed closely around them, another layer of gravel placed above them and the mound placed upon this. Of the remains nearly all the bones were in a good state of preservation, even the bones of the fingers and toes being intact. The bones were evidently of three males and one female and one babe. One of the men appears to have been over six feet high. The skeleton of the woman was, however, of greater interest from the fact that the skull measurements showed that she belonged to one of the lowest types of the human family, and experienced archaeologist claim that in some respects the skull seems to be lower in the scale than the celebrated "Neanderthal" specimen. The distance from the lower portion of the nasal bone to the upper margin of the eye cavities is only four centimeters, and the distance between the eye sockets is only two and three-fourths centimeters.

The forehead is very low and the inner portions of the eyebrow ridges are quite prominent. One of the jaws contained several well preserved teeth. The teeth of the babe were also preserved, but were very small. Its skull was quite thick. One of the male skeletons shows the teeth very much worn, in some instances clear down to the jaw-bone. The earth had been packed so closely around the skeletons that it was with considerable difficulty that it could be penetrated.

There are several other mounds in that vicinity which it is proposed to open, and it is expected that other remains of mound builders will be found. In the mounds near Charles City, which were opened last fall, remains of pottery were found with bones, but nothing of this sort was found at Floyd.

Waterloo Iowa Courier, September 22, 1897
OPENED A MOUND
Skeleton of a Prehistoric Man Exhumed by Relic Hunters
   A party of a relic hunters made an excavation yesterday into one of the mounds which are quite numerous along the Cedar in the vicinity of Waterloo. The mound was 40 feet in diameter and about five feet high in the center. At a depth of 41/2 feet, in the center of the mound fragments of a muman skeleton were found. They consisted of portions of a femur and humerus bones and a section of the base of the skull. It is evident that from the diameter of the bones that they belonged to a man of more than ordinary size, and all that was left of the skeleton can be packed in a cigar box it is probable that the body was laid away many centuries ago. No stone implements or copper ornaments were discovered but several pieces of charcoal were found near the bones and fragments of pottery and arrow heads were found near by.

Kansas

Fort Wayne Sentinel, (Fort Wayne, Indiana) November 28, 1897
Fought Ages Ago
Prehistoric Battlefield Found in Indian Territory.
Thousands of Skulls Dug Up With Arrow Points in Them-An Important Archaeological Discovery.
   Wichita, Kas. Nov. 29-The greatest prehistoric battle and burying ground yet discovered in the United States has just been found near the little town of Redlands. I. T. It lies on the northern border of

the Choctaw Indian reservation and near the Arkansas River.

Prof. Edwin Walters, the archaeologist who discovered it, states that from extensive excavations he has made he believes that nearly 100,000 warriors met death at that point and that the battle occurred 20,000 years ago. He goes a step further and declares the battle was fought between the mound builders and the Maya Toltec race, the latter coming from Yucatan and striving to wrest the Mississippi valley from the mound builders.

The battleground is thirty acres in area and by a series of excavations Professor Walters has satisfied himself that there are nearly 3,000 skeletons to every square acre. His estimate as to the remoteness of the prehistoric conflict is formed by a study of the geological structure of the formation in which the skeletons are found. The bones are buried near the top of deep strata of sand and covered first with a sort of adobe, a formation of the quaternary period, then with alluvial topsoil.

They have been dug out by the carload and almost every skull has from one to five arrow points sticking into it. Sharp arrow points and javelins are also found embedded in other bones of the body and the great number of these instruments of warfare that have been unearthed leave no doubt in the mind of Professor Walters that he has found what was once a field of awful carnage.

The skulls have narrow, retreating foreheads and projecting chins and the skeletons vary greatly in length, some seeming to be those of dwarfs and others of a giant race. The bodies are buried in a circle, feet toward the center and most of them in a sitting posture. At the side of each is found a clay vessel that was evidently filled with food to stay the soul of the departed warrior on his way to the spirit land.

Professor Walters has for many years been studying the mounds and battle grounds of prehistoric races found in the Mississippi valley and declares that the mound builders established a line of defense from Omaha, Neb., south of the gulf of Mexico in their efforts to repel the invasion into their territory of the warlike race of Toltec Maya.

Davenport Daily Republican, June 21, 1898
Scene Of A Big Battle
Interesting Discovery on a Farm in Eastern Kansas-Signs of a Prehistoric Race

J.T. Williamson, a farmer living 13 miles west of Kansas City on the Union Pacific railway, has discovered on his place an ancient burying ground, upon finds in which he vases his belief that his farm was once the scene of a furious Indian battle. It is not an infrequent thing for farmers in counties adjacent to Kansas City to unearth a skeleton or two, and nothing is thought of it, but Mr. Williamson has made an unusual find. From time to time he has dug up skeletons until he estimates that about 30 have been taken out only a small part of his lot. Besides the bones many Indian relics have been recovered. History does not record any great battle in this vicinity, but the finding of so many skeletons is evidence that some kind of warfare existed thereabouts long ago. No one living anywhere in the vicinity can recall any civilized burying ground around there.

In addition to the Indian bones found there is evidence that the Williamson farm was once the scene of the prehistoric race of mound builders. There are four or five mounds in Mr. Williamson's orchard, near his house, which show the art of careful building. One of these, much larger than the rest, was opened several years ago by the relic hunters of the Kansas state university, and it was found to contain some rare things. There was found the skeleton of a person in the mound. It was pronounced by the university experts the bone sof a prehistoric race. It was taken to the university, where it was carefully put together and remains there today. The conclusion that the bones are those of a prehistoric being drawn from the shape of the head, which is something like an egg. It is normal-sized skull, but the forehead shows no traces of development. From the eye sockets the crown of the head slopes back almost to a sharp point, leaving no development whatever of the forehead. The bones are in a fair state of preservation. In the mound the skeleton was found a perfect specimen of pottery, which, when the air struck it, crumbled to pieces. Besides these things two large spear heads were found.

The other mounds that are scattered through Mr. Williamson's orchard have never been molested, but Mr. Williamson expects some day to explore them. He does not expect to find anything of value, but will open them through curiosity to learn what they contain. The discovery of the Indian bones began several years ago, when excavations were made for the foundation of Mr. Williamsons house. At that time 15 skeletons were removed in the digging for the foundation of the various buildings. From time to time one or more turned up in plowing, but another wholesale excavation of dry bones took place last week, when Mr. Williamson's son began regarding part of their lot. They took up within 20 feet of their house, 11 skeletons. The bones were found under not much more than 2 ½ feet of earth. Some of the skeletons were partially petrified and were taken up whole, but most of them fell to pieces when they were picked up. Some of them crumbled into dust when they were touched.

The arrangement of the skeletons showed that some had been buried in confusion, while others were lying in rows and showed evidence of careful burying. There was no wood to show that any sort of casket had been used in burying the dead. Two of the bodies had been buried face downward and one had rested on its side. Nine of them were found in a row and they had been buried with their heads a little to the southeast. Buried with the skeletons were arrow heads, tomahawks, stone pipes, stone axes and a pair of silver earrings.

Several years ago Mr. Williamson unearthed the bones of what was probably a "heap big Indian chief." Part of a fancy burial robe were sufficiently preserved to show that it was a rich silk artistically embroidered. There were silver buttons of ancient pattern on the robe and a string of silver beads found around the neck.

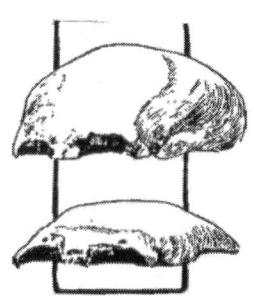

Marion Daily Star, **April 7, 1902**
GLACIAL MAN'S BONES
Important Scientific Discovery on a Kansas Farm
Made While Digging a Tunnel
The Long Looked For Proof That North America Was Inhabited By Man During the Great Ice Period Will Be Furnished It Is Believed, By This Find

The public museum of Kansas City is to be enriched by the addition of the skull and other fragmentary bones of a prehistoric men that were found a few days ago deep in a hillside of a Kansas farm ar a point about two miles in a northeasterly direction from Lansing, says the Kansas City Star. The skull and other bones and their geological environment indicate the skeleton to have been that of a primitive man of the glacial or great ice period centuries ago.

That mankind existed during the glacial period has been established by discoveries in Europe, and while it has been presumed that man also lived in America at the same time, no dubitable evidence of the fact has heretofore been obtained.

When the skull was found, it was not thought to have any scientific value. Several days ago M. C. Long curator of the Kansas City public museum, and Edwin Butts, civil engineer for the Metropolitan Street Railway company, both enthusiastic archaeologist, went to the place of the discovery and secured the fragments of the skeleton and brought them to Kansas City. Both Mr. Long and Mr. Butts are enthusiastic over the discovery. From the appearance of the skull and its position in the earth they are convinced it is that of a glacial man. If this fact be established, it will be the first proof of the kind found on the North American continent. In a short time the skull will be placed on view in the public museum. The facts of the discovery have been communicated to the Smithsonian Institute in Washington.

The find was made on the farm of Thomas and M. Cohncannon. They were digging a tunnel into a grerat hill on their farm with the purpose of using the excavation as a storage place for apples and other fruits. They dug directly into the side of the hill. The skull was found about sixty five feet in. Other bones of the skeleton were beside it. One of the farmers drove a pickax through the skull in loosening it from its stony bed, and later bones fell on it, so that it was broken into half a dozen pieces, but Mr. Long has cemented it together. The skull is that of a man with hardly any forehead. Directly back from the eyes receads the frontal bone. The fragments found shop he had a big jaw. The skull is very thick and strong, and its back part is broad and well developed. The phrenologist avert that this development at the back shows an abnormal nature. But there is no noble dome, no high and rounding forehead, that shows the development of intelligence.

The skull practically intact, a portion of the lower jaw, a part of a thigh bone and several other fragments were found. The bones indicate the man to have been large. The head is small. The orbits for the eyes are close together and appear exceptionally large. Over the orbits are well developed ridges that probably denote perceptive faculties. The bones were found huddled together. They lay partially imbeded in hardpan. A close and exhaustive investigation showed that the various strata of rocks and soils and the "water marks" had never been disturbed vertically and neither had there been any lateral disturbance of the hill. The skeleton evidently had been deposited there before the great mass of rock and soil above and about it. Had mound builders or Indians ever dug deep into the hill they could not have avoided leaving traces of their excavation.

"When we first heard of the find, we deemed it the usual story of a 'mound burial,' said Mr. Long the other day. "our investigation shows beyond all doubt that is a skeleton of a man of the glacial period.. After a most exhaustive investigation, Mr. Butts and I reached the conclusion the skeleton was deposited there during the glacial period or drift. How long ago the ice period was is not definitely known; 50,000 years perhaps; perhaps much longer.

"The evidence is very conclusive that this was not a burial or intrusive deposit, as there was no evidence of any disturbance of the earth. The great depth at which the skeleton was found precludes any idea of a usual burial, and the stratification of the earth both over and under the skeleton shows that the bones lay ther while the mass of soil was deposited over them. Attached to the skull is a kind of stony formation or cement, such as is usually found attached to bones of the mastodon and quite similar to the formation found in the jaws of the mastadon in the public museum."

Mr Long says that the ground around which the skeleton was found shows, conclusive evidence of its glacial formation. Comparison of this skull with photographic illustrations of the skull of the "Man of Spy," a famous skeleton found in a cave in Belgium, shows them to be practically alike.

Kentucky

Counties of Todd and Christian Kentucky,
Historical and Biographical, **1872**

There are numerous mounds in Todd County, but to which of these classes they should be assigned it is difficult to determine from the meager accounts to be gained of them. But one or two have been examined, and these with insufficient care. Skeletons of extraordinary size were found, the skulls of which were passed over the head of the large man, and rested easily upon his shoulders.

Collins Historical Sketches of Kentucky, History of Kentucky, **Lewis Collins, 1874**
Ohio County
Simpson County

Antiquities-A Giant-From a mound on the farm of Eden Burrowes, near Franklin, were exhumed, in may, 1841, at a depth of over 12 feet, several human skeletons. One, of extraordinary dimensions, was found between what appeared to have been two logs, covered with a wooden slab. Many of the bones were entire. The under jaw-bone was large enough to fit over the jaw, flesh and all, of any common man of the present day. The thigh-bones were full six inches longer than those of any man in Simpson County. Teeth, arms, ribs, and all, gave evidence of a giant of a former race. Around his neck was a string of 120 copper beads, and one beads of pure silver, all perfectly preserved. Another skeleton, of smaller dimensions, had around his neck a string of ivory beads, about 100 in number. The string, which had held the beads, was still apparent, though time had destroyed its consistence.

Butler County

Mounds and Cave-On the farm of Judge T. C. Carson, 7 miles below Morgantown, are several mounds-one 8 or 10 feet high, covering between a quarter and half an acre of land. No bones have been found in it; but from a smaller one, a umber of bones belonging to a giant race have been taken-jaw bones which would go over the whole chin of a man, and teeth correspondingly large; the teeth remained sound, but the other bones crumbled on exposure to the air. In Saltpeter Cave, I the Little Bend of Green River, a number of such bones were found.

Harlan County

Antiquities-The first courthouse in Harlan County was built upon a mound in Mount Pleasant-upon, which, in 1808, the largest forest trees were growing. In August, 1838, a new court house was erected upon the same mound, requiring a deeper foundation and more digging-with these discoveries: human bones, some small, others very large, indicating that the bodies had been buried in a sitting posture; several skulls, with most of the teeth fast in their sockets, and perfect; the skull of a female, with beads and other ornaments which apparently hung around the neck. Close by the larger bones was a half gallon pot, superior in durability to any modern ware; made of clay and of pearl winkles pounded of powder; glazed on the inside, and the outside covered with little rough knots, nearly an inch in length.

Allen County

Antiquities-In the west end of the county, about thirteen miles from Scottsville, and seventeen from Bowling Green…

At the west side of the narrow pass, and immediately at its termination, there is a hill similar to the one on the east. Here is to be seen a small mound forty feet in circumference and four feet high. Upon excavating one side of this mound, a stone coffin was dug up two and a half feet long, one foot wide and one foot deep, with a stone covering-the top of the coffin projecting one inch beyond the sides.

Upon opening the coffin, the arm and thigh bones of an infant were found in it. This coffin being removed, others of larger dimensions were to be discovered, but were not removed. Many very large human bones have been exhumed form mounds in this county-some of the thigh bones measuring form eight to ten inches longer that the race of men now inhabiting the country.

Carroll County

Antiquities--…There are a number of mounds in the county, but generally of small size. In 1837, one was examined in which was found the skull and thigh bones of a human being of very large frame, together with a silver snuff box, made in the shape of an infants shoe. On an elevated hill, a short distance from the Kentucky River, in opening a stone quarry, the jaw bone and a large number of human teeth were found; and on the points of the ridges, generally, similar discoveries have been made. About four miles from Carrollton, on the Muddy fork of White Run, in the bed of the creek, on a limestone rock, is the form of a human being, in a sitting posture; and near by, is the form of one lying on his back, about six feet long, distinctly marked.

Collins Historical Sketcher of Kentucky Vol.II, History of Kentucky, 1874, **Madison County, Kentucky**

Ancient cemetery- A race of giants on five high points on Caldwell Campbell's farm, and on a farm of Samuel and Walker Madison, adjoining, 8 miles southwest of Richmond, are burial grounds of pre-historic inhabitants- in all embracing fully 3 acres. On one part, about one and a half acres, have been discovered the skeletons of giants- The femur, tibia, skull, and inferior maxillary bones so large, when compared with the size of the late John Campbell (himself 6 feet 4 inches high) as to indicate a race 7 to 8 feet high. John Campbell slipped the inferior jawbone of one entirely over his own, flesh and all.

New York Times, **February 8, 1876**
THE EARLY AMERICAN GIANT

The public will be unpleasantly reminded of the callous indifference to the future on the part of the prehistoric Americans by the recent discovery of three unusually fine skeletons in Kentucky. A Louisville paper asserts that two men lately undertook to explore a cave which they accidentally discovered not far from that city. The entrance to the cave was small, but the explorers soon found themselves in a magnificent apartment, richly furnished with the most expensive and fashionable stalactites. In a corner of this hall stood a large stone family vault, which the men promptly pried open. In it were found three skeletons, each nearly nine feet in height. The skeletons appear to have somewhat frightened the young men, for, on seeing so extensive collection of bones, they immediately dropped their torch, and subsequently wandered in darkness for thirty-six hours before they found their way back to daylight and soda-water.

Now, it is evident that these gigantic skeletons belonged to men very different from the men of present day. A skeleton eight feet and ten inches in height would measure fully nine feet when dressed in even a thin suit of flesh. The tallest nine-foot giant of a traveling circus is rarely more than six feet four inches high in private life and without his boots, and even giants of this quality are scarce and dear. The three genuine nine-foot men of Kentucky must have belonged to a race that is now entirely extinct, and hence it would be a matter of great interest if we could learn who and what they were.

Many mounds and a henge are still visible in Madison County, Kentucky. This mound is several miles south of Richmond. From *The Nephilim Chronicles, A Travel Guide to the Anceint Ruins in the Ohio Valley*.

History of Kentucky, V. I, 1922
   "Above the mouth of Big Point Creek there is a river bottom extending up the Louisa River about a mile…There mounds were covered with large trees when first seen by white men. The original public highway up the Big Sandy River was laid out to cut the north side of the second mound. In matching this public road the mound was cut and the skeleton of a man of large size was found. It was enclosed in a sort of rude box made by placing flat thin river stones about and over it."

Louisiana

Oakland Tribune, January, 27, 1904
GIANT SKELETONS IN LOUISIANA UNEARTHED
   Skeletons of a race of giants who averaged twelve feet in height were found by workmen engaged on a drainage project at Crowville, near here.
   There were several score at least of the skeletons and they lie in various positions. It is believed they were killed in a prehistoric fight and that the bodies lay where they fell until covered with alluvial deposits due to the flooding of the Mississippi river.
   No weapons of any sort were found and it is believed the Titans must have [...]
The skulls are in a perfect state of preservation and some of the jawbones are large enough to fit around a baby's body.

Minnesota

New York Times, December 25, 1868
Reported Discovery of a Huge Skeleton

From the Sauk Rapids (Minn.,) Sentinel, Dec. 18. Day, before yesterday, while the quarrymen employed by the Sauk Rapids Water Power Company were engaged in quarrying rock for the dam which is being erected across the Mississippi, at that place, found imbeded in the solid granite rock the remains of a human being of gigantic status. About seven feet below the surface of the ground, and about three feet and a half beneath the upper stratum of rock, the remains were found imbeded in the sand, which had evidently been placed in the quadrangular grave which had been dug out of the solid rock to receive the last remains of this antediluvian giant. The grave was twelve feet in length, four feet wide, and about three feet in depth, and is to day at least two feet below the present level of the river. The remains are completely petrified, and are of gigantic dimensions. The head is massive, measures thirty-one and one- half inches in circumference but low in the front, and very flat on top. The Femur measures twenty-six and a quarter inches, and the Fibula, twenty- five and a half, while the body is equally long in proportion. From crown of the head to the sole of the foot, the length is ten feet nine and a half inches. The measure around the chest is ninety-nine and a half inches. The giant must have weighed at least 900 pounds when covered with a reasonable amount of flesh. The petrified remains, and there is nothing left but the naked bones, now weigh 304 pounds. The thumb and fingers of the left hand, and the left foot from the ankle to the toes are gone; but all the other parts are perfect. Over the sepulcher of the unknown dead was placed a large flat limestone rock that remained perfectly separated from the surrounding granite rock.

St. Paul Globe, **August 12, 1896, Minnesota** The skeleton of a huge man was uncovered at the Beckley Farm, Lake Koronig, Minnesota. While at Moose Island and Pine City, bones of other giants came to light.

New York Times, June 30, 1888
Skeletons Of A Former Race

Clear Water, Minn., June 29. -Charles Pinkerton of the town of Cortna, 12 miles from here, in digging for a cellar came across the remains of seven persons in a good state of preservation. They were found in a kind of mound were buried with their heads down, and were 7 to 8 feet in height, and must have been placed there at least 200 years ago, as on top of the mound was the stump of an old elm tree two feet in diameter. From the formation of the skulls they must have been of an inferior race of men. The teeth in the jawbones were mostly sound and not like the teeth of the present race of men.

St. Paul Press, July 29, 1897
Prehistoric Giants Bones of a Brobdingnagian People Found
In Northern Minnesota
Some Recent Remarkable Discoveries in Itasca County
Fine Pottery and Copper Implements- Ponderous Weapons-Queer Hieroglyphics
Special Correspondence of the Globe Democrat

Ely, Minn., August 1. - While Minnesota as a State is getting fairly well along in years, as Western States goes still there are many large tracts over which white men have not traveled to any extent, and about which little is known, except as fragmentary information has been brought in by hunters and guides who have had occasion to pass over the territory. This is particularly true of the extreme northern portion of the State, which until gold was found there was as little known as were the vast wilderness to the west when the Pilgrim Fathers landed at Plymouth Rock.

The steady rush of settlers to this section and of the gold seekers to the region further north has resulted in opening up many thousands of acres of entirely new country and placing before the world a farming territory as rich as any in the United States and a mineral deposit so wonderful and varied in its character that no one may say the end has yet been reached. With the partial settlement of this new country have come many strange finds, showing that at one time in the dark and half-forgotten past the northern section of this State was inhabited by men and animals of which the present generation knows nothing except what scientist and searchers after the curious may say.

The country around Ely is particularly rich in relics of the past, judging from the many finds made by various people since farms have been opened up here. Scarcely a day passes but some new discovery is made, and the relics are particularly interesting to those who have delved into the records of the musty past to any extent. There is evidence to show that at one time this country was peopled by men of tremendous size and by animals in comparison with which the elephants of the circus today would seem like pigmies. There is also evidence to show that the people who then ruled the country were not mere hewers of wood and drawers of water, but possessed skill in various directions.

Cemetery Of The Giants

GIANT SKULL AND MODERN SKULL COMPARED.

Thomas McKinster, who has recently taken up a farm near the mouth of the Little Fork River, in Itasca County, has discovered relics which in point of historic value and interest are far ahead of all previous finds. At one corner of his farm stood a knoll, possibly 150 yards across, covered with heavy growth of pine. On top of this knoll was a huge bird, or rather the outlines of one made of stone not found in any other locality of the State so far as known. In digging into the knoll it was found that the place was once a cemetery, or burying ground of a prehistoric race.

Human bones of great size were found, and all the positions indicated that the original owners thereof had been buried in sitting posture, instead of being stretched out at full length, as is the custom in these days. It was an easy matter to distinguish the bones of the male from those of the female, for they were larger more massive and had preserved their shape better. In some cases the bones fell to pieces at the slightest touch and resolved themselves into a powder, while in others they were soft and spongy, the earth seemingly having entered the cavities formerly occupied by the marrow, and furnished a support for the shells.

COPPER AND STONE IMPLEMENTS.

In one grave, if grave it could be called, were found the bones of what were once probably father and mother and child. The huge bones of the male were about a foot distant from those of the woman and child, which were locked in a close embrace as though the mother had drawn the infant to her bosom just before death came. The larger bones were those of a man who, in life, must have been nearly, if not fully, 9 feet in height, while those of the woman showed that she had been no dwarf, measuring as they did 8 feet 4 inches from the bottom of the bones of the foot to the top of the skull, which was larger, flatter and considerably heavier than the skulls of the people today. The child must have been very young for its bones went to pieces as soon as exposed to the air, and covered the bones of the mother with a fine white powder. By measuring the outlines of the bones of the child it was found that the infant had been nearly 5 feet in height, which is really not so bad for a child of tender years. In the skull of the woman was a big dent, between and slightly above the eyes, showing that she had met death in a violent manner. By the side of the man lay a stone weapon, shaped something like an egg with a depression cut around one end in which a thong might have been twisted. It was of the same kind of stone as the bird on top of the knoll or mound, a hard blue colored granite, almost as close grained as steel.

The Washington Post, **June 17, 1899**
Giant Skeletons In a Cave
Interesting Discovery by a Mexican Sheep Herder in Minnesota
St. Paul, Minn., June 16-Jose Herannda, a Mexican sheep herder in the employ of McLeod Brothers, while rounding up horses in the Sweet Grass Hills, twenty miles north of Columbus, Minn, discovered a large cave, the opening of which has been concealed by heavy underbrush. The cave, 70 feet in length, 35 feet in width, and 10 feet in height, was cut out of solid rock.

In the center, lying side by side, were the well-preserved skeletons of five human beings. These skeletons measure from 7 to 7 1-2 feet in length. Three knife blades evidently made of hardened copper, two bowls hollowed out of granite blocks, two stone hammer heads, and some broken fragments of pottery were also found in the cave.

Evening Tribune **(Albert Lea, Minn.) September 10, 1931**
6 Human Skeletons, Pottery Found on Farm Near Deer River, Minn.
Discovery Leads to Belief That Bands of Prehistoric Mound Builders Once Roamed Northern Minnesota.
Deer River, Minn., Sept 10-Six human skeletons and several pieces of pottery unearthed on a farm 45 miles north of here have led residents to believe that bands of prehistoric mound builders once roamed northern Minnesota.

The skeletons were discovered on the farm of Sam S. Strangeland on the Big Fork river in Gratan township, Itasca county, when Mr. Strangelove began to do dig a root cellar in a mound situated in fromt of his home.

Digging operations on the root cellar had not progressed very far when one of the skeletons was discovered. Four more were uncovered on the same level and a sixth was found about three feet deeper.

All the bodies were buried in a sitting position with the legs doubled under and the head bent downward. Coffins were not used. One skeleton had an exceptionally large jaw bone and the forehead of the skull receded abruptly above the eyes. The other five were of smaller proportions.

Missouri

Scientific American, 1883
"A Tradition of Giants,"
Two miles from Mandan, on the bluffs near the junction of the Hart and Missouri rivers is an old cemetery of fully 100 acres in extent filled with bones of a giant race. This vast city of the dead lies just east of Fort Lincoln Road. The ground has the appearance of having been filled trenches piled full of dead bodies, both man and beast, and covered with several feet of earth. In many places mounds from 8 to 10 feet high, and some of them 100 feet or more in length have been filled with bones and broken pottery, vases of various bright-colored flints and agates. The pottery is of a dark material,

beautifully decorated, delicate in finish and as light as wood, showing the work of a people skilled in the arts and possessed of a high state of civilization. This has evidently been a grand battlefield, where thousands of men and horses have fallen. Nothing like a systematic or intelligent exploration has been made, as only little holes two or three feet in depth have been dug in some of the mounds, but many of the bones of man and beast and beautiful specimens of broken pottery and other curiosities have been found in those feeble efforts at excavation. We asked an aged Indian what his people knew of these ancient graveyards. He answered: "Me know nothing about them. They were here before the red man."

Isis Unveiled, 1877

As we write, there appears in an American paper, *The Kansas City Times,* an account of important discoveries of the remains of a prehistoric *race of giants,* which corroborates the statements of the kabalists and the Bible allegories at the same time. It is worth preserving:"In his researches among the forests of Western Missouri, Judge E. P. West has discovered a number of conical-shaped mounds similar in construction to those in Ohio and Kentucky. These mounds are found upon the high bluffs overlooking the Missouri River, the largest and more prominent being found in Tennessee, Mississippi, and Louisiana. Until about three weeks ago it was not suspected that the mound builders had made this region their home in prehistoric days; but now it is discovered that this strange and extinct race once occupied this land, and have left an extensive graveyard in a number of high mounds upon Clay County bluffs.

As yet, only one of these mounds has been opened. Judge West discovered a skeleton about two weeks ago, and made a report to other members of the society. They accompanied him to the mound, and not far from the surface excavated and took out the remains of two skeletons. The bones are very large-so large, in fact, when compared with an ordinary skeleton of modern date, they appear to have formed part of a giant. The head bones, such as have not rotted away, are monstrous in size. The lower jaw of one skeleton is in a state of preservation, and is double the size of the jaw of a civilized person. The teeth in this jaw-bone are large, and appear to have been ground down and worn away by contact with roots and carnivorous food. The jaw-bone indicates immense muscular strength. The thigh bone, when compared with that of an ordinary modern skeleton, looks like that of a horse. The length, thickness, and muscular development are remarkable. But the most peculiar part about the skeleton is the frontal bone. It is very low, and differs radically from any ever seen in this section before. It forms one thick ridge of bone about one inch wide, extending across the eyes. It is narrow but rather heavy ridge of bone which, instead of extending upward, as it does in these days of civilization, receded back from the eye-brows, forming a flat head, and thus indicates a very low order of mankind. It is of the opinion of the scientific gentlemen who are making these discoveries that these bones are remains of a prehistoric race of men.

History of Dunklin County, Missouri, 1896

The mounds and other ancient earthworks constructed by this people are abundant in Southeast Missouri. Some are quite large, but the greater part of them is small and inconspicuous.

Along nearly all of the watercourses that are large enough to be navigated by a canoe, the mounds are almost invariable found, so that when one places himself in such a position as to command the grandest river scenery he is almost sure to discover that he is standing upon one of these ancient tunnels, or in close proximity thereto. The human skeletons, with skulls differing from those of the Indians that are found in these mounds are usually accompanied by pottery and various ornaments and utensils showing considerable mechanical skill.

Their axes were of stone; their raiment, judging from fragments which have been discovered, consisted of the bark of trees interwoven with feathers, and their military works were such as a people would erect who had just passed to the pastoral state of society from that dependent alone upon hunting and fishing. They were no doubt idolaters, and it has been conjectured that the sun was the object of their adoration. The mounds were generally built in a situation affording a view of the rising sun; when enclosed in walls their gateways were toward the east and, finally medals have been found, representing the sun rays of light. Dunklin County is an especially rich field for the archaeologist. Situated on the farm of C.V. Langdon, one mile south of Cotton Plant, is one of the largest mounds in the county, adjoining are smaller ones.

In the north part, and, in fact, nearly all over the county at comparatively short distances, these mounds are very noticeable. Extra large-sized human bones, skulls, earthen pots, rude ornaments, and various stone implements have been exhumed from many of these mounds

Daily Northwestern, October 30, 1923
Find Bones of a Prehistoric Civilization High Up on Ragged Bluff in the Ozark Mountains

Richland, Mo. July 20 (By Associated Press)-Half way up a rugged bluff of the Ozarks which tower above the Gasconada river about three miles from here, have been found evidence of a prehistoric civilization which aroused the interest of archaeologist of the Smithsonian Institution who will arrive here soon to investigate.

About a year ago, in a large cave which opens to the river, A Steckle, the owner, who was enlarging it to make a resort for tourist, uncovered three human skulls and a number of bones. In addition to pottery and beads. The skulls are unusual in that they do not resemble Indian skulls, but have low receding foreheads and very thick skull bones. The teeth are large, sharp and well preserved.

The bones were uncovered in a bed of ashes directly below a large hole in the stone ceiling, evidently made by the section of fire. More than a foot of earth covered the ashes. Because of this earth, which apparently could have come there only through decomposition, it is believed the race lived 2000 years or more ago.

Morning Herald, Uniontown, Pennsylvania December, 27, 1934
Eight Giant Skeletons Unearthed

Springfield, Mo., Dec. 26-Discovery of eight giant human skeletons arranged in four layers so that each pair formed a cross, in a shallow Indian mound near here apparently shed new light on religious customs of prehistoric tribes who once inhabited the Ozarks.

The Rev. E. P. Newberry, Springfield archaeologist, believed it "highly probable" that the crosses formed by the skeletons were of religious significance.

"The Indian cross," he explained, "has always indicated a place of worship in this particular case it seems a reasonable supposition that the eight skeletons were those of a high priest of some cult, and his attendants.

"It was the custom of some tribes to sacrifice a priest's attendants when the priest died, and to bury them with him just as dead man's horse, dog and food were buried with him."

In the crook of an arm of one of the skeletons was found an ancient stone sacrificial bowl, with markings which were exactly like four strange idols recently discovered by Mr. Newberry in the Coleman cavern north of Springfield.

Other evidence that a race of cultured people lived in this vicinity in prehistoric times, Mr. Newberry said consisted of a vault made of a high type of concrete, found on a farm near Odessa, and dental work in teeth of other skeletons which he has unearthed. American Indians, he explained, never made very good concrete, and never made dental repairs.

Montana

Logansport Reporter, September 3, 1903
GIANT SKELETONS
Prehistoric Relics Unearthed in Montana

   Wonderful finds of fossils and bones of prehistoric animals are being made in the Fish Creek country, Montana, by Professor Marchus S. Farr, and a party of students from Princeton college. The remains of a stone age city have been found in which the bones of animals of great size, along with stone implements of all kinds, many of which are ornamented with gems. In a mound near the creek were found the almost complete skeleton of a man. The bones showed that the man, when alive, measured nearly nine feet in height, and was of powerful build. Nearby was a skeleton of a woman, a trifle smaller in size, and at the foot was a skeleton of an animal that resembled the dog of today except the animal must have been as large as a small horse.

Nebraska

Lincoln Evening News, November 8, 1911
WHO WERE THEY?
Find of Skeletons Puzzles Junction City Farmers

   An ethnologist mystery has been uncovered on the farm of John Noland several miles northwest of this city. In the center of Mr. Nolands's wheat field was a mound which he decided to level, and while doing so he uncovered several human skulls and a large number of bones and teeth, says a Junction City dispatch to the Topeka State Journal.

   One of the skulls and one of the thigh bones, apparently from the same person, were of gigantic stature. The big skull is pierced through the back with several small round holes, apparently such holes as would be made by small shot from a shotgun.

   Persons who have examined the skulls say they do not resemble the ordinary Indian skull and the absence of Indian weapons and utensils would seem to indicate that the bones were not those of Indians. Although a careful search was made, no Indian relics were found within the cairn, although there was a brown powdery substance that might have been wood. The oldest settlers, however say that they do not remember of any white persons having been buried there. The question now is, of what race were those who were buried there?

Lincoln Daily Star, November 30, 1913
Rare Collection To Medical School
Three Skulls Given to University Branch by Dr. Gilder
Types of Three Races Shown by Relics Found in Nebraska Mounds
(Special to the Star)

   Omaha Neb., Nov.29, -A series of three prehistoric skulls, estimated as ranging from 100 years to at least 20,000 in age, and representing the three races of mankind, which have lived in the great Missouri valley since the advent of the human race upon this continent, has just been presented to the new medical college of the University of Nebraska by Dr. R. F. Gilder of Omaha, archaeologist in the field for the university. The three different specimens show upward movement in the human race in its march towards civilization as probably no other collection of prehistoric times does.

Accompanying each skull are implements of chase and war manufactured and used by the contemporaries, although not by the individuals whose crania Dr.Gilder unearthed from where they have laid undisturbed for so many centuries. For the man of 20,000 years ago, whose order of intelligence was the very lowest in the scale, there is an immense war club of stone, so rudely fashioned as to appear to the untrained eye, scarcely more than a rough boulder.

For the mound builder of 2,000 years ago, whose skull is of a decidedly higher grade than that of the " loess man," there are the beautifully polished, clean-cut axes and hammers. And for the American red man of 100 years ago, there are implements showing the magic touch of the master mind of the white man with whom this particular Indian had come in touch, as was shown by the glass beads and metal fringe which still encircled the bony throat.

The oldest skull is technically known as "Nebraska Loess Man No. 8," and is the eighth skull removed from the burial mound known as the "Long Mound," This mound is located a few miles above Omaha and from the huge grave, Dr.Gilder removed twelve skulls, all of an order so low as to be scarcely above the ape in intelligence. Scientist made a careful and minute examination and declared the above this skull had never been disturbed by man, but had been deposited by nature when the hills were made. Geologist declared that this had taken place at least 20,000 years ago and that the age of this people could not be less than that number of years.

"Number-Eight" had almost no forehead at all. There is a supercilliary ridge over each eye as pronounced as the flange on a car wheel, while back of this ridge the skull slopes to the rear of the head. Nature did not fit this man to be the head of a modern trust company, but he was provided with a hcad and skull that would shed missiles like a duck's back turns raindrops. The skulls from this mound have been subject of much interest to archaeologist all over the world and savants from France, Germany, and several other countries have journeyed all the way to Omaha especially to see and study them. Universities in all parts of the world have asked for cast and replicas.

North Carolina

Smithsonian Institute Bureau of Ethnology, 1890-1891

Located on the farm of Rev. T.F. Nelson, in the northwest part of the county, and about a mile and a half southeast of Patterson. It stood on the bottom land of the Yadkin, about 100 yards from the river, and was almost a true circle in outline, 38 feet in diameter, but not exceeding at any point 18 inches in height. The thorough excavation made, in which Mr. Rogan, the Bureau agent, was assisted by Dr. J.M. Spainhour, of Lenoir, showed that the original constructers had first dug a circular pit about 38 feet in diameter to the depth of 3 feet and there placed the dead, some in stone cist and others unenclosed, and afterwards covered them over, raising a slight mound above the pit. A plan of the pit, showing the stone graves and skeletons as they appeared after the removal of the dirt and before being disturbed, is given in figure. 207...No. 16 was unenclosed "squatter" of unusually large size, not less than 7 feet high when living. Near the mouth was an entire soapstone pipe; the legs were extended in a southwest direction upon a bed of burnt earth.

Beehive type tombs within the mound. Drawing of the mound by the *Smithsonian Institute Bureau of Ethnology*, 1890-1891.

Ohio

Ohio's largest mound is located in Miamisburg, Ohio. From, *The Nephilim Chronicles, A Travel Guide to the Ancient Ruins in the Ohio Valley*, 2010.

Athens Messenger, (Athens Ohio) April 21, 1870

In the company with some friends illegible.. to the mound, where we commenced to exhume the skeletons of a race that will remain one of the hidden mysteries forever.

The mound had two circular arches of stone in it. one was about two and a half feet beneath the surface of the mound, and the other about six feet. Immediately under the first arch of stone was found two very large skeletons, in a remarkable state of preservation, and under the second arched wall, other

291

parts of skeletons were found amidst ashes, coals and mussel shells. It seemed apparent that those who had been interred under the deeper arch, had remained there many years, before those under the upper arch were buried. Some of these skeletons were buried with their heads towards the center of the mound, and some with their feet toward the center. Near the thigh bone of one was found a round piece of stone, about six inches long and an inch in diameter, with a half-inch hole through it. It has the appearance of slate-stone, and finely worked off. Parts of seven skeletons have been taken from this mound, and "the end is not yet." But enough about the mounds.

Large mound north of Athens Ohio, in Athen County, Ohio that was part of a henge complex. *The Nephilim Chronicles, A Travel Guide to the Ancient Ruins in the Ohio Valley,* **2010.**

Ohio Democrat, **(New Philadelphia, Ohio) February 24, 1871**
A giant skeleton 8 feet six inches in length, was recently found on the farm of John Buck, in Athens County, Lodi Township, in an old mound.

Athens Messenger, **(Athens, Ohio) July 4, 1878**
A Homer Township Mound Opened
We obtained from Mr. Lewis [...]of Homer Township; a gentleman whose statements may be implicitly relied on, the following information concerning the recent opening of a mound in Homer Township.
The mound in question is on the Jonathon Pedicord farm now owned by Mr. Jas. Carpenter. It is, or was about 30 feet in diameter, and almost 6 feet in height. Induced to do so throughout mere curiosity, Mr. Carpenter hitched his team to a road scraper a few days since, and began the work of moving the earth of which the mound is comprised, and which was so soft and loose that nothing more than a scraper was needed for the work. After about two feet of the top had been taken off, a basin-shaped vault, four feet in depth was discovered. The side of the vault was hard and smooth dressed, and in the bottom was found considerable quantity of ashes, showing that fire had been used in its construction. But the most startling discovery was that of no less than nine skeletons packed in around the rim of this basin. They were placed with their faces downward, and in tiers of three in one place, three in another,

292

and two in another, while the "boss" lay off by himself. Then there was this peculiarity. On top of each skeleton and extending from the head down over the body to the extremities, was a large flat dressed stone, which on being removed showed the skeleton entire. The bones were in a good state of preservation when first exposed to the air, but on an attempt to remove them they parted at the articulations otherwise they are about like those seen in the office of the physicians. One thighbone measured eighteen and a half inches in length. Carpenter has the whole lot piled up like stove wood. Curiosity seekers are beginning to carry away portions of the skeletons. The skull of the "boss," who lay alone is very large. The lower jawbone will go over that part of a living man's corpus, without touching. The teeth found in the skull were sound as a dollar. The neighborhood boys have about all tried their hands in extracting them until not more than two or three are left. But the most singular thing of all is that a perfect horn, pointing forward and downward, grows out of this skull just back of the ear on the left side. Mr. Parsons says, "If you don't believe me, come and see." The horn is about an inch and a half long. Through lack of care, the right side of this skull was mashed to pieces in the place where the horn ought to have been found.

Mr. Parsons told us a good many other interesting things about this wonderful mound, and its dead, and then gave it as his opinion that the skeletons belonged to a race of people of greater antiquity than the ancestors of the red devils who are making such a row up in the Northwest just now.

Humph! We call this the New World. Wouldn't it be better to call it the old one? It is on this continent alone, that such cemeteries as has just been described are found.

There are other mounds in Homer like unto this one just opened, and it is not unlikely that they will be examined in the future, and with more care. By the way, no kind of implements was found save those, which were made out of stone. Of these there were a variety. "Who are these mound-builders?" That's the question.

History of Delaware County, Ohio, 1880

The mounds are mostly sepulchral. One of the most remarkable ever opened in the county, was one on the farm of Solomon Hill, a short distance west of the Girls Industrial Home. We take the following notice of this mound from the Delaware Herald of September 25, 1879. "Saturday we were shown some interesting relics consisting of a queen conch-shell, some isinglass [mica] and several peculiarly shaped pieces of slate, which were found in a mound on the farm of Solomon Hill, Concord Township, Delaware Co., Ohio. The mound is situated on the banks of a rocky stream. The nearest place where the queen conch-shell is found is the coast of Florida; the isinglass in New York State, and the slate in Vermont and Pennsylvania. Two human skeletons were also found in the mound, one about seven feet long, the other a child. The shell was found at the left cheek of the large skeleton. A piece of slate about one by six inches was under the chin.

History of Delaware County, **Ohio,** 1880

On a farm belonging to A. E. Croodrich, in Liberty Township, there is a circular mound, perhaps forty or fifty feet in diameter, which, until it had been largely obliterated by the cultivation of the land on which it lies, was one of the most perfect works of its kind to be seen anywhere. There was another mound on Mr. Croodrich's barn lot a number of years ago, which has been entirely removed. During the process of grading there was found, some distance below the surface, three skeletons in a good state of preservation. One of them was apparently that of a man considerably above medium stature, while the other two were smaller.

Muskingum County, Ohio, Brush Creek Township, **March 3, 1880,**

"To Dr. F. T. Everhart, A.M., Historian:

"Dear sir: On December 1, 1879, we assembled with a large number of people for the purpose of excavating into and examining the contents of an ancient mound, located on the farm of Mr. J. M. Baughman, in Brush creek township, Muskingum county, Ohio.

"The mound is situated on the summit of a hill, rising 152 feet above the bed of the stream called Brush creek. It is about 64 feet in with by about 90 feet in length, having an altitude of 11 feet 3 inches; is nearly flat on top. On the mound were found the stumps of sixteen trees, ranging in size from 8 inches to 2 1/2 feet in diameter.

We began the investigations by digging a trench four feet wide from the east side. When the depth of eight feet had been reached, we found a human skeleton, deeply charred, in close proximity to a stake six feet in length and found inches in thickness, also deeply charred, and standing in an upright position. We found the cranium, vertebrae, pelvis, ad metacarpal bones near, while the femurs and tibula extended horizontally from the stake. At this juncture work was abandoned, on account of the lateness of the hour, until Monday, December 8th, when it was resumed by opening the mound from the northwest. When at the depth of seven and a half feet in the north trench, came upon two enormous skeletons, male and female, lying one above the other, faces together, and heads toward the west. The male, by actual measurement, proved to be nine feet six inches; the female eight feet nine inches in length. At about the same depth in the west trench we found two more skeletons, lying two feet apart, faces upward, and heads to the east. These, it is believed, were full as large as those already measured, but the condition in which they were found rendered exact measurement impossible. On December 22d we began digging at the southeast portion of the mound, and had not proceeded more than three feet whcn we discovered an altar, built of sand rock. The altar was six feet in width and twelve feet in length, and was filled with clay, and of about the same shape that the mound originally was. On the top, which was composed of two flat flag-rocks, forming an area of about two feet in width and six in length, was found wood-ashes and charcoal to the amount of five or six bushels. Immediately behind, or west of the altar, were found three skeletons, deeply charred, and covered with ashes, lying faces upward, heads toward the south, measuring, respectively: eight feet ten, nine feet two, and nine feet four inches in length. In another grave a female skeleton eight feet long, and a male skeleton nine feet four inches long-the female the lowermost, and the face downward, and the male on top, face upward, behind the site of the altar. After proceeding about four feet, we found, within three feet of the top of the mound, and five feet above the natural surface, a coffin or burial case, made of a peculiar kind of yellow clay, the like of which we have not found in the township; consequently, we believe it was brought from a distance. Within the casket were confined the remains of a female eight feet in length, an infant three and a half feet in length, the skull of which was scarcely thicker than the blade of an ordinary case-knife. The skull of the female would average in thickness about one-eighth of an inch, measured eighteen and three-fourth inches from the supra-orbital ridge to the external occipital protuberance; was remarkably smooth; perfectly formed. Within the enclosure was a figure or image of an infant but sixteen inches in length, made of the yellow clay of which the casket was formed; also, a roll of peculiar black substance encased in the yellow clay, twelve inches in length by four inches in diameter, which crumbled to dust when exposed to the air.

We also found what appears to have been the handle and part of the side of a huge vase; it was nicely glazed, almost black in color, and burned very hard. From within a few inches of the coffin was taken a sand-rock, having a surface of twelve y fourteen inches.

(which had also passed through the fire), upon which were engraved the following described hieroglyphics:" [Here a space was left in the note-book for the representation of the inscription found upon the stone; but, for the sake of a true representation, we determined to have photographs made, and make one a part of this report.]

Proceeding north about four feet from where we found the coffin, and within six inches of the top of the mound, we discovered a huge skeleton lying on its face, with the head toward the west. Mr. J. M. Baughman came upon this one accidentally, and, as it fell to pieces, he thinks no one could tell how long it was, but those who saw it unanimously declared it to be the largest of any yet discovered.

We have found eleven human skeletons in all, seven of which have been subjected to fire; and, what is remarkable; we have not found a tooth in all the excavations.

The above report contains nothing but facts briefly told, and knowing that the public has been humbugged and imposed upon by archaeologists, we wish to fortify our own statements by giving the following testimonial:

We, the undersigned citizens of Brush Creek township, having been present and taken part in the above excavations, do certify that the statements herewith set forth are true and correct, and in no particular has the writer deviated from the facts in the case.

[Signed.]   Thomas D. Showers,
                   John Worst all,
                   Marshall Cooper,
                   J. M. Baughman,
                   S. S. Baughman,

St. Joseph Herald (St. Joseph, Michigan,) September 9, 1880
Ancient American Giants

The Rev. Stephen Bowers notes, in the Kansas City "Review of Science," the opening of an interesting mound in Brush Creek Township, Ohio. The mound was opened by the Historical Society of the township, under the immediate supervision of Dr. J. F. Everhart, of Zanesville. It measured sixty-four by thirty-five feet at the summit, gradually sloping in every direction, and was eight feet in height. There was found in it a sort of clay coffin including the skeleton of a woman measuring eight feet in length. Within this coffin was found also the skeleton of a child about three and a half feet in length, and an image that crumbled when exposed to the atmosphere. In another grave was found the skeleton of a man and woman, the former measuring nine and the latter eight feet in length. In a third grave occurred two skeletons, male and female, measuring respectively nine feet four inches and eight feet. Seven other skeletons were found in the mound, the smallest of which measured eight feet, while others reached the enormous length of ten feet. They were buried singly or each in separate graves. Resting against one of the coffins was an engraved stone tablet (now in Cincinnati), from the characters on which Dr. Everhart and Mr. Bowers are led to conclude that this giant race were sun-worshipers.

History of Clermont County, Ohio, 1880

In this connection it might be well to remark that there are several prehistoric cemeteries in this county. The most prominent ones are located near the Miami Township cemetery, on the Cincinnati Turnpike, on the farm of Oliver Perin, in Union Township, and on the farm of Moses Eltsun, Esq., in the same township. In all of these implements are found in connection with the skeletons. The one on the farm of Moses Elstun, Esq., is situated on what is called "Sand Ridge," which runs at right angles with the east fork. In this cemetery the skeletons are found about two feet below the surface, in a cist. On the farm of Daniel Turner, at the mouth of Dry Run, is one, which, as to the number of skeletons found in it, is the largest of any, found so far in the county. It is situated on the brow of the hill, overlooking the east fork valley, at an elevation of two hundred feet above it. This area is about forty feet square, enclosed by flat stones set on edge. This cemetery seems to be a large ditch, in which the bodies have been buried, one on top of the other, to the depth of five feet, and over which is a stratum of earth two feet in thickness.

The forehead is low, and the maxillary bones are unusually large, and so are the femur, which would, in proportion, make a man eight feet in height. One of the largest skeletons noted by the writer was found in the Sand Ridge Cemetery. The skull was in a good state of preservation, together with the teeth; all the rest of the bones were decomposed, with the exception of the femurs, which was unusually large. The cranium, etc., are now in the possession of the Ohio Medical College, at Cincinnati,

History of Clark County, Ohio, 1881

Half a mile north of this fort is a huge mound, the base of which covers about one acre. From this mound many bones have been exhumed, of a race of beings differing greatly from the present, and having no similarity to the red man. A mile west of the fort above mentioned, on the farm of William Allen, is an ancient burying ground of an extinct race. The bones taken from this place are much larger than those of Americans, and, in many respects, give evidence of having belonged to prehistoric people.

History of Champaign County, Ohio, 1881,

The Baldwin Mound-this mound is located on the top of a hill lying between the North and East Forks of Buck Creek at their junction, about eight miles southeast of Urbana... Further excavation disclosed a second skeleton, with head toward the west. The bones of this skeleton were very large and strong, and those of the lower limbs in a remarkable state of preservation; near the hand, and lying across the body, were the flint heads of three spears or arrows. Their position seemed to show that they had been held in the hand by wooden shafts now moldered away. The upper part of the body had been crushed and distorted to a great extent by the pressure above. It had apparently been placed on the left side, and the arrows grasped in the right hand. Removing the earth carefully from this, a third skeleton was seen, its head pointing to the east. This was lying upon its back, and measured form its toes to the top of the head nearly six feet.

Hamilton County, Ohio. "The Past and Present of Mill Creek Valley," 1882

When the Hamilton pike was constructed it cut through a very large mound to the south of the village on the grounds of Mrs. Cummings, near the locality of Station Springs. The West half is still Discernible. Large skeletons were found in it when cut through."

History of Portage County, Ohio, 1885

"Several years ago a burial mound was opened in Logan County, from which three skeletons were taken. The frame of one was in an excellent state of preservation, and measured nearly seven feet from the top of the skull to the lower part of the heel..."

"...In 1850 a mound lying on the north bank of Big Darby about one mile northwest of Plain City, in Union County, was opened and several massive skeletons taken there from. The lower jaw-bones, like those found at Conneaut, easily fitted over the jaw of a very large man, outside the flesh. These bones-and they are usually large wherever found-indicate that the Mound Builders were a gigantic race of beings, fully according in size with the colossal remains that they left behind them."

Other skeletons found in this lateral were those of little children, as indicated by the size of the bones and the thickness of the skulls; some of the bones were from persons of larger stature; at this point the interment ranged in depth from one foot and a quarter to four feet.

Mr .Altick and Mr. Cusick began excavations at the summit of the mound, where a perpendicular shaft was sunk eight feet square, and one foot from the surface in the black leaf mold they found a complete skeleton lying face downward, in horizontal position; however, the bones crumbled when they were lifted from the earth. They excavated another six inches, carefully removing the sand and gravel in order not to injure any deposit they might find; the material removed was screened so that small objects would not escape their notice, and here they came across another skeleton lying face upward, with only six inches separating them. It lay in a sandy mixture, and was in better state of preservation than the first skeleton, and while due precaution was taken in removing it, the bones crumbled as they handled them.

The shaft was then sunk eighteen inches deeper when three more skeletons were unearthed; they were in excellent condition, the bones being firm and hard, due to the greater depth at which they found them. One of the skeleton of a female, one was a child and the other was a male of gigantic stature. As a matter of comparison, Altick held up the femur of the male skeleton by Cusicks leg, and it extended eight inches below his knee; he is six feet in height. The ribs of this skeleton had petrified to a grayish slate color, but none would withstand the contact with the air.

The Stevens Point Journal, **May 1, 1886**
Prehistoric Skeletons
An Ohio County Full of Valuable Relics of the Mound Builders

It is very evident that at an early day in the history of this country this section of Ohio was an important camping ground for the American Indian. There are in this county several burying grounds, and two of them are located five miles west of this city, near Jasper, one on the farm of Mr. William Bush and one on the Mr. Mathew Mark's farm. These burying places are both in gravel banks and were discovered when the banks wer eopened fore the purpose of hauling out gravel. In a conversation with a gentleman who has seen a number of skeletons unearthed at the Mark bank was first opened. Some of these skeletons have been measured and the largest have been found to be nine feet long and over. At one time ten skeletons were exhumed. They had buried in a circle, standing in an erect position, and were in a comparatively well-preserved condition. One remarkable fact about all the skeletons unearthed at these places is the perfect state of preservation in which the teeth are found to be. Not a decayed tooth has been discovered, and this would seem to indicate that these people naturally had excelient teeth or some extraordinary manner of preserving them.

The last skeleton taken up was of ordinary length, but the bones were wonderfully large, and a gentleman who examined them says that the backbone was as large as the backbone of a cow. Some think that this may have been a young fellow who had not yet lengthened out. The opinion is held by not a few that these are not the remains of the common Indian, but that they are the last vestiges of a prehistoric race and extinct race,m as there are several mounds which were undoubtedly constructed by the mound builders located in this and adjoining counties. The writer, in digging in one of these mounds, discovered a portion of a crumbling earthen crock, flint spearheads, ashes, ect. In the Bush gravel bank trinkets and weapons of warfare of various kinds were discovered lying about the skeletons.

Cambridge City, Indiana Tribune, **September 26, 1889**
RELICS IN OHIO MOUNDS
A Gigantic Man Buried Alongside a Panther

Soon after the 1ˢᵗ of March I left for southern Ohio to collect relics to be placed on loan exhibition in the Smithsonian Institute at Washington, say a writer in the Cincinnati *Commercial Gazzette.* During the last two moths eleven large mounds have been opened and their contents taken to museums and placed on exhibition.

These mounds vary in height from eight to thirty feet, are generally conical in shape, and contain all the way from 300 to 10,000 square yards of dirt. They were built by the aborigines in this country hundreds of years ago to serve as burial places for the distinguished dead. They are generally placed near some stream in a vally and not unfrequently on high points of land which commanded a good view of the country, but the larger ones are in the valleys. These mounds are usually composed of clay, sometimes of sand, and often have layers off charcoal or burnt clay in them. These layers are often as brightly colored as if they had been painted.

The first mound opened was on rather high ground, the third river terrace. This mound was 13 feet high, 60 feet wide and 110 feet lone. It took six men eight days to dig through. When about twenty feet from the eastern edge we came upon a thick layer of burned bone. The layer was six inches thick and the width of the trench. The ashes of which it was composed were either resultant from animal or human bones, we could not tell which, but at any rate, whether animal or human, it must have taken thirty or forty skeletons (if cremated) to have made that many ashes.

About five feet above this layer, or nine feet from the summit of the mound, was a skeleton of a very large individuale which had buried by the side of it the bones of a panther. Whether the person had killed the panther and it was buried with him as an honor or whether the panther had killed the individuale I can not say. This much however, can be said, that in forty-three opened no find of this nature has been made. It is therefore quit interesting and important. The skull of the panther was very large, teeth very long and sharp. It would take a moundbuilder of a great deal of nerve to attack a beast of this size if he had nothing but a stone hatchet and bow and arrows to defend himself with. So if he did kill the panther he certainly entitled to a great deal of credit.

Upon opening the large mound last fall, skeletons were discovered. Little attention was given to the bones which soon crumbled. When another mound was opened a few days ago, however, the excavators were struck by the peculiar cranial characteristics. The heads presumably those of men are very much larger than those of present day men. From directly over the eye socket the head slopes straight back, and the nasla bones protrude far above the cheekbones.

The jawbones are so long and pointed that one is struck with their resemblance to those of a monkey. The teeth in the front of the jaw resemble the molars in the mouths of persons today.

Smithsonian Institutes Bureau of Ethnology 12th Annual Report, **1890-1891**
Franklin County

On the level space enclosed by the ditch, 100 feet from the top of the east line of embankment, is a mound (a) 4 feet high and 35 feet in diameter. On the top were numerous flat stones, which it is said had formed graves enclosing skeletons of very large size

Maysville, Ohio Tribune, **June 4, 1890**
PREHISTORIC RELICS
Valuable Discovery in a Mound in Fayette County.

While a party of men, under the direction of T.M. Worthington, were engaged in excavating a large mound on the farm of E.T. Worthington, in the southern part of Fayette county, have made an important find. Two large skeletons measuring seven feet in length had already been exhumed, and May 27th the workman discovered a large bed of black ashes and charcoal. In this bed was found the skeleton of what evidently been a chief. The skeleton lay on its back with the head to the north. Around the neck was a strand of ivory beads, while around the left wrist were two copper bracelets. These bracelets are in excellent state of preservation. They are ten inches in circumference, one third of an inch wide and one-fourth of an ich thick. They are fit together at the ends, but not welded. They appear to be made of a good quality of copper, and were evidently very valuable in their day. Mr Worthington prizes these relics very highly.

The Ohio Democrat, (New Philadelphia, Ohio) November 19, 1891
A Great Find of Prehistoric Remains is Made at Chillicothe, O.
From an Indian Mound
Two Men Working in the Interest of the Worlds Fair Exhume
A Skeleton In Armour
Huge Pearls, Bears Teeth and Other Valuable Relics Also Found-A Second Skeleton Discovered

   Chillicothe, O., Nov. 17-Warren K. Morehead and Dr. Cressen, who have been prosecuting excavations here for the past three months in the interest of the world's fair, have just made one of the richest finds of the century in the way of prehistoric remains. These gentlemen have confided their excavations to the Hopewell farm, seven miles from here, upon which are located some twenty odd Indian mounds. On Saturday they were at work on a mound 500 feet in length, 200 feet wide and 26 feet in height. At the depth of fourteen feet, near the center of the mound, they exhumed the massive skeleton of a man, which was encased in a veritable copper armor. The head was covered by an oval-shaped copper cap. The jaws had copper moldings and the arms were dressed in copper. Copper plates covered the chest and stomach. On each side of the head, on protruding sticks, were wooden antlers ornamented with copper.

   The mouth was stuffed with genuine pearls of immense size, but much decayed by the ravages of time. Around the neck was a necklace of bears' teeth, set with pearls. At the side of the male skeleton was also found a female skeleton, the two being supposed to be man and wife. It is estimated that the bodies were buried where they were found fully six hundred years ago. Messrs. Morehead and Cresson consider this find one of the most important ones they have yet made and believe they have found the king of the mound builders.

The Weekly News (Mansfield Ohio) December 3, 1891

   Near Hamden Junction, O., workmen at the Roscoe mound unearthed a gigantic skeleton encased in a copper covering.

History of the City of Columbus, 1892

   One of the most pretentious mounds in the County was that which formerly occupied the crowning point of the highland on the eastern side of the Scioto River, at the spot where now rises St. Paul's Lutheran Church and adjoining buildings, on the southeast corner of High and Mound Streets in Columbus. Not a trace of this work is left, save the terraces of the church, although if it were yet standing as it stood a century ago, it would be remarked as one of the most imposing monuments of the original Scioto race. When the first settlers came it was regarded as a wonder, and yet it was not spared. As was usual with such works, it was in the form of a truncated cone, its base diameter not less than a hundred feet...All who remember the opening of this mound have a mile of information to add to the story of its demolition.... the father of the late William Platt found a skull so large that it would go over his head.

History of the City of Columbus, **1892**

In William's History of Franklin and Pickaway Counties is a description of some remains of earthworks which occur near Dublin in this county. As these works exist in a much damaged state, the observations made a good many years ago are valuable and here quoted.

"On the banks of the Scioto River, in Perry Township, the Williams History says, "are remains of ancient works which have the appearance of fortification and were undoubtedly used as such by some earlier inhabitants of this county, of whom all trace, further than these forts and mounds, is lost. On the farm of Joseph Ferris, a mile north of Dublin Bridge, are to be seen in a good state of preservation, the outlines and embankments of three forts. One of these is within a few feet of his house and is perhaps eighty feet in diameter inside, with an entrance at the east side. The ditch and embankment are well defined. A short distance northeast of this spot, and within arrow shot of it, is a large fort in square form, and enclosing nearly, or quite, half an acre of ground. Although the tramping of cattle for many years has worn down the embankments, they are several feet high and the ditch, which is inside the works, is now some six feet deep. When the country was first settled this ditch was filled with water, and was a bed of mire, a pole thrust into the ground to a depth of ten feet finding no solid ground beneath. This would tend to show that originally this was a strong place and that the ditch was quite deep. Time has filled it with dead leaves, and refuse matter has assisted in obliterating this work. It is situated on a hill that commands a wide view of the country for a considerable distance in either direction. At a little lower point, and nearer the river, is a small mound. There was also a small mound in the centre of the larger fort, which was opened many years since, and was found to contain the bones of a large man. These crumbled in pieces soon after being exposed to the air. It is possible that by uncovering the ditch of this fort some relics of the extinct race that built these works might be obtained. Search of this kind has generally been turned to the mound, instead of the inner ditches of the fort, where probably was the habitation of the builders. A short distance from this larger fort is a smaller one than that first described. There have been several old works of this kind along the banks of the river between these works and Columbus, but they are mostly obliterated by the cultivation of the land on which they stood."

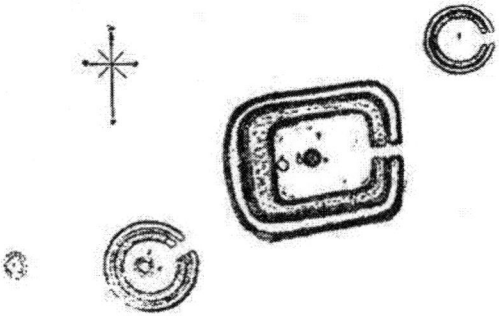

Traces of these earthworks are still visible in a plowed field and the front lawn of a residence.

Where Proctorville now stands was one-day part of a well-paved city, but I think the greater part of it is now in the Ohio River. [Sic] Only a few mounds, there; one of which was near the C. Wilgus mansion and contained a skeleton of a very large person, *with all double teeth,* and sound, in a jaw bone that would go over the jaw with the flesh on, of a large man; the common burying ground was well filled with skeletons at a depth of about 6 feet. Part of the pavement was of boulder stone and part of well-preserved brick.

Burlington Iowa Hawk Eye, **June 7, 1893**
RELICS FROM A MOUND
REMARKABLE DISCOVERIES IN AN OHIO TOWN
A Human Skull Almost Twice the Size of Thise of the Present Day-Looks as if the Lost Race Were Croatians and Phoenicians.

REMOVING THE MOUND.

The work of removing the old Indian mound in Walnut Grove, Martins Ferry, near Bellaire, Ohio, goes on slowly owing to the care exercised that none of the interesting relics to be found in it be lost. Probably the most interesting article taken from the mound is a huge skull, which would seem to indicate that in the days of the mound builders there were giants abroad. The skull is at least twice as large as the normal average of today. This skull is in a good state of preservation. In the collection of relics exhibited in the same place are the cuplike stones, which have been supposed to be cups made by ancient inhabitants. These and other relics are shown in the picture herewith given. Whether these were made by man or are the stones in the process of formation, the outer shell only having become petrified, it would take an expert to decide. Certain it is, however, that similar cups are frequently found in that vicinity in the ground or on its surface, having the appearance of being a natural product and not an artificial one. Some of the spearheads and hatchets found at Martin's Ferry are as perfect as any found anywhere. Bones are found in every part of the

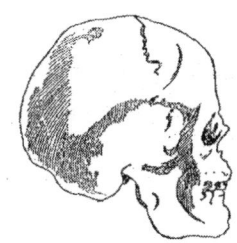

mound, bearing out the theory that such mounds were simply cemeteries instead of graves and monuments at once of the great chiefs. Several corpses have certainly been buried in this mound. The finding of curious precious stones worthy of notice. What seems to be opals, emeralds and crystals nearly like diamonds, have been taken from the mound. Learned savants have held that the proof is complete, from the articles found at Moundsville, W. Va., that moundbuilders were Croations and Phoenicians or both. The presence of precious stones in the vicinity of these corpses, with flint darts, spearheads, stone axes and stones with holes neatly drilled in them, may furnish another clue to the identity of the lost race. The mound at Moundsville is the highest mound found in that part of the country, it being over 100 feet high.

New York Times, **March 5, 1894**
Giants of Other Days
Recent Discoveries Near Serpent Mound, Ohio.
From the *Indianapolis Journal*

Farmer Warren Cowen of Hillsborough, Ohio, while fox hunting recently discovered several ancient graves. They were situated upon a high point of land in Highland County, Ohio, about a mile from the famous Serpent Mound, where Prof. Putnam of Harvard made interesting discoveries. As soon as the weather permitted, Cowen excavated several of these graves. The graves were made of large limestone slabs, two and a half to three feet in length and a foot wide. These were set on edge about a foot apart. Similar slabs covered the graves. A single one somewhat larger was at the head and another at the foot. The top of the grave was two feet below the present surface.

Upon opening one of the graves a skeleton upward of six feet in length was brought to light. There were a number of stone hatchets, beads, and ornaments of peculiar workmanship near the right arm. Several large flint spear and arrow heads among the ribs gave evidence that the warrior had died in battle.

In another grave was the skeleton of a man equally large. The right leg had been broken during life, and the bones had grown together. The protuberance at the point of union was as large as an egg, and the limb was bent like a bow. At the feet lay a skull of some enemy or slave. Several pipes and pendants were near the shoulders.

In other graves Cowen made equally interesting finds. It seems that the region was populated by a fairly intelligent people, and that the Serpent Mound was an object of worship. Near the graves is a large field in which broken implements, fragments of pottery, and burned stones give evidence of a prehistoric village site.

Newark Daily Advocate, **(Newark, Ohio) May 5, 1897**
Licking
Possesses Largest Stone Mound in Ohio
It Overlooks Buckeye Lake-An Interesting Article From Warren K. Moorehead

Prof. Warren K. Moorehead of the Ohio State University, who has made several visits to this county on archaeological investigations, contributes the following to the Popular Science-News on Ohio's Largest stone mound which is located in this county:

"Upon a hill in Licking County, overlooking the reservoir is a famous stone mound. Today it is about ten feet high on the average, and covers a diameter something over 200 feet. In 1811 when first noticed, it was 50 or 55 feet high and about 180 feet broad at the base. No similar stone structure and but few earth mounds could compare with it in size and symmetry.

When the reservoir was constructed some 50,000-wagon loads of stone were hauled from it to "rip-rap" the walls. No one of the ignorant vandals engaged in this demolition had any conception of the importance of the structure. After one or two years of active destruction they reached the bottom of the stones and found eight or ten small earth mounds. It seems that the great structure had been erected simply to cover these. With thoughts of gold and other riches, several of the little tumuli were excavated. In one was found a large skeleton lay in a hollow log. Many beads were strung about the neck, several copper hatchets lay by the hands and copper bracelets encircled the wrists. Important as was the find there is but one slight trace of it remaining at present time.

Years afterward an enthusiastic crank at Newark pretended to have found a stone covered with Hebrew characters and also exhibiting "a picture of Moses" in the central portion of the mound. But all archaeologists immediately set it down as humbug.

In April 1896, the Ohio Archaeological and Historical society explored what remained of the mound very thoroughly. Not much was found, for only too thoroughly has the early explorers done their work of destruction. It was ascertained that the mounds had been erected upon a hard burnt floor and that each one originally contained a skeleton.

The stones weigh from five to twenty pounds and were collected by the aborigines in the neighborhood. Nearby passed the prehistoric trail, leading from Newark's famous fortifications to the stone fort in Perry County. A branch of this led to Flint Ridge, where the ancients procured flint for the manufacture of arrow, spear and lance heads. All through the region are mounds and enclosures."

The Adair County News, (Kentucky) January 1897

An old Indian mound has been opened on the farm of Harrison Robinson, four miles East of Jackson Ohio, and two skeletons of extraordinary size and a great quantity of trinkets have been removed. Some years ago a party of relic hunters, supposed to have been sent out in the interest of the Archaeological society visited the Robinson farm, and after a few days search removed a great collection of stone hatchets, beads and bracelets, which were packed and shipped to an Eastern institute, and until this recent accidental discovery it was supposed that everything had been removed by the relic hunters. It is thought by many that more relics are to be found and preparations are being made for a through investigation.

The Washington Post, December 4, 1898
Dug Up a Skeleton Eight Feet Long

From the New York Journal A remarkable prehistoric skeleton was unearthed the other day by Mr. R. A. Tomlinson on the bank of Owl Creek, a little stream near Londonderry, Ohio. Mr. Tomlinson was engaged digging into a gravel bed, and had penetrated about four feet below the surface when he discovered the bones. The skeleton, which was excellently preserved, was lying at full length on the left side, with its left hand under it. When lifted up the hand was found to hold a dozen darts of the finest workmanship. But it was the size of the skeleton, which amazed those who saw it. When measured it was found to be only about an inch short of eight feet in length, and there can be no doubt that in life the man was fully eight feet in height and probably an inch more than that. The bones were massive, showing that the man was a giant in strength as well as stature. The skull was a third larger than the human skull, and the lower jaw was abnormal in size and thickness. Hundreds of people have viewed the skeleton, and it will doubtless be preserved as a curiosity.

The Washington Post, December 25, 1898
Monster Skeleton Discovered in the Miami Valley
Believed To Be Mound-Builder
Ponderous Jaws, Strongly-marked Orbital Plates, and Queer-Facial Angles Form a Study for the Scientist-Mysterious Stone Implements Found with the Bones Proves that They Belonged to a Man and Not to an Enormous Monkey.

Special Correspondence of The Post. Miamisburg, Ohio. Dec. 25-A discovery of the greatest scientific interest was made when the pick-axes of Edward W. Gebhart and Edward Kauffman, of this place, disturbed the long repose of a skeleton that had been buried for no one can tell how many thousands of years. The body that once proudly strode the earth possessed of the living strength of a giant, with the bones just found for its framework, was surely not of a race of which history has given us any record. The skeleton is that of a human being, who in life must have been of immense size. The face is almost gorilla-like in its angularity; the jawbones are tremendous and the teeth have been pronounced by a local dentist, Dr. Harlan, to be as strong and perfect as any he has ever seen. A number of local practitioners, among them Dr. A. H. Blossom, Dr. Weaver, Dr. Bookwalter, and Dr.

Shuler, have examined the bones and they unanimously agree that the "find" is no less important a one than the skeleton of a prehistoric man, who undoubtedly was on of the Mound-builders, the relics of whose sojourn in the Miami Valley have been so eagerly sought after by archaeologist all over the country. The splendid condition of the teeth has caused the doctors to conclude that the prehistoric man was a root-eater, for no masticator of meats could have retained to an estimated age of the man the perfect molars with which the jaws are fitted. The skull itself is like no skull that any living man carries around on his shoulders, in that it is of extraordinary size and contour. With the massive jaw-bones strongly-marked orbital plates and odd facial angles, the face would present a striking appearance could it be restored by an artist in wax.

Found in a Gravel Pit

The discoveries of the skeleton were digging gravel in a pit about a half-mile from the town when the pick of Edward Gebhart struck the skull. A careful investigation by the two men, from previous discoveries of Mound-builder relics, knew enough of the importance of the find, to at once set about unearthing the skeleton, revealed the rest of the bones, forming an almost complete skeleton. The body had evidently been buried in a sitting posture, for the knees were drawn up and the head had fallen between them. Beside the skeleton was a flat stone, about three inches long and two inches wide, with a hole drilled through one end apparently for the fitting of a handle. The use of this stone is not apparent to the local scientist. Some think it was a charm buried with the body, while others believe it to be the weapon of the dead man, although why such a giant as this prehistoric individual evidently was could carry so unimposing a club in question. It is important that this stone, so clearly fashioned by hand was found in close proximity to the skeleton, for it settles at once all doubts as to whether the great skull originally adorned the body of a man or of a mammoth monkey. At present the bones are on exhibition here. They are serving as a star attraction, for the people are very proud of the fact that once in this neighborhood there lived a race of prehistoric giants who erected the largest mound in the country, and then disappeared, leaving behind them fragments of stone, pieces of pottery, and woven cloth made of bark to show that they once were the owners of the valley. How long this was ago, where the Mound-builders originally came from, what they looked like, whether or not they will be proven some day to have been the long-looked for link between man and the lower animals, are questions that the residents of this locality would like above all things to have some scientist decide. The discovery of the skeleton with its peculiar-shaped skull, will prove of great service in the investigation so far as it concerns the appearance of the Mound-builders.

Athens Messenger, **August 6, 1903**
MOUND BUILDERS' MONUMENTS
The Silent Witnesses of a Gigantic Prehistoric Race, Long Agone.
Wolf's Plains Was the Central Place for the Tribes Who Built the Numerous Mounds- What Became of the Mound Builders?

Wolf's Plains has long been noted for its many mounds, which seem indicative of its having been a central place for the tribes of a people who built them. While a great many of the mounds have been almost obliterated by repeated plowing, yet quite a number remain to give us an idea of their original form and size. It is not known what the exact use of the mounds was, yet the fact that in nearly every one that is opened are found skeletons and various trinkets and copper and stone vessels, seems to throw some light on the subject. Some claim that they were used as places of burial, but they were evidnetly used for a variety of purposes.

The Plain's school house rests on one of these mounds, which has been plowed and worked down to make it suitable for that purpose. During the process of lowering it, numerous trinkets and vessels were unearthed, among which was a particular article, consisting of two rings of copper beads between two pieces of buckskin.

All was in a perfect state of preservation when found, but the air soon changed the buckskin to dust. Charcoal is always present and is as bright as if it had made but yesterday.

A large number of other things were found but some prominent men from a nearby city hastened to the spot and carried all the relics away to place them in museums. These things are always found near the base or bottom of a mound.

A prominent farmer, who not long since passed away and whose name could easily be given, if necessary kept plowing around a mound until he had nearly plowed it down. Near the bottom he unearthed the skeleton of a man who had been buried in an erect position. Charcoal was found near the skeleton, showing that for some reason a fire had been made near the deceased. These remains are always of men of great stature.

Such things as knives, tomahawks, beads, all kinds of copper and stone vessels, skeletons and other objects too numerous to mention have been found in mounds near and on the Plains. Another discovery that might be of interest to those who have not heard of it is that an ancient cemetery or burying ground some distance above where the new shaft is located. When the Hocking Valley railroad was being built it was found necessary to remove a portion of one of the nearby hills, or to make a "cut," and for that purpose a huge steam shovel was employed. While at work they found the ancient cemetery. From the skeletons it is evident that the men were very large. It is said that the bones reaching from the knee to the ground were almost a foot longer than that of the average man of today. These facts, which will be well remembered by many of the older people, indicate that this country was originally inhabited by a race of people of whom we have not the slightest knowledge, except what is contained in their mounds and cemeteries. That this race has lived its day and perished, is rather a sad fact.

Chicago Herald, **April 15, 1904**
FIND SKELELTON OF GIANT
Interesting Relic of Ancient Mound Builders Uncovered in Ohio

A giant skeleton of a man has been unearthed on the Wolverton farm, a short distance from Tippecanoe City, Ohio. It measured eight feet from the top of the head to the ankles, the feet being missing. The skull is large enough to act as a helmet over the average man's head. This skeleton was one of seven found buried in a circle, the feet of all being toward the center. Rude implements were near. The skeletons are though to be those of the mound builders.

Centennial Atlas of Athens County, Ohio, **1905**

A small mound located on the very top of the hill bordering the eastern part of the Wolf's Plains and a little northwest of the house now occupied by Mr. J. Taylor, superintendent of the Johnson Coal Mining company's mine here, was opened by two or three of the citizens in the spring of 1905. They were in search of copper and stone articles and more especially inscriptions. At the bottom of the mound and lying on a huge flat stone was a skeleton apparently of a woman. The lower limbs were crossed. The bones had been much decayed by the action of water. The explorers stated theat the bones were remarkably large. The jaw bone would fit over that of the average man of today and leave plenty of place besides. The forearm bones 5 inches larger than those of the average man. Charcoal was found in three different layers.

***20th Century History of Delaware County, Ohio*, 1908**

"The mound is situated on the banks of a rocky stream...on the farm of Solomon Hill, Concord Township, Delaware County, Ohio. Two human skeletons were found in the mound, one about seven feet long, the other a child. The shell (conch from Florida) was found at the left cheek of the large skeleton."

***The Evening Telegram*, Norwalk Ohio, February 28, 1910**
PRE-HISTORIC MAN IS FOUND NEAR NORWALK
Fitchville Is Exited Over The Finding Of A Giant Skeleton Wich Measured Over Eight Feet In Length.

Norwalk, O., Feb 25-What is believed to be the skeleton of a prehistoric giant, was unearthed in the cemetery in Fitchville. The skeleton of another human being that is thought to have been buried at the same time also was unearthed.

The discovery of the bones was made by Cornelius Springfield of Fitchville cemetery, while he was engaged in digging a grave for the remains of John Laughlin, an aged pioneer resident of Fitchville.

The Laughlin family burial lot in the Fitchville cemetery has been owned by the Laughlin family for the past seventy-five years, and a large number of members of the family have been buried there. The cemetery is located on a high piece of ground and the Laughlin lot is located on the highest point in the cemetery, a sort of a mound.

Springfield had dug down to a depth of about six feet when his spade struck against some hard substance. Scraping away the dirt Springfield was surprised to find that he had unearthed a portion of a human skeleton. Continuing his work, he soon had the entire skeleton unearthed. The skeleton, which was intact and in excellent state of preservation lay on its back, with both arms extended out at right angles with the body, showing that it had never been enclosed in a coffin. In order to unearth the bones of the arms and hands and legs it was necessary to dig into the sides and one end of the grave for quite a distance.

When the skeleton was finally entirely uncovered it was seen to be that of a giant man, which when measured, proved to be about eight feet in length. Springfield is a man of average height, but the hip bones of the skeleton were found to be fully six inches longer than his are. The other bones of the skeleton were proportionately long. The teeth were in sound condition. The cheek bones were wide, and the skull tapered up to a peak, the forehead slanting upward.

After the skeleton of the giant had been removed from the grave Springfield continued has work of digging, and had reached the depth of about eight inches more when he unearthed another skeleton that had been lying directly underneath that of the first. The second skeleton was not as long as the first, being between five and six feet in height.

There is no doubt that the two bodies were buried years before the land there was begun to be used as a modern cemetery, and from the immense size of the skeletons that was first unearthed and the shape of the skull, it is thought that the skeleton must be those of prehistoric human beings.

***History of Seneca County*, Ohio, 1911**

"There were several mounds on the Culver place, form which have from time to time been plowed up bones and ancient crockery. In 1850 one of these mounds was opened and in it was a large skeleton, with a full shaped skull. And among other things a stone pitcher, which seemed to have been made of sand and clay, and smaller vessels filled with clam shells were found therein. These seemed as strange to the Indians as they did to the whites."

306

History of Madison County, Ohio, 1915

"Another burial mound is located on the north bank of Big Darby, about one mile northwest of Plain City, and as this territory originally belonged to Madison county it will be proper to mention it briefly here. It was originally about the same size of the larger mound in Jefferson Township, but is much smaller today. In 1848, a society called 'the Rectifiers" was organized in Plain City, the object at which was the improvements of morals, the advancement of education, benevolence and institutions of charity, and the development of archaeological history. In 1850, the society opened this mound from which they took the remains of some skeletons. The thigh bones were very massive, while the jaw bones were sufficiently large to slip over the face of the ordinary man, demonstrating that the beings to whom they belonged must have been of extraordinary size and proportions."

A Standard History of Ross County, Ohio, Vol., I, 1917

"It is worthy of note that one of the most perfect skulls ever found in the mounds, and one which incontestably belonged to the race who built the earthworks, was discovered in a singularly constructed mound upon the summit of High Hill, which overlooks the Valley of Scioto, and is situated four miles below Chillicothe upon the west side of the river. This skull, described by Professor Morton in his elaborate work, "Crania America," was of unusually large size, and exhibited a facial angle of eighty-one degrees. The internal capacity of the skull was ninety cubic inches-seven inches greater than the mean capacity of the Mongolian skull, three inches greater than the Caucasian, and eight inches greater than that of the American Indian."

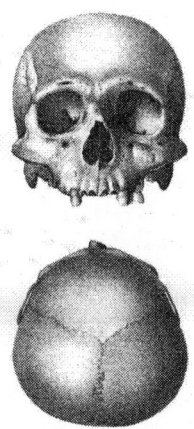

Large Dinaric-like skull with flattened occipant and protruding brow ridge. Skull was uncovered in Ross County, Ohio and presented in *Crania America*.

History of Chillicothe County, 1946

"Near Plainfield, about 1840, Mr. J. D. Workman opened a small earthen mound on his place. He found nothing except several stone relics. Another about two miles below was excavated some ten years later by Wesley Patrick. It contained a few bones belonging to the human skeleton, including the skull, jaw bone and thigh. These were of an unusually large size and indicated the skeleton to be fully seven feet in length."

History of Knox County, Ohio

"In Liberty Township, on the farm once owned by Joseph Beeney, was once a mound of considerable dimensions. It was leveled for a building spot. In it was found a skull of immense size, so large that the largest man in the county could put his head into the cavity with great ease, still leaving unoccupied space."

History of Chillicothe County, **1946**

"Fifty-four years ago a stone mound was opened on a hilltop near the Colonel's house [Colonel Pren Methams]…

A rock pile eighteen feet square and five feet deep, composed of sandstone layers, was removed, revealing a sepulcher floored with a large, flat sandstone, and walled with sandstone slabs. On the floor lay part of a skull, a thigh bone, teeth and a few other fragments of a skeleton. The thigh bone indicated the dead to have been of unusual height, more than seven feet."

Athens Messenger, **August 5, 1970**
Beware of the Super-Size Indian

Let's hope the men who are building highways around Athens aren't superstitious because they just might upsent the resting place of a long dead super-size Indian. And as anyone knows who believes in spirits, there's nothing as nasty as a sleeping sachem, roused from his grave by a bunch of palefaces.

The fellow rests under Harmony Road where he curves eastward near the United Dairy. That's in the

area where the Appalachian Highway will join Route 33 and the Stimson Avenue extension will cross the Hocking River over the structure that will replace the old East Mill Bridge.

Now we really don't know if the guys with bulldozers and drag lines and earthmovers are going to dig up that part of Harmony Road, but if they do, then let them be warned. Resting under the curve is believed to be an Indian whose companions were eight feet long, and he might be larger.

Back in 1905 a county highway crew was working on Harmony Raod when a large mound stood in their way, so they began excavating the obstacle and soon uncovered a burial place containing 32 skeletons. Thurman Knox was the foreman and he related the story years later to the late C. H. Harris of The Messenger, telling him the skeletons were more than eight feet tall.

Over the entire mound was a foot thick layer of plastic clay in which gravel was mixed. This formed a protective covering over the entire mound, making it impervious to rainfall. Supporting the roof were clay columns of the diameter of a barrel and made in sections about as long as a barrel and set on top of each other.

Knox told Harris that when thew bones of the 32 men were uncovered all but one disintegrated to dust when exposed to air. One skeleton of a man between eight and nine feet tall was recovered almost intact. The bones of the forearm, partially eaten away, were still several inches longer than those of a larger man. The jawbone filled with perfect teeth was so large it fitted over the jaw of a large man like a mask.

The excavation at East Mill went down 15 feet and in the center of the skeleton group was found a hard clay box-like structure about six feet wide and 12 feet long. It was never opened and now lies under the center of the road near the dairy. [...]

Oregon

Hammond Times, (Indiana) June 30, 1939
DISCOVER INDIAN RELICS ON FARM

Salem, O.-(I N S)- Discovery of ancient skeletons and priceless relics in an Indian mound at North Benton, northwest of Salem, by two Alliance, O. mail carriers, has brought hundreds of visitors to the scene and attracted the attention of expert archaeologist.

The two amateur archaeologist, Roy Saltsman and Willis Magrath, made the excavation on the farm of John Malmsberry.

After examining the mound, Richard G. Morgan, state archaeologist, declared that the work of the two Alliance men was the most important archaeological discovery in this section of the state in recent years.

Morgan said that the remains were those of the Hopewell Indians, who supposedly migrated to America across the Bering Straights long before the advent of the white man. He estimated the age of the findings at more than 2,000 years old.

One skeleton uncovered was that of a man, apparently a chief, estimated to have been nearly seven feet tall. Whose skull was 25 inches in circumference.

Other findings included flint arrows, the stones of three sacrificial alters, spear heads, flake knives and beautifully wrought objects of copper.

Tennessee

The Natural and Aboriginal History of Tennessee, John Haywood, 1768

On the farm of Mr. John Miller of White County are a number of small graves, and also many large ones. The bones in which show that the bodies to which they belonged, when alive, must have been seven feet high and upwards.

About the year 1814 Mr. Lawrence found, in Scarborough's Cave, which is on the Calf-killer River, a branch of Cany Fork, about 12 or 15 miles from Sparta, in a little room in the cave, many human bones of monstrous size. He took a jawbone and applied it to his own face, and when his chin touched the concave of the chin bone, the hinder ends of the jawbone did not touch the skin of his face on either side. He took a thighbone, and applied the upper end of it to his own hip joint, and the lower end reached four inches below the knee joint.

Mr. Andrew Bryan saw a grave opened about 4 miles northwardly from Sparta, on the Calf-killer Fork. He took a thighbone, and raising up his knee, he applied the knee-joint of the bone to the extreme length of his own knee, and the upper end of the bone passed out behind him as far as the width of his body. Mr. Lawrence is about 5 feet 10 inches high, and Mr. Bryan about 5 feet 9.

Mr. Sharp Whitley was in a cave near the place where Mr. Bryan says the graves opened. In it were many of these bones. The skulls lie plentifully in it, and all the other bones of the human body; all in proportion, and of monstrous size.

Human bones were taken out of a mound on Tennessee River, below Kingston, which Mr. Brown saw measured by Mr. Simms. The thighbones of those skeletons, when applied to Mr. Simms thigh, were an inch and a half longer than his, form the point of his hip to his knee; supposing the whole frame to have been in the same proportion, the body it belonged to must have been seven feet high or upwards.

Colonel William Sheppard, late of North Carolina, in the year 1807 dug up, on the plantation of Col. Joel Lewis, 20 miles form Nashville, the jawbone of a man, which easily covered the whole chin and jaw of colonel Lewis, a man of large size.

Mr. Cassady dug up a skeleton from under a small mound near the large one at Bledgoe's Lick in Sumner County, which measured little short of seven feet in length.

Human bones have been dug up at the plantation where Judge Overton now lives, in Davidson County; four miles south westwardly form Nashville, in making a cellar. These bones were of extraordinary size. The under jawbone of one skeleton very easily slipped over the jaw of Mr. Childress, a stout man, full fleshed, very robust and considerably over the common size. These bones were dug up within traces of ancient walls, in the form of a square of two or three hundred yards in length, situated near an excellent, never failing spring of pure and well tasted water. The spring was enclosed within the walls. A great number of skeletons were found within the enclosure, a few feet below the surface of the earth. On the outer side were the traces of an old ditch and rampart, thrown up on the inside. Some small mounds were also within the enclosure.

At the plantation of Mr. William Sheppard, in the county of Giles seven and a half miles north of Pulaski, on the east side of the creek, is a cave with several rooms. The first is 15 feet wide, and 27 long, 4 feet deep; the upper part of solid and even rock. Into this cave was a passage, which had been so artfully covered that it escaped detection till lately. A flat stone, three feet wide and four feet long, rested upon the ground, and inclining against the cave, closed part of the mouth. At the end of this, and on the side of the mouth left open, is another stone rolled, which filling this also, closed the whole mouth. When these rocks were removed and the cave opened, on the inside of the cave were found several bones-the jawbone of a child, the arm bone of a man, the skulls and thighbones of men. The whole bottom of the cave was covered with flat stones of a bluish hue, being closely joined together, and of different forms and sizes. They formed the floor of the cave. Upon the floor the bones were laid. The hat of Mr. Egbert Sheppard, seven inched wide and eight inches long, but just covered and slipped over one of the skulls.

About ten miles form Sparta, in White county, a conical mound was lately opened, and in the center of it was found a skeleton eight feet in length. With it was found a stone of the flint kind.

At Fort Chartres was found a human skull of astonishing magnitude. A jawbone was taken form the mound near Natchez, which the gentleman who saw it could with ease put over his face; also a leg bone, which from the ground reached three inches above the knee. Many other instances might be enumerated to establish the position that a race of men of much larger bulk than any in America of this day, formerly resided upon the Cumberland River and its waters, and upon the Tennessee and its waters, and below them upon the Mississippi, as well as upon the rivers of north of Cumberland and in some parts of Virginia.

The Natural and Aboriginal History of Tennessee, by John Haywood, 1823

At the distance of about four miles southwest of Sparta on the waters of the Cany Fork, are the remains of an ancient fortification, containing about five acres, perfectly square, the walls being composed entirely of dirt, as appears from the present state of its ruins. Here is a great burying place. The human skeletons discovered here are remarkable for their gigantic stature. From all that can now be discovered, this must have been a race of men averaging at least 7 feet in height. Such men, it is probable, never grew in the tropical climates. No instance is recollected of giants between the tropics. Some were planted by the Scythians in Palestine, when in a very distant age they penetrated as far as to the confines of Egypt, and built the city of Scythopolis. But such men never came from between the tropics. The skeletons now under consideration were some of the ancient Scythians who, sown to the Christian era, terrified the nations which they invaded, by their enormous bulk.

Logansport Reporter, July 3, 1903
GIANT SKELETONS FOUND AT RIPLEY
Indian Mound Near Tennessee Town Yields Remains of Gigantic Men

Ripley, Tenn., April 23. (AP)-Skeletons of three gigantic men buried by a forgotten race have been unearthed by a fisherman digging in an old Indian mound near here. Tribal finery in which they were interred was recovered intact. One of the skeletons bore ivory beads and a long ivory ornament. The other was decorated with copper beads and designs of bone and mica.

Two were found near the surface. Further down the largest of the skeletons was discovered in a sitting position on a carpet of ashes. Pottery, one piece containing the bones of an infant, was found nearby. It was in a fine state of preservation.

Indian mounds abound in this section but hitherto none had yielded skeletons the size of those found by the fisherman.

He has offered them to the Tennessee Historical Society at Nashville.

12 Annual Report of the Bureau of Ethnology to the Secretary of the Smithsonian Institution
1890-1891
Roane County Tennessee

Underneath the layer of shell the earth was very dark and appeared to be mixed with vegetable mold to the depth of 1 foot. At the bottom of this, resting on the original surface of the ground was a very large skeleton lying horizontally at full length. Although very soft, the bones were sufficiently distinct to allow of careful measurement before attempting to remove them. The length from the base of the skull to the bones of the toes was found to be 7 feet 3 inches. It is probable therefore, that this individual when living was fully seven and a half feet high. At the head lay some small pieces of mica and green substance, probably the oxide of copper, though no ornament or article of copper was discovered.

Utah

History of Muskingum County, Ohio, 1882

"From an interesting account of certain mounds in Utah, communicated by Mr. Amaza Potter in the 'Eureka Sentinel' of Nevada, as copied by the Western Review of Science and Industry. I make the following extracts:

"The mounds are situated on what is known as the Paysm farm, and are six in number, covering about twenty acres of ground. They are from ten to eighteen feet in height, and from five hundred to one thousand feet in circumference."

"The explorations divulged no hidden treasure so far, but have proved to us that there once undoubtedly existed here a more enlightened race of human beings than that of the Indians who inhabited this country, and whose records have been traced back hundreds of years."

"While engaged in excavating one of the larger mounds, we discovered the feet of a large skeleton, and carefully removing the hardened earth, which was embedded, we succeeded in unearthing a large skeleton, without injury. The human frame-work measured six feet six inches in length, and, from appearances, it was undoubtedly that of a male. In the right hand, was a large, iron or steel weapon, which had been buried with the body, but which crumbled to pieces on handling. Near the skeleton, was also found pieces or cedar wood, cut in various fantastic shapes and in a state of perfect preservation; the carving showing that the people of this unknown race were acquainted with he use of edged tools. We also found a large stone pipe, the stem of which was inserted between the teeth of the skeleton. The bowl of the pipe weighs five ounces, and is made of sandstone, and the aperture for tobacco had the appearance of having been drilled out."

"We found another skeleton, near that of the above mentioned, which was not quite as large, and must of been that of a woman. There was a neatly carved tombstone near the head of this skeleton. Close by, the floor was covered with a hard cement, to all appearances, a part of the solid rock, which, after patient labor and exhaustive work, we succeeded in penetrating, and found it was the corner o a box, similarly constructed, in which we found about three pints of wheat kernels, most of which was dissolved when brought in contact with the air. A few of the kernels found in the center of the heap looked bright, and retained their freshness on being exposed. These were carefully preserved, and, last spring, planted, and grew nicely. We raised four and a half pounds of heads from these grains. The wheat is unlike any other raised in this country, and produces a large yield. It is the club variety; the heads are very long, and hold very large grains."

He has offered them to the Tennessee Historical Society at Nashville.

West Virginia

History of Kanawha County 1876

"In this village [Clifton] three cellars have been dug by citizens, and in each case an entire human skeleton was exhumed. A square ground embracing about ten acres in that portion of the village fronting the river, seems to have been set apart for a cemetery… Mr. Marshall Hansford, while digging a post-hole in his yard a few years ago, found, about eighteen inches below the surface, nine pieces of sheet copper, several inches square, and rolled very thin. In digging his cellar he found the skeleton of a large sized man, and a great variety of bones and birds, bears and other wild animals. As a proof that these skeletons, relics and the like, were remains of an ancient race, I need only to inform the reader that not long before these discoveries were made, the earth above them was literally covered with stalwart sycamores, which Mr. Hansford informed me were fully five hundred years of age."

History of Preston County, West Virginia, 1882

That they were large is established by their bones; that there were giants among them is proved by the wonderful large bones found interspersed through the mounds, showing that some of them were about seven feet high, while the majority was from five feet eight inches to six feet in height.

Sandy Creek Mound

From this mound, the writer obtained a strange skull out of the top layer of bones. Digging down, we came upon several skulls in the bottom layer, but could not get them out, as they crumbled to pieces in our hands; finally the top of one was secured, and where the sutures meet on the top of the Caucasian head, they were prevented in this head by a small bone of about one inch in length by one-half inch in width, of a peculiar shape. All the other skulls possessed this same peculiar bone. The top of the skull secured and the others that crumbled, showed the heads of the race to have been long and narrow, with low foreheads and long narrow faces.

Bureau of Ethnology, 5th Annual Report, 1883-4

"Below the center of No. 7 (see plate), sunk into the original earth, was a vault about 8 feet long, 3 feet wide, and 3 feet deep. Lying extended on the back in bottom of this, amid the rotten fragments of a bark coffin, was a decayed human skeleton, fully 7 feet long, with head west. No evidence of fire was to be seen, nor were any stone implements discovered, but lying in a circle just above the hips were fifty circular pieces of white perforated shell, each about 1 inch in diameter and an eighth of an inch thick."

Bureau of Ethnology 12th Annual Report, 1894
Kanawha County

No. 11 is now 35 by 40 feet at the base and d 4 feet high. In the center 3 feet below the surface, was a vault 8 feet long and 3 feet wide. In the bottom of this, among the decayed fragments of bark wrappings, lay a skeleton fully seven feet long, extended at full length on the back, head west. Lying in a circle above the hips were fifty-two perforated shell disks about an inch in diameter and one-eighth of an inch thick.

Bureau of Ethnology 12th Annual Report, 1894
Kanawha County

Mound 19, the one farthest to the east, is 60 feet in diameter and 5 feet high. It was found to contain a rude vault of angular stones; some of them as much as two men could lift. This had been built on the natural surface and was 8 feet long, 4 wide, and 3 high, but contained only the decaying fragments of a large skeleton and a few fragments of pottery.

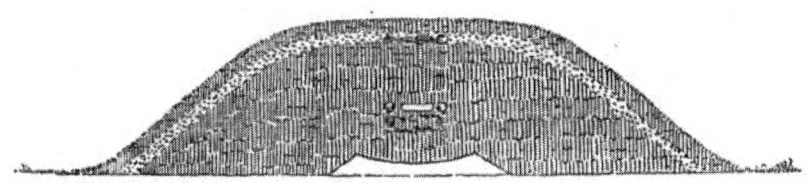

Bureau of Ethnology 12th Annual Report, 1891
Kanawha County

Mound 31 measured 318 feet in circumference, 25 feet high, and 40 feet across its flat top. (See Fig. 302) A 10-foot circular shaft was sunk from the top and trenches run in from the side. The top layer consisted of 2 feet of soil, immediately below, which was 1 foot of mixed clay and ashes. Below this, to the bottom, the mound was composed of earth apparently largely mixed with ashes, placed in small deposits during a long period of time. Three feet below the top were two skeletons, one above the other, extended at full length, facing each other and in close contact. Above but near the heads were a pipe, celt, and some arrow or spear heads. Ten feet below these were two very large skeletons in a sitting position, facing each other, with their extended legs interlocking to the knees. Their hands outstretched and slightly elevated, were placed in a sustaining position to a hemispherical, hollowed, course-grained sandstone, burned until red and brittle. This was about 2 feet across the top, and the cavity or depression was filled with white ashes containing fragments of bones burned almost to coals. Over it was placed a somewhat wider slab of limestone 3 inches thick

Mound 31 is still visible in a Charleston, West Virginia park. *The Nephilim Chronicles, A Travel Guide to the Ancient Ruins in the Ohio Valley,* **2010.**

Bureau of Ethnology, 12th Annual Report, **1890-1891**
Mason County

"In an old cultivated field stream with mussel shells, are one large and several small mounds. In all those which were explored there was a layer of skeletons on the natural surface, and two, or sometimes three, other layers above them to a height of 5 feet. The appearance of the mounds justified the statement of Mr. Couch and others that at least one more layer had been removed during fifty years of steady cultivation. The skeletons were well preserved many of them large, in a prostrate position, with no particular arrangement."

"Five miles above the mouth of the Kanawha, on the south side, on the farm of Charles E. McCulloch, is the largest mound in this section. Unlike most of the large mounds, it is not on the river bottom, but on a sloping terrace nearly a hundred feet higher, and after long cultivations are still 20 feet high and fully 300 feet in circumference... A circular shaft 11 feet in diameter was sunk down through the center to the bedrock a foot below the base of the mound. A rock heap at the top had been made in a depression evidently caused by the caving in of a vault. This rock heap had been disturbed by parties who found a very large skeleton with some stone weapons. Beneath it sandstone slabs as heavy as a man could lift were scattered through the shaft, and at the bottom enough of them standing and lying at all angles to have covered the vault.

Cabell County

"About 1 mile west of Barboursville, on a hill nearly 500 feet above the Guyandolte, overlooking that stream for a long distance and offering a fine position for defense, is a group of mounds...

"At 10 feet from the south edge of No. 5 were two medium-sized skeletons, a lance head by the right side of each. These were lying at the foot of the hard, conical core, instead of reclining upon it. About 2 feet below the top of this ancient moundlet or core, and 4 feet from the top of the modern one built over it, were one very large and two ordinary sized skeletons, all having skulls above the ribs as though buried in a sitting posture facing each other."

Coshocton County

"Just east of Col. Methan's residence (Jefferson Township, near Warsaw), on a high point overlooking the valley for 3 or 4 miles, was a mound about 5 feet high, made of flat stones, in layers one over another, with the spaces between (where they did not fit closely) filled with broken stone. This had been built up over a stone box-grave containing a skeleton 7 feet long and a few relics."

History of Preston County, West Virginia, 1882

That they were large is established by their bones; that there were giants among them is proved by the wonderful large bones found interspersed through the mounds, showing that some of them were about seven feet high, while the majority were from five feet eight inches to six feet in height.

The Washington Post, June 23, 1908
Giant In Ancient Mound
Curious Relics of Prehistoric Times is Found in the Tomb

Special to the Washington post

Huntington, W. Va., June 22-The municipal authorities of Central City, four miles west of here, three weeks ago ordered the removal of a prehistoric mound from Thirteenth Street. To-day twelve feet above the base of the mound a gigantic human skeleton was discovered. It is almost seven feet in length, and of massive proportions.

It was surrounded by a mass of rude trinkets. Eight huge copper bracelets were discovered. These when burnished proved to be of purest beaten copper and a perfect preservation. Rude stone vessels, hatchets, and arrow heads were found with the skeleton.

A curiously inscribed totem was found at the head of the skeleton. The Smithsonian Institution will be notified of the discovery.

Daily Kennebec, June 26, 1908
Giant In Ancient Mound
Curious Relics of Prehistoric Times id Found in the Tomb

_Huntington, W. Va., June 25-The municipal authorities of Central City four miles west of here, three weeks ago ordered the removal of a prehistoric mound from Thirteenth Street. Today twelve feet above the base of the mound a gigantic human skeleton was discovered. It is almost seven feet in length, and of massive proportions.

It was surrounded by a mass of rude trinkets. Eight huge copper bracelets were discovered. These, when burnished, proved to be of purest beaten copper and a perfect preservation. Rude stone vessels, hatchets, and arrow heads were found with the skeleton. Smithsonian Institution will be notified of the discovery.

The Daily Mail, Hagerstown Maryland December 3, 1908
BONES OF GIANTS
Skeletons Found Believed To Be Those Of Mound Builders

Friendly W. Va., Dec 2-Prof E. L. Lively and J.L Williamson have made an examination of the giant skeletons found by children playing near that town. The femur and vertebrae were found to be in remarkable state of preservation and showed the persons to be of enormous stature.

The skeletons ranged in height from 7 feet 6 inches down to 6 feet 7 inches. The skulls found are of peculiar formation. The forehead is low and slopes back gradually, while the back of the head is very prominent, much more so than the skulls of people today. The legs are exceedingly long and the bones unusually large.

The findings of the skeletons has created a great deal of interest and the general impression is that the bones are the remains of the people who built the mounds, the largest in the country being located at Moundsville, Marshall county.

Raleigh Herald, **December 15, 1916**
GRAVE OF A GIANT

Workmen last Tuesday, while grading for the Fairmont & Wyatt railroad on the B. W. Shian farm near Pine Bluff, Marion county, unearthed the skeleton of what appears to be a giant of olden times. The skeleton was found in the mound opposite the residence of B. W. Bogges. The skull and several of thre large bones were in a good state of preservation. The skull being something near the size of a two-gallon bucket, with the low forehead and a long under-jaw. The ribs, so far as can be determined appear to be about three times the size of that of the average man of present time. One thigh bone which is well preserved, measures four and one half feet in length which shows this was a person of great height.

Small bones which appear to be those of the hands and fingers would indicate hands of enormous size. The feet appear to be in proportion with the other parts of the body.

There has been quite a little speculation as to what that old mound contained, but up to the time the railroad commenced cutting it open no one ever had the nerve to open it. Some citizens of that section claimed that it was just an ordinary mound, such as is found in so many places in Ohio, Indiana and numerous other states, built by the mound builders, but now as the mound has been opened and nothing found but a single skeleton all have come to the conclusion that the mound was nothing more than the grave of the giant, who must have been the king of his tribe.

Charleston Daily Mail, **October 22, 1922**
"Skeletons in Mound"

One of the most interesting of the five state parks is Mound Park, at Moundsville from which that city derived its name. Probably no other relic of pre-historic origin has attracted as wide study among archaeologists as the Grave Creeks mound which has given up skeletons of the ancients who constructed it.

Aside from the mammoth tumulus, itself 69 feet high and 900 feet in circumference, there were originally no fewer than seven mounds situated in the broad plain at the point. None was nearly equal to the one now standing, and the locations of most of the smaller ones are now lost to all excepting a few.

Archaeologists investigating the mound some years ago dug out a skeleton said to be that of a female because of the formation of the bones. The skeleton was seven feet four inches tall and the jawbone would easily fit over the face of a man weighing 160 pounds.

That the women of that ancient day were not unlike the woman of today in their liking for finery was evidenced by the articles that were found beside the skeleton of what centuries ago was a "flapper." Seventeen hundred ivory beads, 500 seashells of an involute species and five copper bracelets were found in the vault. The beads and shells were about the neck and breast of the skeleton while the bracelets were upon the arms.

There was also taken from the mound the skeleton of a man eight feet tall. There were no ornaments beside it. These skeletons were sent to the Smithsonian Institution in Washington.
"Human Bones Found"

At many places near the mound human bones of large size have been found and relics in large number and great variety have been picked up. Many beads found nearby were of porcelain-like substance, and a stone image was found representing a human figure sitting in a cramped position with face and eyes projecting upward and hair knotted in the back of the head. The features of the figure, especially the nose, were distinctly Roman. It is thought to have been a god. The figure disappeared, and its whereabouts are now unknown, but, until recent years, it was among the relics in the mound museum inside the huge pile, which was discontinued some years ago owing to the decay of the walls.

One interesting feature of the excavating was the formation of the ground composing the mound. It resembled the surrounding soil and was sandy until a depth of about eight feet was reached when blue spots were noticed. These increased upon approach to the center until they were so closely laid as to give the soil a clouded appearance. Examination showed that the spots contained bits of bone and ashes which led the investigators to the belief that the entire mound had been built of cremated bodies which builders piled about and upon the vault of the chief and his queen. Others maintain that the mound was the burying place of a chief and his queen and that the mound was constructed by earth taken from a large and regularly shaped basin at no great distance from the mound and piled up a shovel full at a time.

Among the interesting finds made in one of the small mounds were several stone tubes. These were made from a fine lead-blue steatite 12 inches long, 1½ inches in diameter at one end, and 1¼ inch at the other end. The tubes were bored in the manner of a gun barrel to within a short distance of the larger end, and a small aperture was left. The use of these tubes has never been learned but the workmanship is anything but rude.

Charleston Daily Mail, May 2, 1926
SKELETONS IN MOUNDS NEAR TOWN
The mound indicated by the cross in the diagram accompanying this article was 312 feet in circumference and 25 feet high. A second growth of timber was then on the mound, the decaying stumps of the first growth being still present. Sinking a shaft in the mound, a large vault was soon disclosed which contained numerous human bones and two entire skeletons. Four feet below the deposit and just below the original level of the ground were found six circular oven shaped pits three feet in diameter and three in depth. Those unearthed were in a semi-circle and it was assumed that the pits extended all the way around the mound. The pits contained a dark substance taken from the remains of indian corn.

The mound indicated at the extreme left of the same row was called the alter mound and was taken to be a connecting link between the mounds of this region and those of Ohio. It was 318 feet in circumference. At the depth of two feet the shaft disclosed a foot layer of clay and ashes, in which two entire skeletons lay horizontally, one immediately aboce the other. The upper and larger one lay with the face down and the lower with the face up, indicating a double burial.

The usual plan of sinking a shaft was followed. Earth and irregular sandstones, some of which made a load for two men, covered a vault seven by four feet in dimensions. In the vault was found a large and much decayed skeleton, wanting the head, which could not be found. A rough spear head was found with the skeleton. The Indian was seven and a half feet tall. It was enclosed in a bark coffin, and was placed on the back with the head towards the east. Six heavy copper bracelets were on each forearm.

Charleston Gazette, **June 15, 1930**
Salem Professor Discovers Huge Skeletons in Mounds
Dr. Sutton Believes Tribe of Giants Once Inhabited Doddridge County Section; Data on Exploration Will Go to Smithsonian Institution.

SALEM, June 14-Excavation of two mounds near Morganville, in Doddridge county, about 11 iles west of here revealed what Prof. Ernest Sutton, head of the history department of Salem college, believes is valuable evidence of a race of giants who inhabited this section of West Virginia more than 1,000 years ago.

Professor Sutton revealed tonight that he had been excavating the two mounds for the past several months. Skeletons of four mound builders indicating they were from seven to noine feet tall have been uncovered. Professor Sutton believes they were memebers of a race known in anthropology as Siouan Indians.

The best preserved skeleton was found enclosed in a casting of clay. All the vertabrae and other bones excepting the skull were intact. Careful measurement of this specimen indicated it was a man seven and a half feet tall.

The Kanawha Spectator, **Vol. I, 1953**

"The two largest mounds excavated and explored by Colonel Norris were on opposite sides of the Kanawha. One of them was situated on what was then a farm, belonging to the estate of Colonel Benjamin H. Smith, where Dunbar is now; and the other was on the Creel farm, now part of South Charleston. This mound still remains in a small park between MacWorld and Seventh Avenues at 'D' Street. At the time of Norris' explorations there were so many of these mounds in this section, indicating a once ancient community, that Norris suggested the title of Great Kanawha City as a suitable name 'for this ancient hive of people.'

"The excavations made in the Dunbar mound revealed it to be a double storied structure. The exploration was made by first sinking a vertical shaft through the center of the mound, down to and slightly below the original surface of the ground. This mound was about 175 feet in diameter and 35 feet high-as high as a modern three-story house. Within the mound were found successive layers of skeletons, some of them sepulchered in a stone vault, and those nearer the bottom in a large wooden vault. Some of the beams of the latter were of walnut, and were 12 inches in diameter.

"About half way down to the bottom the earth was mixed, for a depth of three or four feet, with ashes. One skeleton, still enclosed in a coffin made of bark, was in a better state of preservation than the others.

"Within the large wooden vault, near the bottom layer of earth, lay the principle figure, a huge skeleton measuring seven and a half feet in length and nineteen inches between the shoulder sockets. This figure lay prone, the head pointing toward the east. Around this skeleton were four others. Dr. Hale, who watched some of Colonel Norris' excavations, states that the irregular positions of these four skeletons indicated that they had been placed in a standing position, at each of the four corners; and that their irregular heaps suggested to some who saw them 'the possibility that they may have been buried alive, to accompany their great chief to the happy hunting grounds and land of spirits.'"

The End of the Allegewi Hopewell

"There was a sign from the sun, the like of which had never been seen and reported before. The sun became dark and its darkness lasted for 18 months. Each day, it shone for about four hours, and still this light was only a feeble shadow. Everyone declared that the sun would never recover its full light again."
John Ephesus, 535 A.D.

The year 535 marked great changes across the globe. In South America, Teotihuacan vanishes, the Roman Empire falls, the Persian empire crumbles and in North America it is the end of the Adena Hopewell Empire. 535 A.D. was a time of plague, famine and the beginning of the Dark Age in Europe. Why? According to to David Keys in, *Catastrophe: A Quest for the Origins of the Modern World,* 1999, these events were caused by the volcanic Indonesian island of Krakatoa exploding and sending the world into what would be the equivalent of a "nuclear winter."

The Delaware Indians kept an account of their history in what is called the *"Bark Record"* also known as the *"Walam Olum."* This history was preserved on painted sticks, translated by Rafinesque from the original symbols and the Algonquin words written along with them by an interpreter who understood both. The movement to the south of the Delaware and the Algonquin people also coincides with David Keys theories that there was a cataclysmic event that occurred in 535 A.D; effecting the weather for years after the explosion of the island of Krakatao. The *Walam Olum* chronicles the Algonquin Nations movements from the north (Canada) to the south into lands inhabited by the Adena Hopewell. What event caused the whole Algonquin Nation to move south? The *Walam Olum* says, "It freezes where they abode, it snows where they abode, it storms where they abode, it is cold where they abode." Did the explosion on the Isle of Krakatoa and the ensuing global cataclysm precipitate the Algonquin nation's move to the south? Did those eighteen months without sunlight cripple the Allegewi Hopewell empire to the point where they too moved south, abandoning the earthworks in the Ohio Valley, or where they defeated in battle by the the Algonquin tide that appeared in overwhelming numbers from the north?

The Prehistoric Aborigines of Minnesota and Their Migrations, N. H. Winchell, 1908

"It will be anticipated, from what has been said thus far, that the original mound-builder dynasty in the Ohio Valley was destroyed by an incursion of hostile people belonging to the Algonquin stock. It will burden of the rest of this paper to establish that great prehistoric event, and to show what effect it had on Minnesota.

Dr. Cyrus Thomas is to be accredited with the most thorough investigation of the aboriginal earthworks of the country. Under the direction of the Bureau of Ethnology he has established some important generalizations and has traced out some of the movements of the tribes that were concerned in the war which resulted in the expulsion of the original mound-builders from Ohio and the contiguous regions. Suffice it to say here that he considers that the evidence shows a movement, at least an extension, of the earliest mound-builders from the region of eastern Iowa, southeastern Minnesota, and southwestern Wisconsin, across Illinois and Indiana into Ohio. He shows that these people were driven out toward the east and southeast. He traces this retreat, which may have required several hundred years for its completion, with the most patient and convincing research, and arrives at the conclusion that when the whites came upon the scene the defeated and expelled people were known as Cherokee, living in western North Carolina and eastern Tennessee, and were still building mounds

But this line of persistent agression from the northwest to the southeast, resulting in the expulsion of the Cherokee from the upper part of the Ohio Valley, was not the whole of the great war, though it is part that has been established by evidence like that adduced by Dr. Thomas. It can hardly be questioned that such an incursion would have had disastrous effect on the mound builders of the whole Ohio Valley, and that they were driven out at the same time by the same hostile force."

Cyrus Thomas based some of his findings from the accounts of John Heckwelder, a Moravian Missionary, living with the Delaware or Lenni-Lenape in Pennsylvania, who gave the first printed account of the migrations and subsequent hostile incursion of the Delaware into the Allegewi Hopewell homeland.

The History, Manners and Customs of Indian Nations Who Once Inhabited Pennsylvania and the Neighboring States, John Heckwelder 1822

"Their object was the same with that of the Delaware's; they were proceeding on to the eastward, until they should find a country that pleased them. The spies, whom the Lenape had sent forward for the purpose of reconnoitering, had long before their arrival discovered that the country east of the Mississippi was inhabited by a very powerful nation, who had many large towns, built on the great rivers flowing through their land. These people (as I was told) called themselves Talligewi. Colonel John Gibson however, a gentleman who has a thorough knowledge of the Indians, and speaks several of their languages, is of opinion that they were not called Tallegewi, but Allegewi, and it would seem that he is right, from the traces of their name which still remain in the country, the Allegheny river and mountains having indubitably been named after them. The Delaware still call the former *Allegewi Sipu,* the River of the Allegewi. We have adopted, I know not for what reason, its Iroquois name, Ohio, which the French had literally translated into La Belle Riviere, The Beautiful River. A branch of it, however, still retains the ancient name Allegheny.

Many wonderful things are told of this famous people. They are said to have been remarkably tall and stout, and there is a tradition that there were giants among them, people of a much larger size than the tallest of the Lenape."

The Allegewi name still survives in the Allegheny Mountains and River. Further, the Ohio River's ancient name was *Allegewi Sipu* or the river of the Allegewi. Allegewi mounds are found within the Ohio River drainage or on the Ohio River itself which supports this connection. The Delaware also say that there were giants among the Allegewi.

A series of circular fortifications were constructed from Kosciusko County in north central Indiana across the southern tier of Lake Erie and Ontario and east along the St Lawrence River. Many of these forts have been attributed to the Iroquois, but there are also indications of an earlier Allegewi presence who may have allied with the Iroquois in northern Ohio and east along the southern tier of the Great Lakes. Were these Allegewi forces sent north to stem the tide of the Algonquin invasion?

There is evidence of Allegewi in apparent large numbers south of the Great lakes. Robert Converse writes in *The Archaeology of Ohio,* 2003 : "Strangely, nearly all the large classic Flint Ridge Adena spears have been recorded as isolated surface finds in northern Ohio. Although these typical cultural objects-spears, cache blades, pendants and gorgets occur as burial accompaniments in the mounds of the Adena homeland in southern Ohio and the Ohio River Valley, they are not commonly found as surface artifacts there. At no place in the entire Adena area do they occur in the numbers seen in northern Ohio."

In addition to the large amounts of Allegewi arrows and spearheads found in Northern Ohio and along the southern tier of the Great Lakes, there are accounts of mass graves containing large skeletons.

Migrations of the Lenni Lenape or Delawares
American Antiquarian, Vol., 19, 1897

"The relation, geographically, of the Iroquoian family to the Algonquins may, it is presumed, be taken as an indication that the former preceded the latter in the possession of the eastern territory, whether we adopt the one theory or the other, in reference to the general course of migration, Dr. Daniel Wilson in his paper on "The Huron-Iroquois of Canada" (Royal Society of Canada, 1884) takes the view in regard to the comparative ages of these two groups in this region. As a stream meeting an obstruction it cannot overwhelm, divides and circles about it, so it would seem that the Algonquin tide, finding the firmly planted Iroquois an obstruction it could not sweep away, flowed around them, filling the unoccupied spaces. What was the general course of the Algonquin tide? As there are few, if any, scholars of the present day who claim that this course was northward, in prehistoric times, except along the limited space of the New England coast, we may dismiss this view from consideration. Mr.

Gallatin, who studied the languages of this family with special care, expresses the opinion in his "Synopsis of the Indian Tribes," that the northern Algonquins were probably the original stock or family, In this northern division he includes the tribes dwelling north of the Great Lakes.

One of the oldest and most important traditions of this family is that of the Lenni Lenape or Deleware Indians, recorded by Heckwelder, but given more fully in the well known *"Bark Record"* or *Walam Olum.* This begins with a creation myth, then gives an account of their wanderings, the passage over some important stream or water way, their war with the Talligewi, in which they were aided by the Talamantans (Hurons), and final settlement on the banks of the Delaware, from which they obtained their modern name. It is now generally admitted that the *Nemassipi* ("Fish River" or *"Messusipi"*), of the tradition could not have been the Mississippi, as Heckwelder supposed, but the St. Lawrence in some part of its course, most probably in one of the links connecting the lakes, as, for example, Detroit River. The correctness of this opinion cannot be better shown than in Dr. Hales words, which we quote as follows:

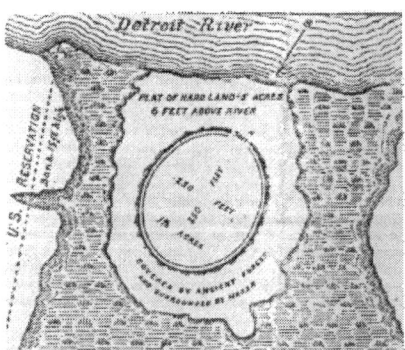

Iroquoian earthen fortification located on the Detroit River.

"The country from which the Lenape migrated was *Shinaki,* the 'land of the fir trees,' not in the west, but in the far north, evidently the woody region north of Lake Superior. The people who joined them in the war against the Allegewi or Tallegwi, as they are called in this record were Talamatans, a name meaning 'not of themselves,' whom Mr. Squire identifies with the Hurons, and no doubt correctly, if we understand by this name the Huron-Iroquois people, as they existed before their separation. The river which they crossed was the *Messusipu,* the 'Great River,' beyond which the Allegewi were found, 'possessing the east.' That this river was not our Mississippi is evident from the fact that the works of the Mound-builders extended far to the westward of the latter river, and would have been encountered by the invading nations, if they had approached it from the west, long before they arrived at its banks. The "Great River" was apparently the upper St. Lawrence, and most probably that portion of it which flows from Lake Huron to Lake Erie, and which is commonly known as the Detroit River. Near this river, according to Heckwelder, at a point west of Lake St. Clair, and also at another place just south of Lake Erie, some desperate conflicts took place. Hundreds of slain Allegewi, as he was told, were buried under mounds in that vicinity."

Called the "Great Mound" at the mouth of the River Rouge and Lake St. Clair.

This precisely accords with Cusick's statement that the people of the great southern empire had almost penetrated to Lake Erie at the time when the war began. Of course in coming to the Detroit River from the region north of Lake Superior, the Algonquins would be advancing from the west to the east. It is quite conceivable that, after many generations and wanderings,they may themselves have forgotten which was the true *Messusipu* or 'Great River.' of their traditionary tales"

It will be seen from this that Dr. Hale places the starting point in the "far north--the woody region north of Lake Superior:" an opinion with which we can justly agree. His further remark that "in coming to the Detroit River from the region of the north of Lake Superior the Algonquins would be advancing from the west to east," is noticeable, considering his theory heretofore mentioned. Dr. Brinton passes to the far east in his interpretation of this tradition. "Were I," he remarks, "to reconstruct the ancient history from the *Walam Olum*, as I understand it, the result would read as follows: "At some remote period their ancestors dwelt far to the northeast, on the tidewater, probably at Labrador, They proceeded south and west, till they reached a broad water full of Islands and abounding in fish, perhaps the St. Lawrence about the Thousand Isles. They crossed and dwelt for some generations in the pine and hemlock regions of New York, fighting more or less with the Snake people and Talega (Sioux and Allegewi), agricultural nations living in stationary villages to the southwest of them, in the area of Ohio and Indiana."

Snakes is the term that was used by the Algonquins to describe the Lakota up to historic times. In this ancient record of the *Walam Olum*, we can assume that the Snake Land is the territories held by the Hopewell Sioux. Today, the Sioux wish to be called Lakota, Dakota because S*ioux* is an Algonquin word that means *enemy* or *snake*. The Algonquin word for enemy was originally *nadouessi* which became *nadouesioux* by the French and finally *sioux* by the English. While the label of, "snake" could apply to any people that Algonquians were at war with, the Sioux and Algonquians fought inheritable wars against one another, and even when they had made contact with Europeans, the Algonquians called the Dakota the *Nadouessi*. While this name has been interpreted as being disparaging to the Lakota Dakota, its roots may be derived from the snake/sun worship practiced by the Hopewell and the snake effigies that are found in abundance within their early homelands of the Midwest.

American Antiquarian, Vol., 19, 1897

" The rigorous climate of their original home, and its geographical position, are clearly set forth in the following verses, with the original pictographs from the *Bark Record* or *Walam Olum.*

"It freezes where they abode, it snows where they abode, it storms where they abode, it is cold where they abode.
At this northern place they speak favorably of mild cool lands with many deer and buffaloes.
In that ancient country, in that northern country, in that turtle country, the best of the Lenape were the turtle men."

"The direction in which they started and traveled is also clearly stated."

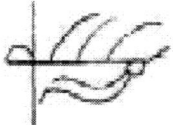

"To the Snake Land, to the east they went forth, going away, constantly grieving."

"It is difficult to understand how this course would take them from Labrador to the Thousand Islands It is true it is said":

"The fathers of the Bald Eagle and the White Wolf remain along the sea, rich in fish and mussels."

Floating up the streams in their canoes, our fathers were rich, they were in the light when they were at those islands."

"This however, would apply with far greater consistency to Hudson Bay, or even Lake Winnepeg than the ocean coast, as in following the rivers south, or southeast, they would be moving up stream. Moreover, it is twice expressly stated that Snake Island was "to the east."

In attempting to explain the tradition we should follow it as closely as possible consistent with other data. The tenor of the Bark Record indicates a movement southeastward and is at variance throughout with the idea that they came from the coast of Labrador or from the Atlantic shore at any point. If we will bear in mind the fact that if they started from the shores of the Hudson Bay, this great water would be to them the "sea" the great water. until they came into the vicinity of the true ocean, the difficulty of explaining the references to the sea will vanish. Add to this the the generally accepted tradition of the Indians of New England as given by Roger Williams, that they came from the southwest, and we would, according to the theory which brings them from Labrador, carry them completely around a circle.

"Having turned their course toward the Snake Land, Snake Island, it seems that, on their way, they crossed on ice during the winter some broad water, which from the language does not appear to have been a river, to which they applied the name, "sea."

"Over the water, the frozen sea, They went to enjoy it (Snake Island) On the wonderful slippery water, On the stone hard water they all went, On the great sea, the mussel-bearing sea."

"There are reasons for believing this refers to a winter of unusual severity as the Shawnee tradition mentioned hereafter claims that the water was rendered solid by the power of their magicians.

Having passed water they come to the "land of the spruce pines." What water is here alluded to is of course a question difficult to decide satisfactorily. However, I am inclined to the opinion that it was somewhere about Lake Huron, and that Snake Island was in the same locality. The events and movements which follow appear to be explained more consistently with the geography, by supposing that a crossing into the northern end of the lower peninsula is here alluded to, than by assuming this crossing to have been farther east. They had not yet advanced as far south, or east, as the country of the Talega, and hence must have been north and west of Lake Erie.

They come now to the "land of the spruce pines,"--northern part of Michigan--where they dwelt for an indefinite period. Here they fought with the *Akowini*, "Snake People" or Snake tribe; a weak people who hid themselves in the "swampy vales," which are by no means uncommon in northern Michigan. Again they decide to move on."

"The Snake land was at the south: the great Spruce-Pine land was toward the shore:
To the east was the Fish land, towards the lakes was the Buffalo land"

This seems to apply very well to central Michigan. The Snake land was southward in Indiana and Ohio; the Spruce-pine land was that which they had just abandoned in the northern part of the peninsula; assuming Detroit river to be the *Nemesipi* (Fish river), we can readily locate the "Fish land," the "Buffalo land," which was "toward the lakes," is descriptive of the prairie region of northern Illinois and northwestern Indiana around the southern end of Lake Michigan.

Snake land was reached, and after "much warfare south and east," full possession was obtained. Here they remained during the reigns of ten chiefs, probably not less than one hundred nor more than two hundred years. Here they first learned the use of maize.

"Shiver-with cold was Chief, who went south to the corn land.
After him Corn-Breaker was Chief, who brought about the planting of corn"

This implies of course, that immediately south of them were people who cultivated corn, possibly the

327

Alligewi, with whom they afterward warred, but more likely some other tribe.

At this point in the narrative there are some puzzling statements difficult to reconcile with each other and with the general trend of the story.

"After him the Salt-man was Chief: after him the Little-One was Chief
There was no rain and no corn, so they move further seaward.
At the place of caves in the buffalo land, they at last had food, on a pleasant plain"

The mention of "Salt-man," "seaward," and "place of caves in the buffalo land--on the pleasant plain,"--in such close connection, lead to the supposition that they relate to the same period in the migration. It is possible, however, that the reference is made to incidents in the history of the different tribes of this family group, which appear at this period of the narrative to have spread themselves over northern Indiana, the eastern portion of Illinois and the northwestern border of Ohio, a portion lingering in southern Michigan. As it is legitimate to infer that the Shawnee formed the chief off-shoot going south, it is not straining a point to suppose that the salt springs on the Saline river in southeastern Illinois had been discovered. On the other hand, the figure of the "caves in the buffalo land" bear a remarkably close resemblance to tents or wigwams. It is noticeable in this connection that in a preceding verse (III, 1) it is said that "the Lenape of the turtle, (turtle clan) were close together, in hollow houses, living together (or in a town) there.

The translator, in the vocabulary under the original word,'*wolokgun*', gives the following, Cane house; '*walak*,' hole; '*walken*,' he is digging a hole." The word signifying "a cave" (*waloh*) appears to be derived from the same root. According to Father Zenobius, who accompanied La Salle in his first expedition through Illinois, the Indians in the northern part of what is now this State, "made their cabins of mats of flat rushes, sewed together double." Hennepin speaks of the same kind of covering to their cabins, which he says is so well sewed as to be impervious to wind, rain and snow. He says they make their cabins in the form of a tent. These facts taken together, and the symbol referred to, lead to the belief that the word "caves" in this passage of the Record signifies cabins, or wigwams.

That they were still west of the Allegewi is evident from what follows: hence the statement that they moved "farther eastward" or eastward, being far from the sea, which is the real meaning, is consistent.

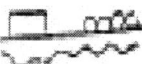

"They settled again on Yellow River, and had much corn on stoneless ground"

It is evident that they, or the main body to whom the tradition especially relates was now moving eastward from the western limits of their route, and was passing over some of the points touched on the westward march. If Dr. Brinton be correct in his supposition that this river is a small stream in northwestern Indiana, a tributary of the Kankakee, there is perfect agreement with the route of the migration so far as we have traced it. His note on the passage is as follows: "*Wisawana*, the Yellow River. There is a small river so called in the state of Indiana, a branch of the Kankakee called on Hough's 'Map of the Indian Names of Indiana,' We-tho-gan, a corruption of '*wisawana*.' When the Minsi made their first migration west, about 1690, they directed their course to this spot, where they were found by Carlevoix in 1721."

The cause of this eastward movement appears to be explained by the following verse:

"White-Fowl was chief: again there was war north and south.
The Always-Ready One was chief: he fought against the Snakes.

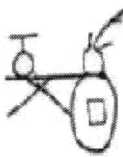

The Strong-Good-One was chief: he fought against the northerners.
The Lean-One was chief: he fought against the Tawa (Ottawa) people.

The Opossum like was chief: he fought in sadness, and said, they are many, let us go together to the east to the sunrise.

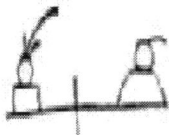

They separated at the Fish River: the lazy ones remained there.
Cabin-Man was chief: the Allegewi possessed the east.
Strong Friend was chief: he desired the eastern land."

"They were pressed by foes on both northern and southern flank; their old enemies, the Snakes, again warred upon them. The figures indicate that the "northerners" and "Tawa people" mentioned were on the west, now the rear, of the Lenape. As we have seen above, on leaving the land of the spruce pines, "Snow Bird went south, and with him probably much the larger body, from whom the Shawnee split off and remained in the southwest when the others retraced their steps toward the east. "White Beaver went east" leading the smaller body and probably stopping along the west bank of the Detroit River and Lake St. Clair. It was toward these, the western band now moved determined to proceed toward the east, south of the lakes. Some of the eastern band refused to join them, and "they separated at Fish River, the lazy ones" remaining there, where they had probably found food abundant.

Scarcely had the march toward the east begun before the Allegewi were encountered."

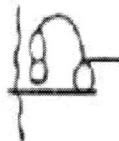

**"Some passed on east: the Allegewi ruler killed some of them.
All say in unison,'war, war!"**

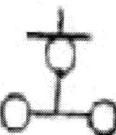

**The Talamantan, friends from the north come, and all go together.
Sharp One was chief: he was the pipe-bearer beyond the river."**

It appears from the language and the figure (IV,52,) that there was a rubicon here, a river which the Allegewi considered a boundary of their territory, whether the Maumee or some other stream farther east or south, can be decided by conjecture only, nor is it important in the present investigation.

This contest with the Alligewi, though doubtless long and sanguinary, is told in a few brief lines.

**"They rejoiced greatly that they should fight and slay the Allegewi towns.
The Stirrer was chief: the Allegewi towns were to strong.
The Fire-Builder was chief: they all gave to him many towns.**

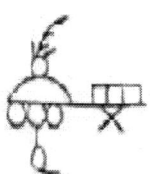

**The Breaker-in-Pieces was chief: all the Allegewi go south
He-Has-Pleasure was chief: all the people rejoice.
They stay south of the lakes: the Talamatan friends north of the lakes."**

"Their confidence in their ability to overcome the Allegewi appears to have been, at first, a little disappointed as the towns, for a time resisted their attacks directed by *Pimokhasuwi* (Stirrer) The next chief however, if we may judge by his name, as translated by Dr. Brinton-Tenchekenut (Fire Builder)- used the torch as a more effective weapon (Dr. Hale translates it ("Open Path"). If we are justified in supposing that the square, circular and other ancient works of Ohio indicate these towns, this would imply that the walls were surmounted by stockades or woodwork of some kind. Be this as it may, the Lenape appear to have been successful, and the Allegewi were driven south, finding at last a permanent resting place in the mountains of western North Carolina and eastern Tennesse. The Lenape remain south of the lakes and the Talamantans return to their country on the north side. The figure (IV,61) appears to refer to a single lake, presumably Lake Erie, and indicates that the Lenape occupied or controlled the entire country south, from end to end. At least the Talamantans symbol stands above (north of) the middle of the lake, and Lenape symbols below (south of) each end. The friendship which had existed between these two nations was not of long duration."

Tales of the war that resulted in the expulsion of the mound builders from the Ohio Valley continued into historic times. A legendary, bloody conflict that occurred at the Falls of the Ohio, near Clarksville, Indiana was passed down from generation to generation. The following articles also describe a light skinned people, that is descriptive of the Sioux , who have historically been fair skinned and have been described by other tribes as "white."

George Catlin commented on the physical appearance of the Mandan Indians in, *Letters and Notes on the Manners, Customs, and Conditions of the North American Indians.* "A stranger in the Mandan village is first struck with the different shades of complexion, and various colors of hair which he sees in a crowd about him; and it is at once almost disposed to exclaim that "these are Indians." There are a great many of these people whose complexions appear as light as half-breeds; and amongst the women particularly, there are many whose skins are almost white, with the most pleasing symmetry and proportions and features; with hazel, with grey, and others with blue eyes."

Catlin also described the Osage in the same publication. "Osages were the tallest men in North America, and other contemporary observers agreed. Very few of the men, at their full growth, are less than six feet in stature and very many of them six and half, and others seven feet."

History of Marietta and Washington County Ohio, 1902
The history of this West is a long history of war, from the earliest days even to our own century. This territory between the Alleghenies and the Mississippi is one of the greatest battlefields in the world. It is certainly the oldest and most renowned in our America. The first of our race to enter it looked with

331

wondering eyes upon the monstrous earthen forts of a prehistoric race whom we have named the relics they left behind the "Moundbuilders." Of this race the Indians knew nothing, save what the legends handed down by their fathers who told of a race of giants which was driven out of the Central West, and sent flying down the Ohio and Mississippi to reappear no more in human history.

Prehistoric Men of Kentucky by Col. Bennett A. Young 1910

Col. James Moore, of Kentucky, was told by an old Indian that the primitive inhabitants of this state had perished in a war of extermination waged against them by the Indians; that the last great battle was fought at the Falls of the Ohio, (Clarksville Indiana); and that the Indians succeeded in driving the aborigines into a small island below the rapids, 'where the whole of them were cut to pieces'. The colonel was assured that the evidence of this event rested upon facts handed down by tradition, and that he would have decisive proofs of it under his eyes as soon as the waters of the Ohio became low. When the waters of the river had fallen, an examination of Sandy Island was made, and 'a multitude of human bones was discovered'. There is a similar confirmation of this tradition in the statement of General George Rogers Clark, that there was a great bury-ground on the northern side of the river, but a short distance below the Falls. According to a tradition imparted to the same gentleman by the Indian Chief Tobacco, the battle of Sandy Island decided finally the fall of Kentucky, with its ancient inhabitants when Colonel McKee commanded in the Kanawha, (says Doctor Campbell), he was told by the Indian Chief Cornstalk, with whom he had frequent conversations, that Ohio and Kentucky (and Tennessee also is associated with Kentucky in prehistoric ethnography by Rafinesque) had once been settled by white people who were familiar with arts of which the Indians knew nothing; that these whites, after a series of bloody contest with the Indians, had been exterminated; that the old burial places were the graves of an unknown people; and that the old forts had not been built by Indians, but had come down from ' a very long ago' people, who were of a white complexion and skilled in the arts'.

Washington Post, October 27, 1912
Curse of Yellow Hair
Recent Murder Recalls Strange Indian Legend of Prehistoric White Race
On the Ohio
St. Louis Globe-Democrat

The last connecting link with a prehistoric race was destroyed when George Kelly murdered his poor old grandmother and then killed himself at Jeffersonville Ind., a few months ago. The aged woman had $75, and the eighteen-year old boy got it, spent it and then took his own life when his brother accused him of having committed the crime.

The victim was the widow of Valentine Kelly, who was run over and killed by a train many years ago, but she was known among the savants of Indiana as Mary Kelly, who direct descendant of Black Hawk Stewart, a famous Shawnee Indian Chieftain, whose title dated back to the conquest of the land from a prehistoric race that inhabited it.

The little farms that lie close to the banks of the Ohio Falls are to this day fertilized with the bones of these people, and the only clew to their identity was a fragment of song that Mrs. Kelly remembered to have heard her mother sing. Mrs. Kelly told the writer it had been handed down from generation to generation for hundreds of years and that she believed it to be true. In fact, there is much to this day to bolster up this belief.

At the time of the permanent peace established by General George Rogers Clark, Black Hawk, who was one of the most ferocious of all the Indian chieftains, washed the war paint from his face, buried the hatchet and resolved to devote his talents to the arts of peace. By an arrangement with General Clark, a deed of title from the United States Government was secured for him to a plot of land on the falls, and on the very land for 300 years the tepees of his forefathers and stood. He was born there and has bones are buried there. The land never passed out of the family, and it is still held under the original title. This, it should be explained, was not the Black Hawk who figured in the war in Northern Illinois.

"But there is a curse on the place." said Mrs. Kelly to the writer, who knew her very well in the long ago, when her memory was much better than it was in her later years.

"Yellow Hair cursed it, and none of my people ever die a natural death. One after another I have seen them go, and I have always wondered if it will extend to me."

"If there is anything in it," said old Valentine Kelly, her husband, "it will reach me, too."

"The next night he was walking on the railroad track when a train hit him and killed him. Several of the family has been drowned in the water of the falls, and now Mrs. Kelly is dead at the hands of her beloved grandson, who also slew himself. A few years ago the old house erected by Black Hawk himself, when he determined to adopt the ways of the pale face, was destroyed by fire of a mysterious origin.

There was apparently no way for it to have caught fire, and as she sat in the roadway at the front gate, viewing the smoldering ruins, Mrs. Kelly said, solemnly:

"It is the curse of Yellow Hair."

And her sons believed her and the neighbors believed her-and it may have been as she said.

For three miles the Beautiful River (Ohio in the Indian tongue) makes a bend between Jeffersonville, Ind., and Louisville, Ky., and rushes westward with a terrific roar. Inspired by a fall of about 25 feet. In the center of the cataract is what has long been known as Corn Island. On the Indiana side the big eddy whirls past Wave Rock, the graveyard of many proud steamboat. In low water the place is dotted with the dismantled hulks. And just below the whirlpool lies the Kelly property. There is a big spring bubbling out of the side of the path that leads down to the rocky shore that is said to have been dug by Yellow Hair. To the right of it, going up the bank, is a graveyard, where hundreds of prehistoric people lie buried, and to the left is the Kelly farm, on the river edge of which are 50 tombs of the same mysterious people. The first cemetery is undoubtedly that of the common people. They were of medium stature, and were all buried facing the rising sun. Their bones fertilize the cornfields of the farm of Edward Commines on land that was originally settled by William Beach. Occasionally a skull or a portion of a skeleton is dug up by the plow, but the matter-of-fact farmer tosses it back and the next furrow covers it from sight. Every man who has ever owned the Commines land has met with a violent death. Commines's father was killed by a train a few years ago.

The other cemetery contains the bones of 50 dead Kings. The tombs are made of rough hewn stone and the occupants were all men, not one of whom was less than six and one half feet high. They were buried in sitting posture, with their faces turned toward the rising sun and their weapons must have been buried with them, evidently placed in their laps. But the peculiar coincidence is that the left temple of each had been crushed in by some blunt instrument. Whether it was as religious rite or a precaution against burying them alive is a matter of surmise. The writer, who opened one of the graves with Prof. Green, the eminent geologist and at one time State Geologist of Indiana, believes it was a religious rite. The school history of Kentucky says when the first white settlers arrived at Louisville they found piles of human skeletons on Corn Island and some are found there now. To the early settlers it appeared that there had been a great battle fought and that one tribe had been entirely wiped out. All of the skeletons were those of people of medium stature, save one, that of a man, and he must have been seven feet high. On the banks of the falls to this day are found thousands of Indian arrows and spear heads, with an occasional battle ax, and once a stone owl was found that had probably been fashioned by one of the prehistoric people. This description represents the concrete facts and is the corroborative evidence of the weird tale told by Mrs. Kelly and her ancestors in their mystic chant of the vanishing of a strange race of people. The story had better be given in her own words to the writer of this narrative.

"When I was a wee bit of a girl," said Mrs. Kelly, "my mother sang me to sleep with the words of this song. It was a sort of a chant in the Indian tongue, and I do not remember it all. Translated so you will understand it, it was to the effect that a white people lived here on the falls and that they were mighty. A tall Chief with yellow hair ruled over them and four ages they fought off the red men and held the fisheries of the falls and the hunting grounds for their own. The sun was the god they worshipped, and he appeared to have blessed them with peace and plenty. Yellow Hair our people called the Chief, who was a giant. The Chiefs or Kings must have maintained the great stature by intermarrying in the royal family, probably killing all the females except just enough to perpetuate, the race. My mother thought they saved the best developed girls for the wives of the Chief in order to perpetuate the governing race. I did not ask her why she formed this opinion, and it may have been part of the legend. But our people had long viewed the land from afar and they determined to possess it. The Chief at that time was Hawk Wing, the line through which I come. He sent spies to make overtures to the strange white people and they visited Yellow hair and told him the Shawnees wanted to share with them the fisheries and the hunting grounds. Yellow Hair listened to their statements and then told them that there was just enough for the white people and that he and his people preferred to live by themselves. Then the Ambassadors of the Shawnees said that if the white people would not submit peacefully to having them for neighbors they would slay them and take their possessions. At this Yellow Hair laughed disdainfully and said the sun god would destroy his enemies with fire from heaven and that every man who took part in such a bloody and unprovoked massacre would die a violent death and that the curse would have the effect as long as one of the offending race remained on earth.

But Hawk Wing had faith in the Great Spirit, that he and his tribe worshipped, and he collected his warriors and set out for the home of Yellow Hair. In some way, the scouts of Yellow learned of there near approach, and he and his people leaped into their canoes and went to Corn Island. The dangerous whirlpools and the treacherous eddies, with which they were familiar, they thought would protect them from the less skilled Shawnees. But they did not know Hawk Wing. He and his braves had been accustomed to the water from infancy and they were almost as much at home in the torrent as Yellow Hair and his people. So that night while Yellow Hair was peacefully sleeping in fancied security. Hawk Wing and his braves were making canoes and getting ready for battle. Just as the sun was breaking through the murky sky of the east the canoes of Hawk Wing reached the shores of the island. Yellow Hair and his people were awakening from sleep and were falling on their knees in prayer to their sun god. They were in this position when the yells of my people burst upon them. Many were slain as they knelt, but Yellow Hair was a warrior, and though taken by surprise, he seized his battle-ax and valiantly defended his subjects. With his single-hand he slew more than a score of our people. Then when he was weary from fighting Hawk Wing confronted him. Behind Yellow Hair were his wives and children kneeling in prayer and in front of him were Hawk Wing and his warriors. The two chieftains sprang at each other with their battle-axes. My ancestor was used to war and familiar with all the tricks. As a result, after a terrible encounter, during which both were covered with wounds, Yellow Hair sank exhausted and hawk Wing's battle ax was buried in his brain.

"Maddened by the conflict, Hawk Wing turned upon the kneeling women and children and slew them. He and his men kept up the slaughter until not one of the white race remained. Every single one of them had been killed and the scalp lock of Yellow Hair dangled at the belt of Hawk Wing. Till his death he kept it and it was buried with him.

"Then the Shawnees took possession of the houses and lands of the vanquished people and the Kelly's are the last of the victims, for the Shawnees have all gone to the happy hunting grounds, and they have but a remnant of the original blood in them.

"There is one other little bit of information I can give you on the subject, but I do not know how I learned it. On the island in the falls is a small cave, which was once known as 'Yellow Hair's Bath,' but which is now always referred to as the 'Crystal Bath,' It is said Yellow Hair bathed in this every day after he prayed to the sun. The cave is of solid stone and a small stream of water trickles through the top, making a natural, shower bath, where the fisherman to this day often bathes."

"Finally, the last of the habitations of the strange people was torn down and 300 years later, when General Clark came here and found Black Hawk in possession, nothing remained save the bones of the murdered people on the island.

One after another I have seen my people killed in some manner and misfortune has stricken them from the face of the earth. Do you blame me for thinking that the curse of Yellow Hair is upon us?"

Valentine Kelly, who was a Spiritualist, told the writer that he was once standing in a shed near the royal tombs when a gigantic white man with yellow hair peered in at the window. He said he saw him, as clearly as could be, for it was broad daylight and he could not have made a mistake. However, Mr. Kelly was a firm believer in ghost and hobgoblins, and it may be that he did not actually see Yellow Hair, but he believed to the time of his death that he had seen him. He permitted Prof. Green and the writer to open two of the graves on his farm, but stopped further excavating, as he said the scientist would soon dig up the best part of his farm if he permitted them to do so. But there were originally 50 of the tombs and now more than 40 remain. The high water washed away some of them, and two were opened by man.

One of the best-known archaeologists of Indiana, Dr. W. F. Work, of Charlestown, Ind., found seven similar stone tombs 13 miles from the scene, and he noticed that the left temple of each dead man was crushed in and that the bones were those of men of gigantic stature. Dr. Work spent much time in exploring the habitations of the cliff dwellers of Arizona and has written much on the subject. He believes Yellow Hair's people were the Mandan Indians. Orlando Hobbs, also an archaeological authority of Indiana and a man known widely for his learning and research, holds this opinion.

There is a rich field for science on the falls of the Ohio, and may be that when the distant fields are thoroughly explored those at home will be given the attention they deserve. In this connection it may be stated, by way of parenthesis, that adjoining the farm of the Kelly's are 1,000 acres of land that are still in Virginia, although it is surrounded by Indiana and cut off from the state to which it belongs by Kentucky. Yet Virginia gave this land to George Rogers Clark and his heirs forever with-out taxes in reward for his services in ridding the section of the Indians." And it is not on the map of Indiana, through a mistake in drawing the outlines. It is governed by three trustees, one appointed by Clark County, another by Floyd County, and the third perpetuates himself by naming some one who is to succeed him when he dies. But this is another story.

History of Muskingum County Ohio, 1882

Many centuries ago the inhabitants of America, who were the authors of the great works in the Mississippi Valley, were driven south by an army of savage warriors from the north. The Indian traditions locate their original home north of the Great lakes. In the process of time some of their people migrated to the river Kanawag, (St. Lawrence). After many years a foreign people came by sea and settled south of the lakes. Then follow long accounts of wars and fierce invasions from the north.

Mississippii as a Province, Territory and State
J.F.H. Claiborne, 1880

The Choctaws preserve a dim tradition that, after crossing the Mississippi, they met a race of men whom they called the Na-hon-lo, tall in stature and of fair complexion, who had emigrated from the sunrise. They had once been a mighty people, but were then few in number, and soon disappeared after the incoming of the Choctaws. This race of men were, according to tradition, tillers of the soil and peaceable.

The following passages are from The Book of Mormon, where I found some accounts of the prehistoric Nephites and their war with the Lamanites. Early Mormon leaders believed there was a connection between the Nephites and the mounds and earthworks in the Ohio Valley. I found that Nephi, their leader, being large in stature and erecting earthen fortifications, an interesting similarity with the *Walam Olum* and tales by the Lene Lanape of the Allegewi having giants amongst them. Their leaders name was Nephi, that is reminiscent of the Nephilim, is there a connection?

Book of Mormon, Chapter 2:1
And it came to pass in that same year there began to be a war again between the Nephites and the Lamanites. And not withstanding I being young, was large in stature; therefore the people of Nephi appointed me that I should be their leader, or the leader of their armies.
Nephi, 4:31
And now I, Nephi, being a man large in stature and also having received much strength of the Lord, therefore I did seize upon the servant Laban, and held him, that he should not flee.
Alma 48:8
Yea, he had strengthening the armies of the Nephites, and erecting small forts, or places of resort; throwing up banks of earth round about to enclose his armies, and also building walls of stone to encircle them about, round about their cities and the borders of their lands; yea all about the land.

The following section contains legends and commentary by people who had spent time with the Siouan tribes in the west and chronicled their migrations from the Ohio Valley.

Migrations of the Hopewell Sioux

Siouan Tribes And The Ohio Valley, American Anthropologist 1943, John Swanton

"When Europeans came in contact with the Dakota tribes they were living in three main divisions with several detached tribes. The largest group was located west of the Mississippi from Lake Winnipeg in Manitoba south to the mouth of the Arkansas River. The second largest group was in the Piedmont of Virginia and the Carolinas with settlements that extended to the coast and into the Appalachian Mountains. The third group was found on the Pascagoula River in Mississippi and along the gulf coast were the Biloxi; and on the Yazoo River, the Oto. The Winnebago Indians lived on the western shores of Lake Michigan at present day Green Bay, Wisconsin, and were separated in later days from the Mississippian tribes by Algonquin Indians.

Linguistically, the tribes in North and South Carolina were the most aberrant of all of the Siouan speech. The speech of the Virginia Sioux is closer to that of the western dialects. The conclusion is that at one time the Virginia Sioux and the Mississippian Sioux was split into two groups."

John Lawson writes in *The History of Carolina* 1860, concerning the eastern Sioux, "When you ask them whence their forefathers came, that first inhabited the country, they will point westward, and say, "where the sun sleeps our forefathers came thence. Thus the western tribes say they came from the east and the eastern tribes say they came from the west and in between is the Ohio Valley."

Swanton continues, " In brief, traditions among the western Siouans, J.O Dorsey, the Siouan specialist, wrote of the Omaha in 1882: "indicated a former home in the east toward the country of the eastern Siouans and in the case of some tribes, notably the Quapaw, residence on the Ohio, while traditions among the eastern Siouans pointed to a home toward the west in the direction of the western Siouans. These traditions would not seem to vary then from what one might have expected."

Excursion through the Slave States 1840, G. W. Featherstonhaugh

Maj. George C. Sibley, son of the famous Dr. John Sibley, who in 1834 furnished G. W. Featherstonhaugh with the following information he had obtained from an old Osage chief: "The old chief further said...that the tradition had been steadily transmitted down from their ancestors, that the Whashash (Osage) had originated and emigrated from the east in great numbers, the population being to dense for their hunting-grounds; he described the forks of the Allegheny and Monongahela river, and the falls of the Ohio, where they had dwelt some time, and where large bands had separated from them and distributed themselves in the surrounding country. Those who did not remain in the Ohio country, following its waters, reached St. Louis, where other separations took place, some following the Mississippi up to the north, others advancing up the waters of the Missouri."

George Catlin, *North American Indians* 1860, discussing the Mandan Indians, saying: "There are other, and very intersesting traditions and historical facts relative to a still prior location and condition of these people, of which I shall speak more fully on a future occasion. From these, when they are promulgated, I think there may be a pretty fair deduction drawn, that they formerly occupied the lower part of the Missouri, and even the Ohio and Muskingum, and have gradually made their way up the Missouri to where they are now."

In 1908, N.H. Winchell added in the, *"The Popular Science Monthly," "The Sioux and Iroquois Legends, Prehistoric Aborigines of Minnesota and Their Migrations."*

"The Osage and perhaps the Omaha, who belong to the Dakota stock, and who have a tradition which is confirmed by other traditions, that they once lived east of the Mississippi in that very region, [southern Ohio]. With this understanding it is, I repeat, a remarkable fact that, aside from the Muskogee earthworks of the gulf coast, which have distinctive characters, only the Dakotan and Iroquois stocks can be shown either by history or tradition to have been characteristic mound builders.

This legend is found amongst several of the Dakota tribes, and even amongst the later Algonquin who returned westward to the Mississippi Valley. The Osage, Omaha, Mandan, Kansa and Akansea, and Ponca. These tribes concur in saying that they formerly dwelt in the Ohio and Wabash valleys, and that they moved down the Ohio Valley, where they were separated into two divisions at the mouth of the Ohio River, some of them going down the Mississippi and some of them up the same river.

It is due to the research of the Late J. V. Brower that the Dakota tribes of Minnesota have proved to belong to the so-called mound-builder dynasty.

"There is also a remarkable series of effigy mounds in central and southern Wisconsin, which extended across the Mississippi into Minnesota and Iowa. As to the prevalence of serpent worship, we have shown that there were serpent effigies in Ohio, Illinois, Minnesota, Wisconsin and Dakota, and that all these were situated along the line of migration, which according to tradition of the Dakotas, was followed by their ancestors on reaching their later seats on the Mississippi and Upper Missouri Rivers." We may conclude from this that the Winnebago were not only effigy builders, but they were serpent worshipers, and that these various serpents were their work."

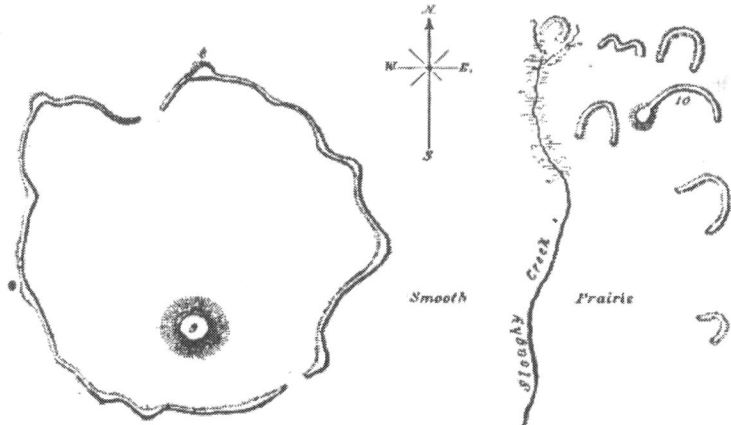

Earthen serpent effigies surrounding a Hopewellian burial mound, along with bird-like and serpent effigies on the eastern side of the creek; similar effigies are also found in southern Ohio. Note the earthwork at the fork of the creek that resembles the snake swallowing the sun disc at the Serpent Mound in Adams County, Ohio. Mounds and earthworks are located in in Pipestone County, Minnesota. Diagram is from *The Mound Builders*, Shetrone, 1941.

Aboriginal Religions in America, Stephen Peet, 1905: "The region in Ohio where the serpent effigies are the most prominent was once the dwelling place of a tribe of hunters who are known to have migrated from their original seats east of the Alleghenies, following the buffalo in their retreat westward, namely, the Dakotas or Sioux, and it is quit likely that the name of the snake people, which tradition has preserved, was the one given to them. Confirmatory of this is the fact that the serpent effigies are found all along the track taken by the Dakotas in their migration westward to their present seats. One was discovered by the writer on the bluff near Quincy, Illinois, another on the bluff neat Cassville, Wisconsin, another on the ridge near Lake Wingra, near Madison, Wisconsin, another near Mayville, still another, discovered by Prof. J. A. Todd, on the bluff called Dakota. And the fact that carved animal pipes, resembling those in Ohio, have been found in the mounds in Illinois and Iowa, the most interesting of which has the serpent coiled around the bowl exactly as the one found in the fort called Clarks works."

Stephen Peet writes in *Ancient America Vol. II,* 1892 "The effigy mounds may have been built by the Winnebago who fled northward when the Lenape crossed the Mississippi. They are thought to have remained in southern Wisconsin during the Lenape-Alligewi war, and in the end were there to welcome the fugitives back when they returned. This may explain why the Winnebago are isolated in the west from the other Dakota tribes. It is also worthy to note that the Winnebago is oldest dialect of the western Dakota and are called "grandfathers" by the other tribes."

As to the prevalence of serpent worship, we have shown that there were serpent effigies in Ohio, Illinois, Minnesota, Wisconsin and Dakota, and that all these were situated along the line of migration, which according to tradition of the Dakotas, was followed by their ancestors on reaching their later seats on the Mississippi and Upper Missouri Rivers."We may conclude from this that the Winnebago were not only effigy builders, but they were serpent worshipers, and that these various serpents were their work."

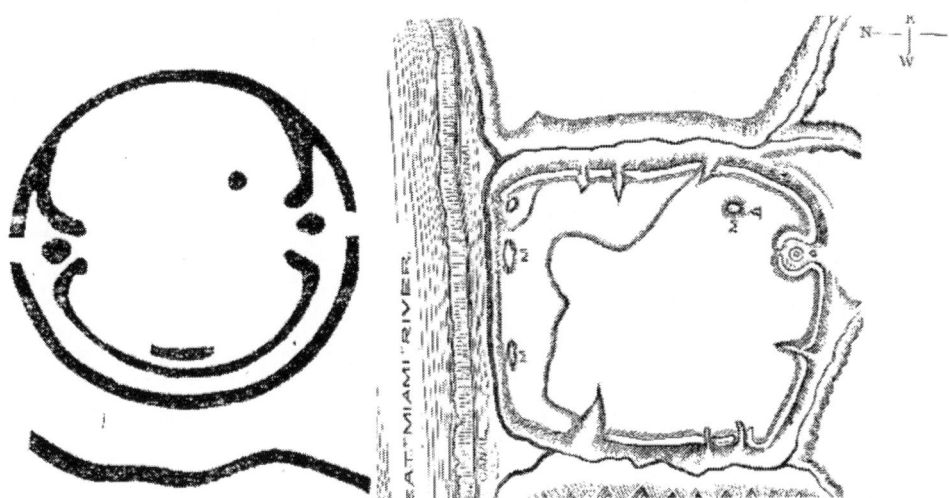

To the left is an earthwork that was located in Dade County, Missouri originally published in the *American Antiquarian* in July, 1878, showing a burial mound within a circular enclosure with two serpents whose heads are poised toward a central circular mound, symbolic of the sun. To the right is an earthwork in Butler County, Ohio and originally published in *"The Mound Builders"* by J. P. McLean in 1879 that is of similar iconic design.

In an unpublished work written in June 2001, archaeologist and ethnohistorian Alan R. Woolworth dealt with the question of the tribal affiliation of the effigy and burial mounds in Minnesota in *"Who Built the Prehistoric Burial Mounds in Minnesota and the Dakotas."*

"Available information demonstrates that the burial mounds in most of southern Minnesota were created by Siouan speaking Indians who had originated in the Ohio River Valley and eventually migrated into this region.

The ancestral homes of many Siouan-speaking Indians such as the Teton, Yankton, Yanktonai, Assinboine and Santee Dakota peoples were in the Ohio River country. Also, in this general area were the Dhegiha and Chewere Siouan speakers. Perhaps all of these groups migrated down the Ohio and Mississippi rivers to the general vicinity of Cahokia and then travelled northward to Minnesota region.

Chewere Siouans composed of the Winnebago, Iowa, Oto and Missouri groups, had likewise migrated down the Misssissippi River to the large fortified Azatlan site in the south central Wisconsin and later moved westward into the Minnesota area."

The era of the Allegewi Hopewell had ended. The Lakota-Dakota Sioux, Cherokee, Iroquois would rise again in a new tradition of mound building called the Mississippian, (800 A.D.-1500A.D). In this era, people were still buried in mounds of earth but the geometric earthworks of the Adena and Hopewell were replaced by great platform mounds where the houses of the Chiefdoms and Sun Priests resided.

Evidence of a Siouan presence in southern Indiana was documented in the, *Archaeological Notes on Posey County,* Wm. R. Adams, Indiana Historical Bureau, 1949 "Griffin has recently stated the possibility of a proto-historic occupation of the Ohio Valley from the Wabash to the Mississippi by the *Arkansas* and mentions the Murphy Site as "most strongly" suggesting such an occupation. Earlier Eli Lilly had made the suggestion that Murphy Bone Bank, and the Indian "citadel" at Merom on the Wabash in Sullivan County, were indicative of a Siouan relationship, equating the three with the Oneota culture."

"Moorehead called attention still earlier to the anomalous nature of the materials from Murphy. He rightly compared the ceramics with "Missouri Arkansas" forms and might have included Kentucky and Tennessee in the same temporal reference. He also stated the pipes were early Siouan."

Physical remains of the Mississippian Sioux, within the geographic area of the guide are still visible at the platform mounds at Marietta, Angel Mounds in Evansville and at the mouth of the Wabash. One of three large mounds at Vincennes, Indiana was a platform mound implying that the remaining two, may also prove to be Mississippian Sioux in origin.

According to the Lene Lenapi Alqonquins, the Adena were the the Allegewi; a name that still survives today in the Allegheny River and Mountains. The Hopewell were a confederation of the kindred tribes of the Sioux, Iroquois and Cherokee. Mounds and earthworks found in the travel guide are located in the southern tier of Michigan, northeast Indiana and northwest Ohio were constructed by the Point Peninsula Iroquois, who at first shared many Allegewi traits.

Iroquois in western Indiana and southwest Michigan were influenced by Hopewell Sioux residing in southern Illinois and Missouri. Archaeologist call these Siouan tribes, Havanna Hopewell. Ft. Ancient mounds that once were numerous in southern Indiana and Ohio were constructed by the Shawnee.

The Shawnee migration south was chronicled in the ancient *Bark Record* or *Walam Olum* of the Lene Lenape, or Delaware. This movement to the south may have been precipitated by the island of Krakatoa exploding; sending the world into "nucular winter" for a duration of about two years. Evidence of this sanguin battle was chronicled in county histories in the form of mass graves containing large skeletons that are associated with the Allegewi. Further evidence of an Adena presence north of their mound sites are the number of surface finds of their spear heads.

The Sioux nation and all of their tribes, concur in a migration out of the Ohio Valley, some going north to Wisconsin and Minnesota and other s going south into Missouri and Arkansas. Whether this migration was precipitated by the explosion of Krakatoa and the climatic change that followed or was the result of the incursion of the Algonquins or a combination of both is yet to be determined.

Of greater importance is that we know what Native Americans built the mounds and earthworks found within this guide. Under the provisions of the *Native American Graves Protection Act,* 1991, all of the mounds in this guide should be protected from any further excavations by universities. The current trend in archaeological excavations is directed toward the ceremonial centers. These Native American sacred sites should be given the same Federal or State protection as the burial mounds.

Generations ago, we took these people's land; let's not be the generation that takes their dead.

Printed in Great Britain
by Amazon.co.uk, Ltd.,
Marston Gate.